Clinical Signs in Humans and Animals Associated with Minerals, Trace Elements, and Rare Earth Elements

Clinical Signs in Humans and Animals Associated with Minerals, Trace Elements, and Rare Earth Elements

Mike Davies
Author - Mike Davies, BVetMed CertVR CertSAO FRCVS;
RCVS Specialist in Veterinary Nutrition (Small Animal Clinical Nutrition)

Academic Press is an imprint of Elsevier
125 London Wall, London EC2Y 5AS, United Kingdom
525 B Street, Suite 1650, San Diego, CA 92101, United States
50 Hampshire Street, 5th Floor, Cambridge, MA 02139, United States
The Boulevard, Langford Lane, Kidlington, Oxford OX5 1GB, United Kingdom

Library of Congress Cataloging-in-Publication Data
A catalog record for this book is available from the Library of Congress

British Library Cataloguing-in-Publication Data
A catalogue record for this book is available from the British Library

ISBN: 978-0-323-89976-5

For information on all Academic Press publications
visit our website at https://www.elsevier.com/books-and-journals

Publisher: Charlotte Cockle
Acquisitions Editor: Anna Valutkevich
Editorial Project Manager: Lena Sparks
Production Project Manager: Niranjan Bhaskaran
Cover Designer: Christian Bilbow

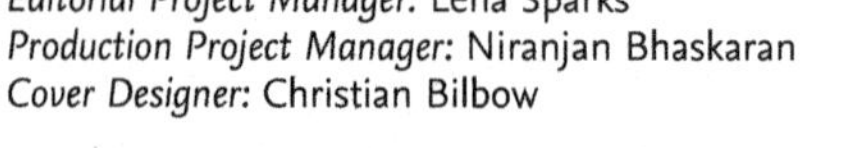

Typeset by STRAIVE, India

Dedication

Dedicated to my partner Lana Kyte for her unbelievable patience and support during the many months it took me to write this book, to Professor David Gardner at the University of Nottingham for embracing some of my suggestions and clinical research into minerals, and finally to the memory of two of my longstanding friends and colleagues who always encouraged me in my work Keith Plunkett and Dr. Jennifer Poland OBE.

Contents

Author biography

Mike Davies is an experienced veterinary clinician having worked in private practice, industry (pharmaceutical, computer, and pet food companies), and in academia (Veterinary Schools at the Universities of London and Nottingham). He has postgraduate qualifications in Veterinary Radiology, Small Animal Orthopedics, and Clinical Nutrition. He is a Royal College of Veterinary Surgeon's Specialist in Veterinary Nutrition (Small Animal Clinical Nutrition) speaks internationally and has published widely in the areas of geriatrics and clinical nutrition, including most recent papers on minerals in UK pet foods and the mineral status of dogs with the developmental skeletal disease.

Preface

Written for medical and veterinary clinicians, researchers, academics, public health officials, farmers, and scientists this unique book is a quick-access reference to clinical signs associated with deficiencies, toxicities, imbalances, or environmental exposures to minerals, trace elements, and rare earth elements. It includes data for humans, other primates, companion animals, horses, rabbits, reptiles, rodents, ruminants, poultry, fish, and selected Zoo species. The book aims to increase awareness to improve diagnosis and to encourage further investigation based on comparative data. The subject material is divided into three sections so the information can be accessed by clinical signs, specific elements, or by species.

Introduction

Historical perspective

The important role of minerals in the maintenance of health has been known for centuries with evidence that domesticated animals were supplemented with salt (NaCl) as early as 40 BC. Between the 1930s and 1950s, there was an explosion in the generation of scientific information about minerals but, perhaps surprisingly, our knowledge about some important trace elements (e.g., As, B, Pb, Li, Ni, Si, Sn, and V) only began in the 1970s. A review is available in McDowell (1992a).

At the time of writing this book, mineral-associated disease is still common in humans and animals. According to the World Health Organization (WHO) more than two billion people suffer from micronutrient deficiency globally. Absolute figures are not available for other species but mineral-associated disorders are common in grazing domesticated animals, mineral-related developmental skeletal diseases such as metabolic bone disease are common in pets, and mineral-related diseases are frequently diagnosed in captive animals in zoological collections.

Role of minerals in the body

Minerals have several important roles in the body (Underwood and Suttle, 1999) including:

- Structural components of organs
- Cofactors in enzyme systems (metalloenzymes)
- Hormone synthesis
- Maintenance of osmotic pressure
- Maintenance of acid–base balance
- Membrane permeability
- Tissue irritability
- Cell replication and differentiation

Nutrients that have one or more specific roles in the body are called type 1 nutrients and include the minerals copper, iodine, and iron. Type 2 nutrients are required for numerous general metabolic processes and include the minerals magnesium and zinc.

The presence of an essential nutrient in a food does not mean it is entirely bioavailable to the animal. For example, in simple stomached species if a mineral is present as phytate or oxalate it may not be accessible (as with phosphorus and calcium, respectively) however ruminants have a microbiome that produces enzymes (e.g., phytase) that can breakdown these compounds and make the minerals available for absorption. (Underwood and Suttle, 1999).

Despite a lot of research our understanding of the biological roles of both macrominerals—required in greater than 100 ppm (ppm) and microminerals (trace elements) – required in less than 100 ppm is still incomplete.

There are 39 minerals of importance to human and animal health:

Aluminium (Al), Antimony (Sb), Arsenic (As), Barium (Ba), Bismuth (Bi), Boron (B), Bromide (Br), Cadmium (Cd), Calcium (Ca), Chlorine (Cl), Chromium (Cr), Cobalt (Co), Copper (Cu), Fluorine (F), Germanium (Ge), Iodine (I), Iron (Fe), Lead (Pb), Lithium (Li), Magnesium (Mg), Manganese (Mn), Mercury (Hg), Molybdenum (Mo), Nickel (Ni), Phosphorus (P), Potassium (K), Rubidium (Rb), Selenium (Se), Silicon (Si), Silver (Ag), Sodium (Na), Strontium (Sr), Sulfur (S), Tin (Sn), Titanium (Ti), Tungsten (W), Uranium (U), Vanadium (V), Zinc (Zn).

About 17 of these are regarded by the National Research Council (2005) as being *essential nutrients* for humans (Ca, Cl, Cr, Co, Cu, F, I, Fe, Mg, Mn, Mo, P, K, Se, Na, S, Zn), Vanadium (V) is *probably* a required nutrient, five are *possibly* essential (As, B, Ni, Ru, Si) and 16 are considered to be *nonessential* (Al, Sb, Ba, Bi, Br, Cd, Ge, Pb, Li, Hg, Ag, Sr, Sn, Ti, W, U). However, there are differences between animal species, so nonessential nutrients for one may be essential for another. The status of nonessential elements is kept under constant review and the classification of some of these may change in the light of new scientific evidence. However, agreement on essentiality can be controversial so in the USA Cr is considered essential for humans, in the EU it is not!

There are 17 so-called rare earth elements (REE): Cerium (Ce), Dysprosium (Dy), Erbium (Er), Europium (Eu), Gadolinium (Gd), Holmium (Ho), Lanthanum (La), Lutetium (Lu), Neodymium (Nd), Praseodymium (Pr), Promethium (Pm), Samarium (Sm), Scandium (Sc), Terbium (Tb), Thulium (Tm), Ytterbium (Yb), and Yttrium (Y). They are often found in minerals with Thorium (Th), and less commonly Uranium (U).

Rare earth elements are broadly grouped into "light" lanthanides (La, Ce, Pr, Nd, Sm, Eu, and Gd) and "heavy" (Y, Tb, Dy, Ho, Er, Tm, Yb, and Lu) classes (Wells and Wells, 2001).

The National Research Council (2005) considers that some REE may be essential nutrients for some animals. Most clinical reports of health problems due to exposure to RERs come from occupational exposure in humans and the health risk to animals from exposure to even large amounts of REE is considered low. However, some REEs have been linked to cancer, respiratory problems, tooth loss, and death. (Massari and Ruberti, 2013). REEs are now in widespread use, and the associated health risks to people and animals have been reviewed (Rim et al., 2013) as well as the impact of the global RE industry (Koltun and Tharumarajah, 2014). There is growing concern about the health risks identified due to rare earth metal accumulation in marine life in two recent systematic reviews (Blinova et al., 2020; Malhotra et al., 2020).

Some other mineral-associated elements and their compounds such as metals that may cause health problems are also included in this book: Beryllium (Be), Cesium (Cs); Gallium (Ga), Gold (Au), Indium (In), Niobium (Nb), Palladium (Pd), Tellurium (Te), Thallium (Tl), and Zirconium (Zr).

I have not included a treatise on the health risks and clinical signs associated with exposure to the naturally occurring radioactive isotopes of these elements where they exist—but these are summarized in Appendix 3.

Human foods

For many years the WHO has had dedicated advisory committees focused on the Global health risks of malnutrition, including deficiencies and toxicities relating to minerals and trace elements. In the last decade, there has been a great improvement in the labeling of human foods and many governments have issued advice to citizens about the risks of excessive mineral intake—notably salt (NaCl). However, it is still not easy to find out exactly what minerals and trace elements are in a food, and the online community is full of disinformation about what is "good" or "bad" for health. As humans eat such a wide range of foods and extracting a detailed and accurate dietary history is so difficult, it can be very difficult to determine whether an individual's food intake is adequate or not.

Animal foods

In the USA and EU all animal foods must be labeled as being "complete" or "complementary". If labeled "complete" they must contain all the essential nutrients required by the target species and carry an Analysis that specifies the ingredients list and chemical analysis. In Europe, they carry an "as fed" or "typical analysis" which lists the percentage composition of each group of nutrients, for example, 10% protein. In the USA, the ingredients are listed as a guaranteed analysis which specifies a minimum or maximum levels for each type of nutrient, for example, minimum of 10% protein. None of the animal food labeling regulations require the content for individual minerals or trace elements to be listed however total mineral content is expressed as "ash" or "inorganic material."

Legislation governing animal foodstuffs also includes regulations to protect human and animal health and for some nutrients and nonessential minerals or heavy metals, there are legal maximum limits set.

Animal food can become contaminated with minerals or heavy metals in several ways:

1. Minerals and metals accumulate in plant and animal matter so toxic levels can be in food ingredients
2. Contamination can occur during manufacturing, home preparation, storage (from storage containers), accidental

Animals can acquire toxic levels of minerals or metals by ingestion (drinking, licking, or eating), contaminated environmental sources including artifacts like batteries, coins, cage metals, painted wood, linoleum, metal toys. In cats and dogs, the most common toxicities are due to arsenic, cadmium, lead, and zinc.

The presence of low levels of potentially toxic metals in commercial pet foods is common and some of these elements will accumulate in the body over time. To confirm a cause and effect relationship in an animal showing clinical signs there must be confirmation of toxic tissue levels (e.g., blood, kidney, liver) from the animal as well as a presence in the food. Even then the high levels of an element may not be responsible for ill health. For example, domesticated cats fed fish-based foods accumulate arsenic in their kidneys over time, however, this does not necessarily cause harm (Alborough et al., 2020).

Deficiencies

Inadequate intake of an essential nutrient can lead to deficiency disease. This may be caused by inadequate amounts of the nutrient in the food or interference with bioavailability—digestion, absorption, transportation, or metabolism.

To me, it is unbelievable that we are in the 21st Century and deficiencies of Ca, Co, I, Fe, Mg, and Se are all still considered to be common in humans. These problems are not limited to the developing World as a systematic review found that Se status of humans was poor across the European Union and Middle East (Stofaneller and Morse, 2015). Nutritional secondary hyperparathyroidism (aka. metabolic bone disease) resulting in bone under-mineralization with consequential skeletal problems including deformity and spontaneous fractures is common in dogs, cats (domestic and captive big cats), and reptiles fed an all meat ration, which lacks several essential nutrients including calcium and has a large inverse Ca: P ratio.

With improvements in education and the will of the medical and veterinary professions with support from governments and global stakeholders such as the WHO and Charities, it should be possible to eliminate mineral and trace element-associated deficiencies due to poor dietary intake—and this should be a priority.

Toxicities

Exposure to high amounts of an element through dietary intake or environmental exposure can cause toxic effects that are harmful to health, and for many elements, sufficiently high intake can result in death.

Acute toxicities due to high dose exposures are relatively easy to recognize, as are dermal, hair, ocular, or respiratory tract signs associated with exposure to caustic or irritant substances by direct skin contact or inhalation of mists or fumes. Clinical signs associated with chronic exposure to low doses, during which time the element may accumulate in the body, are often insidious in onset and much more difficult to identify and confirm.

Toxicity can result from intake of large amounts of an element in food, water, or air, but also accumulation in tissue due to impaired excretion via the liver (in bile) or kidneys (in urine) and other mechanisms (sweating and expiration). Many of the elements included in this book are used in some form in medicine and iatrogenic toxicity needs to be considered. For example, dental implants may contain alloys based on metals (gold, platinum, palladium, or silver), nickel, or copper all of which can be toxic.

The health risk to humans and animals can be graded as *low risk* for most (36) of these elements (Al, Sb, Ba, Bi, Ce, Cr, Co, Dy, Er, Eu, Gd, Ge, Ho, I, La, Li, Lu, Mg, Mn, Nd, Ni, Pr, Pm, Ru, Sm, Sc, Ag, Sr, Tb, Sn, Tm, Ti, W, U, Yb, Y). A *moderate* (*medium*) *risk* to health results from exposure to eight of these elements (As, B, Br, Ca, Fe, P, K, Zn), and 11 present a *high risk* to health (Cd, Cu, F, Pb, Hg, Mo, Se, NaCl, S, V).

Whether an individual develops toxic signs depends upon the form that the element is in (elemental or as a compound), the dose of exposure, duration of exposure, route of exposure, and genetic predisposition.

Carcinogenicity, embryotoxicity, mutagenicity, and teratogenicity

Elements included in this book may have by themselves (e.g., heavy metals) or through their naturally occurring radioactive isotopes, a carcinogenic, embryotoxic, mutagenic, or teratogenic effect because they cause damage to cells.

Carcinogenicity: Can induce cancer in tissues. There are many papers in which carcinogenic changes can be induced in tissues by giving extremely high doses of a substance but these effects may not be likely to occur in the real world. Indeed, the over the interpretation of studies conducted in rodents using high doses has been questioned (Ames and Gold, 1990). It is important that all the available data has been scrutinized properly by independent scientific experts such as the International Agency for Research on Cancer (IARC) and the US National Toxicology Program (NTP) so that conclusions about carcinogenicity are valid.

Mutagenicity: Causes permanent, transmissible changes in the genetic material (chromosomes and genes) of cells or organisms (amount and structure).

Embryotoxicity: Crosses the placenta and damages a developing embryo in utero.

Teratogens: produce physical or functional changes in developing embryos or fetuses. Severity depends on chemical form, the dose of exposure, the stage of development that the embryo/ fetus is in at the time of exposure. Ionizing radiation such as that emitted by natural radioactive substances can damage the embryo due to cell death or chromosome injury.

Species differences

There are thought to be over nine million different species of animal split into vertebrates and invertebrates. It is beyond the scope of this book to attempt to cover all species, and in any event, there is very little known about the specific mineral, trace element, and rare earth effects for most species. So, I have included information about humans, domesticated species, laboratory species, and selected wild species.

Humans usually self-select their food intake and so accurate and detailed records should be available, except for people with impaired recall such as those with dementia.

Domesticated and captive animals are usually fed a controlled ration by the owner/farmer/keeper and, unless the animal has access to additional food—such as domesticated cats that hunt prey, or dogs that are regular scavengers—it should be possible to determine whether dietary intake is adequate and safe.

Whilst there are many basic similarities between species, through evolution, they have adapted to suit their niche environment food supply and can have very different gastrointestinal tracts, metabolic and physiological processes.

Exotic/Zoo species

Very little is known about most of the dietary requirements for essential minerals and trace elements or the effects of rare earth exposure in Zoo species. However, for some, there are published National Research Council Guidelines, and there are excellent Care Manuals, which include nutritional advice, published by the Association of Zoos & Aquariums (ASA), freely accessible online at https://www.aza.org/animal-care-manuals.

In the wild many species can self-regulate mineral intake to meet their needs, so for example in Gorillas (Rothman et al., 2008) juveniles select different food materials and so consumed more minerals (Ca, P, Mg, K, Fe, Zn, Mn, Mo) per kilogram of body mass than adult females and silverback males.

The highly specific (extreme) dietary intake of Koalas (eucalyptus) and Giant pandas (bamboo) are well known, and these species have evolved physiological and metabolic strategies to be able to digest and utilize their staple ration to survive on these ingredients and, in the case of koalas, to avoid the toxic effects.

Even within species there can be major differences, for example, some captive Tapirs are prone to develop iron overload whereas others are not, and in the wild they do not get this condition.

Historically, keepers of exotic species have assumed that their nutritional requirements are similar to species that share similar gastrointestinal anatomy, but care is needed when extrapolating data from different species because they have different physiological and biochemical profiles. So, a mineral may be an essential nutrient for one species but not for others, as an example several nutrients are considered essential for Goats (As, Br, Cd, Li, Pb, Ni, Si, Sn, V) which are not currently considered essential in other small ruminants.

For any individual the dietary requirement for an essential nutrient will vary depending on:

1. Chemical form. For example, copper as the oxide is not bioavailable to cats.
2. Impaired digestion if there is a lack of digestive enzymes or bile in simple stomached species due to pancreatic insufficiency or liver disease, or abnormal microbiome in herbivores such as ruminants and horses.
3. Impaired absorption in disease like inflammatory bowel disease or following surgical removal of sections of the gastrointestinal tract.
4. Impaired transport in the body due to lack of carrier substances, e.g., ferritin in iron deficiency, albumin in protein deficiency.
5. Many factors can affect normal metabolism impairing utilization of nutrients.
6. Impaired excretion in reduced renal or hepatic function.
7. Within the body physiological status can affect nutrient utilization including: age, breed, ethnicity, epigenetics, sexual status, and species.
8. Extraneous factors that affect nutrient bioavailability include: nutrient-nutrient interferences (e.g., fiber, phytate, minerals) and drug-nutrient interactions (Table 1).

Mineral and trace element requirements may vary from normal in the presence of disease (NRC, 2018).

Lack of awareness in the professions

In my experience, clinicians are often unaware of the many minerals, trace element, and rare earth element-related diseases and clinical signs that exist, as listed in this publication. Another problem is that clinicians may not have time to fully assess

TABLE 1 Some drug-nutrient interactions.

Drug interference	Element affected
Antihypertensive drugs	Folate
Nonsteroidal antiinflammatory drugs (NSAIDs)	Iron
Hypoglycaemic drugs	Calcium, cobalamin
Antacids (proton-pump drugs)	Calcium, cobalamin, magnesium, iron, zinc

After NRC. Examining special nutritional requirements in disease states: proceedings of a workshop (2018). National Academies Press; 2018. http://nap.edu/25164.

the nutritional status or they may shy away from discussing nutrition-related topics with patients/clients. I know that both of these scenarios are common in the veterinary profession resulting in missed and underdiagnosis of mineral-related disorders and poor owner education.

Primary care medical and small animal veterinary clinicians rarely obtain a detailed dietary history for their patients. As a result, nutritional errors may not be included in their differential diagnosis list and even referral clinicians often fail to consider diet when investigating refractory or complex clinical cases. At an International Veterinary Conference, I attended in 2019 an eminent Professor was speaking on the paraneoplastic syndrome. It was an excellent presentation until he showed his slides of the causes of persistent hypercalcemia. There was no mention of hypervitaminosis D—which can be fatal due to soft tissue calcification and at the time of the presentation Ca toxicity due to hypervitaminosis D was a problem in dogs because of accidental over-supplementation of some pet foods distributed worldwide.

Human food intake is dictated by many factors including cultural and family tradition, religious beliefs, social peer pressure, perceived taboos, emotions, learned likes and dislikes (texture, flavor, animal/plant origin), food availability in shops, and cost. The nutritional status and diet-related health status of low-income families are known to be worse than wealthy members of society.

There is huge interest in food, massive coverage in the media-sadly with often conflicting claims about what is good and bad nutrition, yet rarely do individuals tailor their food intake based on scientific guidelines, indeed there has been growing uptake of extreme diets including most recently the arrival of vegan restaurants and products in supermarkets. Individuals in developed parts of the World eat such a wide, varied ration that it can be very difficult to identify nutritional problems unless they are eating an extreme form such as veganism.

There has been a huge increase in owners feeding raw meat foods to dogs and cats even though there is no scientific evidence to show that they need to be fed raw food and the health risks to the pets, their owners, and in-contact animals humans are well documented. At the other end of the spectrum, there are a growing number of vegetarian and vegan foods for pets.

In production animals, local soil and surface water composition will affect crop analysis and animal consumption rates. Without supplementation or strict restriction of intake, (as appropriate) deficiency or toxicities can and do easily occur.

In humans, most reports of nutrition-related diseases also originate from countries where basic food ingredients have an extreme content, for example, high or low mineral content in the soil, which affects mineral content in crops producing disease in livestock and humans consuming the food generated from the land. In the 21st century, we now live in a global market and our meats and cereals bought locally could have originated from anywhere in the World and they do contain variable nutritional content.

Lack of awareness about the role of minerals, trace elements, and rare earth elements in disease undoubtedly means that there are gross under-recognition and under-reporting of clinical cases. Many common diseases are likely to be associated with inappropriate nutrition for example in all cases of developmental skeletal disease in which there is impaired endochondral ossification it is highly likely that mineral imbalance is involved. At the University of Nottingham, we found that a very large number of pet foods in the UK declared to be "complete" were actually deficient in minerals including copper, and some of those also had excessively high amounts of other minerals, e.g., calcium that would compete with any copper present—making the deficiency-associated disease a high probability (Davies et al., 2018).

Behavioral problems such as learning difficulties and poor cognitive performance in both children and adults may have a nutritional (mineral) basis which should be evaluated fully and ruled out before neurological system-modifying drugs, with potential adverse effects, are prescribed long term.

Maintaining adequate mineral and trace element intake is an essential step in the successful management of patients with severely reduced gastrointestinal function either because of disease or surgical removal (NRC, 2018) as illustrated by one study of children with intestinal failure on parenteral nutrition—97% had anemia, 20% had iron deficiency anemia, 56% Cu deficiency, 40% Fe, 35% Se, and 31% Zn deficiency (Yang et al., 2011).

Limitations in standards of scientific evidence

The accepted approach to determining the validity of scientific data is summarized by evidence pyramids such as this (courtesy of Dale Hattis) (Fig. 1).

Ideally, all proposed cause-and-effect relationships between intake or exposure and clinical signs would be based on systematic reviews and meta-analysis of multiple, high-quality randomized, controlled studies with data that can be directly compared. Unfortunately, this is not the case as most studies in this field are case–control, cross-sectional, or case studies, often involve low numbers of individuals, lack statistical power, and use different methodologies and outcome measures.

The situation is further complicated by the confounding, sometimes complex interrelationships between different minerals and other dietary components, concurrent exposures, and species differences in gastrointestinal functionality and metabolism. This all means that for most mineral-disease associations the scientific evidence is only weak-moderate, and also direct transfer of information across species is not always valid.

Lack of high-quality randomized controlled studies is a big issue when conducting a review of nutrition-related publications and further studies are desperately needed to determine the true nutrient requirements for many species. In human nutrition, the GRADE framework has now been adopted by many organizations including the WHO to assess nutritional studies. In the UK the Scientific advisory committee on nutrition (SCAN) advises the government. Some basic information is still needed, such as the need to establish a "no observed adverse effect level" (NOAEL) for macrominerals such phosphorus in domestic species (e.g., cats).

Further scientific research is essential to determine the true significance of reported mineral-related associations and effects in body systems, but at the same time, there is, rightly, a societal desire not to employ live animals in scientific research. So, progress in this field will rely increasingly on clinical case reports and epidemiological studies.

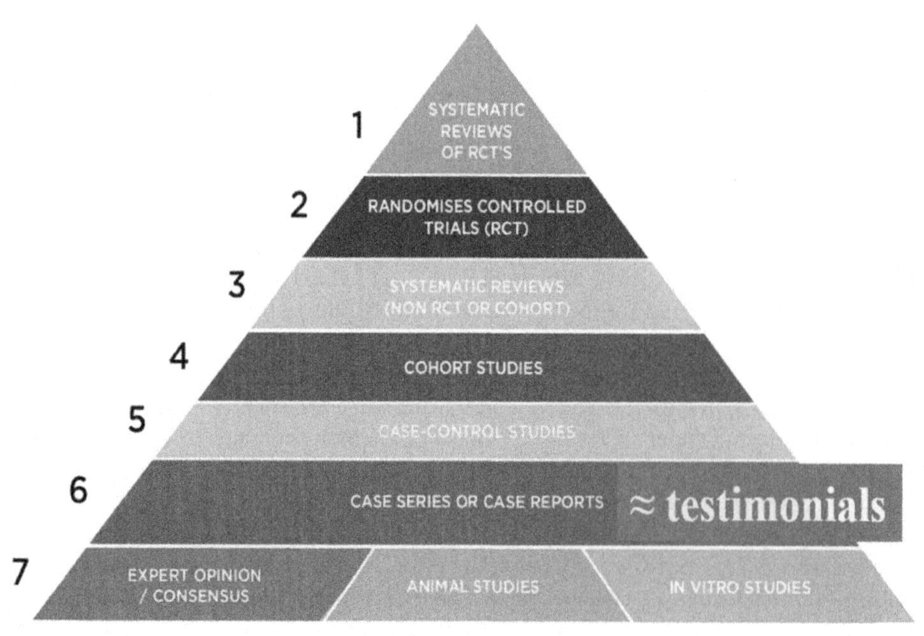

FIG. 1 Evidence pyramid.

Objectives of this book

This book aims to provide medical and veterinary clinicians, researchers, academics, and other stakeholders such as Public Health scientists and farmers with a comprehensive but easy to access reference tool to identify when minerals, trace elements, or rare-earth elements (or their compounds) may be involved in the development of clinical signs due to deficiency of essential nutrients or toxic effects of excessive intake or exposure.

There is an old saying "If you don't know about something you won't look for it, and if you don't look for something you won't find it" (Anon). One of the main aims of this book is to act as a sentinel to flag up known or suspected mineral-disease relationships to increase awareness in the clinical, public health, and scientific research communities to stimulate investigation and reporting of cases and to encourage more sharing of comparative data across the medical, veterinary, and research communities.

In this book, I have chosen not to list recommended daily intakes for every species because of the wide variation in requirement depending on physiological status (activity, genetic variation, growing, lactation, pregnancy, and advancing age), as well as confounding factors such as other dietary components, the effect of drugs or the presence of clinical or subclinical disease. In addition, recommended daily intakes are often controversial, vary in different geographical regions, and are constantly under review. Readers wanting to know the specific requirements for a nutrient in a species should consult with their local official agency.

Limitations of this book are that references are only included if they were available in electronic search engines (CABI, Google Scholar, and PubMed) in English so foreign language studies are not reported; descriptive terminology used by some authors may differ from those used by others and some authors may have reported only a limited number of outcome measures and clinical signs rather than reporting all possible signs. For example, some toxicities may be associate with renal failure, but not all authors report that azotemia was measured.

I look forward to hearing from readers about how this book may help or be modified further to help achieve its objectives.

Mike Davies
Oxfordshire
1st October 2020

Key features

- The first book that compiles the clinical signs associated with deficiencies, toxicities, imbalances, or exposures to minerals, trace elements, and rare earth elements.
- Includes data for humans and a wide range of species, providing valuable comparative information.
- Provides concise accessible summary tables of clinical signs and the species in which they are seen; of the individual elements and what species have been reported to show deficiencies or toxicities and, for each species deficiency or toxicity-related signs that they show for each element.
- The reader is provided with relevant references and other useful information.

Section 1

Clinical signs

In this section, clinical signs are listed in alphabetical order and known associations with elements or their compounds are tabulated. Species are listed alphabetically after humans. Further details can be found in Section 2 in which the minerals, trace elements, and rare earth elements are listed alphabetically, and Section 3 which lists species.
Associations between clinical signs and minerals, trace elements, or rare-earth elements are taken from the published literature available on electronic search engines (CABI, Google Scholar, and PubMed), published bibliographies in textbooks, and scientific papers. The absence of a species from a list of clinical signs does not mean that the species does not develop those signs if the individual is deficient or exposed to toxic levels of the element—the association may simply not have been observed, not reported, or was not found in the scientific literature.

Key to tables

Clinical signs: For the purposes of this section, the term "clinical signs" includes variations from normal detected from clinical history, physical examination, diagnostic tests (for example imaging or laboratory tests), or postmortem examination.
Comments: A column for comments is included in tables when points of clarification or explanation are required.
DD: Some differential diagnoses that could be considered.
Element: The element or one of its compounds.
?: Suspected but not confirmed.

8-Hydroxy-2′-deoxyguanosine (8-OHdG)—High levels

A free radical biomarker for oxidative stress and carcinogenesis.

Element	Species	Comments
Indium toxicity *as indium tin oxide (ITO)*	Humans	In leukocytes and urine

DD: *In diseases*: Alzheimer's disease, atopic dermatitis, cancer, chronic hepatitis, cystic fibrosis, diabetic nephropathy, diabetic retinopathy, Huntingdon's disease, hyperglycemia, inflammatory bowel disease, metabolic syndrome, oxidative stress, pancreatitis, Parkinson's disease, and rheumatoid arthritis; *Following exposure to* asbestos, benzene, chromium fumes, cigarette smoke, diesel exhausts, manganese fumes, toxic metals, styrene, toluene, radon, vanadium fumes, and zylene. (Reference: www.doctorsdata.com Accessed 30th September 2020.)

Abdominal cramps

Nonspecific term for discomfort or pain anywhere in the abdomen. Also called "stomach ache."

Element	Species
Cadmium poisoning	Humans
Gadolinium toxicity	Rats
Magnesium toxicity	Humans
Mercury toxicity	Humans

Clinical Signs in Humans and Animals Associated with Minerals, Trace Elements, and Rare Earth Elements. https://doi.org/10.1016/B978-0-323-89976-5.00004-9

Phosphorus toxicity	Humans
Samarium toxicity	Mice
Vanadium toxicity	Humans

Abdominal distension

Swelling of the abdomen.

Element	Species
Selenium toxicity	Birds

DD: cysts, organomegaly; fluid accumulation (ascites, blood, and chyle), tumors.

Abdominal pain (colic)

Painful discomfort in the abdomen. Also called colic or "stomach ache."

Element	Species
Arsenic toxicity (acute)	Dogs, various
Barium toxicity	Humans
Bismuth toxicity	Humans
Bromine toxicity	Humans
Calcium toxicity	Humans, reptiles
Copper toxicity	Various
Europium toxicity	Humans
Iron toxicity	Humans, cattle
Lanthanum toxicity	Humans
Lead toxicity	Humans (children and adults), dogs
Lithium toxicity	Humans
Mercury toxicity	Humans, dogs, horses
Nickel toxicity	Humans
Phosphorus toxicity	Cattle (mild)
Selenium toxicity	Humans
Sodium toxicity	Humans, dogs
Sulfur toxicity	Humans
Thallium toxicity	Humans
Zinc toxicity	Humans, dogs
Zirconium toxicity	Humans

DD: abdominal wall muscle strain, aneurysm (abdominal aorta), appendicitis, cancer, cholecystitis, cysts, diverticulosis, ectopic pregnancy, endometriosis, food allergies, food poisoning, gallstones, gas, gastrointestinal disease (Crohn's disease, irritable bowel disease (IBD), obstruction, ulcerative colitis), gastrointestinal ischemia, gastroesophageal reflux (heartburn), hernias, nephroliths, pancreatitis, urinary tract infection.

Abnormal/unusual thoughts

Element	Species
Mercury toxicity	Humans

DD: mental illnesses.

Abortion

Spontaneous expulsion of a nonviable fetus before completion of pregnancy.

Element	Species	Comments
Aluminum deficiency	Goats	Anke et al. (2005a)
Arsenic deficiency	Goats	
Arsenic toxicity	Humans	Water supply contamination
Bromine deficiency	Goats	
Cadmium deficiency	Goats	
Cadmium toxicity	Various	
Copper deficiency	Cats, sheep	
Iodine deficiency	Humans, cattle, sheep	
Iron deficiency	Humans, camels	With low Zn
Lead toxicity	Sheep	
Lithium deficiency	Goats	
Manganese deficiency	Cattle, goats	
Mercury toxicity	Humans, monkeys	
Molybdenum deficiency	Goats	
Phosphorus deficiency	Cattle	
Rubidium deficiency	Goats	*Not confirmed as an essential nutrient*
Selenium deficiency	Cattle	
Selenium toxicity	Cattle	
Vanadium deficiency	Goats	
Zinc deficiency	Humans, guinea pigs, nonhuman primates	*Humans:* with low Fe

DD: hormonal abnormalities, infections, ionizing radiation exposure, toxic agents, uterine abnormalities.

Abasia

Lack of motor coordination in walking.

Element	Species
Bismuth toxicity	Humans

DD: midbrain, thalamic, lenticular, and frontal lobe lesions or psychogenic causes. Neurological diseases (cerebellar damage, Guillain-Barré syndrome, hydrocephalus (normal pressure), Parkinson's disease, stroke).

Acetonemia

A metabolic state characterized by the presence of ketone bodies in the blood. Also called ketosis.

Element	Species
Fluorine toxicity	Cattle (adult)

DD: lactation in cattle.

Achlorhydria

Lack of hydrochloric acid in gastric secretions.

Element	Species
Zirconium toxicity	Humans

DD: autoimmune disorders, hypothyroidism, *H. pylori* infection, medications (antacids, hydrogen receptor inhibitors, proton pump inhibitors), surgery (gastric bypass).

Achromotrichia

Absence of pigment in hair, fur, or feathers.

Element	Species	Comments
Copper deficiency	Humans, buffaloes (*Vitiligo*), cats, cattle, dogs, guinea pigs, poultry, rabbits, rats, sheep	Vitiligo: lack of pigment in skin
Iron deficiency	Foxes, mink, poultry (rare), rats	
Molybdenum toxicity	Cattle, sheep	Due Cu deficiency
Zinc deficiency	Rabbits	Dark to gray
Zinc toxicity	Nonhuman primates	Due Cu deficiency

DD: genetic factors, other nutritional factors: amino acid deficiency (phenylalanine to tyrosine, vitamin (biotin, cobalamin, vitamin C, and vitamin D), pituitary disease, thyroid disease).

Acne

Inflammation of the skin glands and hair follicles.

Element	Species
Lithium toxicity	Humans
Thallium toxicity	Humans
Zinc deficiency	Humans

DD: high testosterone levels, medications (e.g., corticosteroids and lithium).

Acrodynia

An allergic reaction in children causing: pink discoloration of hands and feet; local swelling; intense pruritus (itching); insomnia; irritability; photophobia. Also called: erythredema, pink disease, Swift's disease.

Element	Species	Comments
Mercury toxicity	Humans	Pain and pink discoloration and swelling fingers and toes

Acroparesthesia

Sensation of burning, tingling, pricking, or numbness in the extremities.

Element	Species
Selenium toxicity	Humans

DD: carpal tunnel syndrome, Fabry disease (inherited disorder).

Acute airway obstruction

Element	Species
Sulfur toxicity	Humans

DD: allergic reaction, asthma, bronchoconstriction, foreign object obstruction, infections.

Acute kidney injury (AKI)/acute renal failure

Sudden damage to the kidneys resulting in loss of function.

Element	Species
Aluminum toxicity	Dogs
Antimony toxicity	Humans
Barium toxicity	Humans
Bismuth toxicity	Humans
Mercury toxicity	Humans, cattle, horses, sheep
Selenium toxicity	Humans, sheep
Uranium toxicity	Humans
Zinc toxicity	Dogs

DD: cardiac failure, contrast media, dehydration, glomerulonephritis, hypovolemia, infections, liver failure, nephrotoxic drugs, sepsis, vasculitis.

Acute tubular necrosis

Death of tubular epithelial cells that form the renal tubules of the kidneys.

Element	Species
Zinc toxicity	Humans

DD: hypotension, nephrotoxic drugs, toxins.

Adrenal hypertrophy

Increased adrenal cell size without an increase in cell numbers.

Element	Species
Sodium deficiency	Poultry
Zinc deficiency	Mice, Pigs

DD: ACTH-secreting pituitary adenoma (Cushing's disease); ectopic ACTH from a nonpituitary tumor.

Adrenal insufficiency (Addison's disease; hypoadrenocorticism)

Inadequate production of steroid hormones by the adrenal glands.

Element	Species
Nickel toxicity	Humans

DD: adrenalectomy, adrenal gland damage (amyloidosis, cancer, hemorrhage), autoimmune disease, infections (AIDS, TB).

Agitation

A state of nervous excitement or anxiety.

Element	Species
Lithium toxicity	Humans

DD: alcoholism, anxiety, autism, bipolar syndrome, brain tumors (rare) depression, hormonal disorders (hypothyroidism), mood disorders, neurological disorders.

Aggressive behavior

Hostile or violent behavior.

Element	Species	Comments
Calcium deficiency	Squirrels	
Iodine toxicity	Dogs	In raw pet food
Mercury toxicity	Humans	Excessive anger

DD: social pressures, mental disease.

Alanine transaminase (ALT)—Elevation

An enzyme found in many organs and plasma but especially the liver. Used as a biomarker for liver disease.

Element	Species	Comments
Aluminum toxicity	Rats	
Cerium toxicity	Mice, rats	After IV injection
Gadolinium toxicity	Rats	

Gallium toxicity	Mice	
Iron toxicity	Sheep	
Lanthanum toxicity	Humans, rats	*Humans:* Dialysis patients
Molybdenum toxicity	Rats	
Selenium deficiency	Horses	

DD: alcoholic liver disease, alpha1-antitrypsin deficiency, drug-induced liver disease, hemochromatosis, hemolysis, hepatic lipidosis, drug-induced liver injury, hepatitis (autoimmune, infections (hepatitis B, hepatitis C, leptospirosis), muscle disease, thyroid disease, Wilson disease).

Albuminuria (proteinuria)

The presence of albumin (a small protein) in urine. Used as a marker for kidney damage. Also called proteinuria.

Element	Species
Cadmium toxicity	Humans, rats
Molybdenum toxicity	Rats

DD: diabetes mellitus, hypertension.

Aldolase—Reduced levels

An enzyme found mainly in liver and muscle that is involved in glucose metabolism.

Element	Species
Lithium deficiency	Goats

DD: causes of damage to muscle (myocardial infarction) or liver (cirrhosis, hepatitis).

Alkaline phosphatase—Activity reduced

Alkaline phosphatase is an enzyme found in several tissues including the liver, skeletal muscle, small intestine, and bone. Dogs have a corticosteroid-specific isoenzyme.

Element	Species
Zinc deficiency	Horses

DD: hypophosphatasia (rare genetic disease affecting bones and teeth), malnutrition.

Alkaline phosphatase—Increased blood concentrations

Alkaline phosphatase is an enzyme found in several tissues including liver, skeletal muscle, small intestine, and bone. Dogs have a corticosteroid-specific isoenzyme.

Element	Species	Comments
Aluminum toxicity	Rats	
Beryllium toxicity	Rats	Develop rickets
Boron deficiency	Poultry (chicks)	If chicks are Vit D deficient
Calcium deficiency	Nonhuman primates	*Lemurs*
Cerium toxicity	Rats	
Iron toxicity	Sheep	
Lanthanum toxicity	Humans, rats	*Humans:* dialysis patients
Manganese deficiency	Humans	
Molybdenum toxicity	Sheep, rats	*Sheep:* Blood ALP *Rats:* Small intestine and renal isoenzymes BUT liver ALP is lower
Phosphorus deficiency	Various: Nonhuman primates, fish	

DD: bone tumors, cancer, in dogs high corticosteroids, during skeletal growth, liver conditions (bile duct obstructions. gallstones), skeletal muscle disease.

Alkalosis—Metabolic

A metabolic condition in which the pH of tissue is above the normal range (7.35–7.45).

Element	Species
Chlorine deficiency	Humans, cattle
Potassium deficiency	Humans

DD: alcoholism, decreased blood carbon dioxide (diabetic ketoacidosis, kidney disease), dehydration, increased blood bicarbonate (anorexia, dehydration, diabetic ketoacidosis, chronic obstructive pulmonary disease, vomiting), hypokalemia (some antibiotic use, diuretic overdose, diarrhea, folic acid disease, renal disease and excessive losses in urine, vomiting), chronic kidney disease.

Allergic reaction

When the immune system overreacts to a harmless substance (called an allergen).

Element	Species	Comments
Gadolinium toxicity	Humans	After injection for MRI scan
Molybdenum toxicity	Guinea pigs	Contact sensitivity
Nickel toxicity	Humans	Allergic contact dermatitis (common)
Titanium toxicity	Humans	To dental implants

DD: other causes: drug reaction, other food ingredients, insect bites/stings, mold, pet dander, plant toxin, pollen.

Alopecia

Hair loss.

Element	Species	Comments
Copper deficiency	Rabbits	
Dysprosium toxicity	Guinea pigs	Epilation
Iodine deficiency	Cats, cattle, dogs, pigs	
Iodine toxicity	Horses (foals)	
Lutetium toxicity	Humans	Transient
Magnesium deficiency	Guinea pigs, rabbits	
Mercury toxicity	Sheep	Wool loss
Molybdenum toxicity	Rabbits	Due Cu deficiency
Selenium deficiency	Nonhuman primates	*Squirrel monkeys*
Selenium toxicity	Humans, pigs	
Sodium deficiency	Dogs, gerbils	
Tellurium toxicity	Rats	*Experimental*
Thallium toxicity	Humans	
Thyroid deficiency	Dogs	
Thyroid toxicity	Horses (foals)	
Zinc deficiency	Humans, cattle, goats, horses (foals), mice, nonhuman primates (squirrel monkey), rabbits, rats, sheep	
Zinc toxicity	Nonhuman primates	*Due galvanized cages*
Ionizing radiation	Various	Exposure

DD: allergic reaction, hormonal, parasites (demodex, ringworm, sarcoptic mange).

Alveolitis

Inflammation of the alveoli in the lung.

Element	Species	Comments
Zinc toxicity	Guinea pigs	Inhalation

DD: repeated exposure to animal or vegetable dusts, molds.

Alzheimer's disease

An irreversible, progressive brain disorder resulting in loss of memory and brain function.

Element	Species	Comments
Aluminum toxicity	? Humans	?Al content is higher in brains of people with Alzheimer's disease. Cause and effect relationship **not** confirmed

Amblyacousia

Impaired (dullness) of hearing.

Element	Species	Comments
Tin toxicity	Humans	Organotin toxicity

DD: anything that impairs hearing.

Aminoaciduria

Urine contains abnormally high amounts of amino acids. Normally 95% of amino acids that pass through the glomerulus are reabsorbed in the renal tubules.

Element	Species
Bismuth toxicity	Humans

DD: congenital amino acid metabolism disorders (e.g., phenylketonuria), liver diseases.

Aminolevulinic acid dehydratase (ALAD)

ALAD is a protein coding gene.

Element	Species	Comments
Lead toxicity	Birds	In blood, brain, liver
Palladium toxicity	Mice	

DD: alcoholism, dehydration, fasting, infections, some drugs (? oestrogens, progesterone), stress (physical and psychological).

Amyloidosis

Accumulation of abnormal protein (amyloid) in tissues.

Element	Species
Palladium toxicity	Mice

DD: chronic infectious or inflammatory disease (rheumatoid arthritis, Crohn's disease, or ulcerative colitis), dialysis-related, multiple myeloma (immunoglobulin light chain amyloidosis).

Amyotrophic lateral sclerosis

A disorder that affects neurones responsible for voluntary movements.

Element	Species
Aluminum toxicity	Humans

DD: inherited defect, other neurological diseases.

Anemia

A low number of red blood cells, or low hemoglobin concentration in the red cells.

Element	Species	Comments
Aluminum toxicity	Humans	Especially in patients with renal failure. Microcytic hypochromic anemia (not Fe deficiency) in patients with Al osteomalacia and encephalopathy
Arsenic toxicity	Humans	Due low Fe
Cadmium toxicity	Humans, rats	*Humans:* Due low Fe
Cobalt deficiency	Humans, cattle, goats, pigs, sheep	Humans: megaloblastic pernicious anemia Ruminants: normochromic, normocytic anemia
Cobalt toxicity	Cattle, goats, sheep	
Copper deficiency	Humans (infants), buffaloes, camels, cats, cattle, dogs, goats, mice, pigs, poultry, rabbits, rats, sheep	
Copper toxicosis	Various	
Fluorine toxicity	Cattle	
Gallium toxicity	Humans, mice	Microcytic
Iron deficiency	Humans, birds, cats, cattle (calves), dogs, elephants, fish, foxes, horses, marine mammals, mice, mink, nonhuman primates, pigs (common), poultry, rabbits, rats, reptiles, sheep (experimental)	Microcytic-hypochromic anemia
Lead deficiency	Rats	Microcytic, hypochromic
Lead toxicity	Humans (adults, children), birds, cattle, reptiles	Microcytic, hypochromic
Magnesium deficiency	Guinea pigs	
Molybdenum toxicity	Cattle, poultry (chicks), rabbits, rats	Due Cu deficiency
Nickel toxicity	Dogs	
Palladium toxicity	Rats	
Phosphorus deficiency	Cats	Hemolytic anemia
Selenium toxicity	Dogs	
Tin toxicity	Humans	
Uranium toxicity	Humans	
Zinc deficiency	Humans, ferrets	
Zinc toxicity	Humans, birds, dogs, reptiles	*Dogs:* hemolytic *Humans:* sideroblastic anemia, due to copper deficiency

DD: bone marrow disorders, vitamin deficiencies—cobalamin, folic acid, hemorrhage, inherited disorders (e.g., thalassemia or sickle cell disease), malabsorption.

Anesthesia

Temporary loss of sensation.

Element	Species	Comments
Selenium toxicity	Humans	peripheral

DD: exposure to anesthetic agents or analgesics.

Androgenic receptor protein expression reduced

Element	Species
Aluminum toxicity	Rats

Angiedema

Swelling of the tissue under the skin or mucous membranes.

Element	Species
Bismuth toxicity	Humans

DD: allergies (food-eggs, milk, nuts, shellfish), insect bites or stings (bees, wasps), latex, medicines (antibiotics, NSAIDs—aspirin, ibuprofen); medicine effects (nonallergy) [angiotensin-converting enzyme (ACE) inhibitors—enalapril, lisinopril, perindopril, ramipril; ibuprofen and other NSAIDs, angiotensin-2 receptor blockers—ARBs].

Angular deformity of long bones

Impaired endochondral ossification resulting in abnormal formation of long bones (bowed legs) often due to premature closure of growth plates (e.g., ulnar) with the continued growth of adjacent bone plate (e.g., radius)—common in dogs.

Element	Species	Comments
Calcium deficiency	Birds, camels, cats, dogs, foxes, goats, horses, nonhuman primates, reptiles, squirrels	*Waterfowl:* carpal rotation *Lemurs:* (Tomson and Lotshaw, 1978). *Squirrels:* limbs, spine, tail
Calcium toxicity	Dogs	Radius curvus
Copper deficiency	Humans, camels, cats, cattle, pigs, poultry (chicks)	*Cats:* Deformed carpi *Pigs:* Crooked forelimbs Cattle: Rickets *Poutry:* Deformed metatarsals
Iodine deficiency	Humans, dogs	
Manganese deficiency	Birds, dogs, goats, guinea pigs, pigs, rabbits, rats	
Molybdenum toxicity	Horses, rabbits, rats	*Horses:* rickets
Phosphorus deficiency	Birds, dogs, mink	
Zinc deficiency	Cattle	Bowing of hindlegs

DD: trauma to growth plate(s) resulting in premature closure.

Anhydremia

A deficiency of water in the blood.

Element	Species
Sodium toxicity	Cattle

DD: burns, diabetes insipidus, hemorrhage, edema.

Ankylosis

Immobility of a joint due to fusion of the bones.

Element	Species	Comments
Fluorine toxicity	Cattle	Joints and spine

DD: injury (trauma), postjoint infection, advanced osteoarthritis, surgical fusion.

Anorexia (decreased food intake; inappetence)

Absence or insufficient food intake to meet nutritional needs.
Self-imposed anorexia is common in people which induces severe calorie malnutrition, starvation resulting in anemia, bradycardia, cardiac arrhythmias, dehydration, fractures, heart failure, increased risk of infection, kidney failure, menstruation ceases, osteomalacia, osteoporosis, suicide risk, loss of thermoregulation, low white blood cell count (after https://www.bulimia.com/topics/medical-issues/).

Element	Species
Antimony toxicity	Humans
Arsenic deficiency	Goats
Arsenic toxicity	Fish
Beryllium toxicity	Humans
Boron toxicity	Humans, cattle
Bromine deficiency	Goats
Cesium toxicity	Humans
Cadmium toxicity	Fish, various:
Calcium deficiency	Birds, camels, fish, reptiles, squirrels
Calcium toxicity	Cats, cattle, pigs, poultry
Chlorine deficiency	Humans, cattle, rats
Cobalt deficiency	Humans, cattle, sheep
Cobalt toxicity	Cattle, sheep
Copper deficiency	Humans, buffaloes
Copper toxicity	Fish
Florine toxicity	Cattle, horses, pigs
Iodine deficiency	Cattle
Iodine toxicity	Cattle, sheep
Iron deficiency	Humans (children), cattle (calves), pigs, reptiles

Iron toxicity	Cattle, sheep
Lanthanum toxicity	Humans
Lead toxicity	Humans, birds, cats, fish, reptiles
Magnesium deficiency	Humans, cats, cattle, goats, dogs, fish
Magnesium toxicity	Goats
Manganese deficiency	Sheep
Manganese toxicity	Sheep
Mercury toxicity	Humans, cats, ferrets, marine mammals (seals), mink, pigs, sheep
Molybdenum toxicity	Cattle, rabbits
Nickel toxicity	Cattle, dogs, fish, mice, monkeys, pigs, poultry (*chicks, ducks*), rabbits, rats
Palladium toxicity	Rabbits, rats
Phosphorus deficiency	Humans, buffaloes, cattle, dogs, goats, horses, poultry, sheep
Potassium deficiency	Cats, cattle, fish, goats, horses, pigs, sheep, rats
Selenium deficiency	Dogs, horses
Selenium toxicity	Humans, cattle, deer, dogs, nonhuman primates (*Macaques*), pigs, rats
Sodium deficiency	Cats, cattle, goats, horses, marine mammals, pigs, poultry, rats, sheep
Sodium toxicity	Cattle
Sulfur deficiency	Cattle, goats, sheep
Sulfur toxicity	Cattle
Tellurium toxicity	Humans
Thallium toxicity	Humans
Tin toxicity	Humans (*Organotin compounds*)
Vanadium toxicity	Sheep
Zinc deficiency	Humans, cattle, fish, horses, nonhuman primates, pigs, poultry, rabbits, rats, fish
Zinc toxicity	Birds, dogs, fish, reptiles
Zirconium toxicity	Fish

DD: inability to eat due to facial or upper alimentary tract disorders (facial injuries, foreign bodies, linear foreign bodies around the base of tongue (string in cats), major organ system disease, mental disorders, pain, pyrexia, tainted food).

Anosmia

Lack of sense of smell.

Element	Species
Antimony toxicity	Humans
Cadmium toxicity	Humans
Lanthanum toxicity	Rats

DD: nasal congestion (allergy, influenza, sinusitis), poor air quality, nasal polyps, nasal surgery, trauma.

Antler abnormalities

Developmental abnormalities.

Element	Species	Comments
Calcium deficiency	Deer	Impaired growth, delayed velvet shedding
Phosphorus deficiency	Deer	Impaired growth, delayed velvet shedding

DD: trauma, other nutritional problems (inadequate protein/vitamin intake).

Antral stenosis

Narrowing of the antrum part of the stomach adjacent to the pyloric canal which leads to the duodenum—first section of small intestine.

Element	Species
Zirconium toxicity	Humans

DD: ingestion of caustic materials.

Anuria

Absence of urine production.

Element	Species	Comments
Bismuth toxicity	Humans	
Iron toxicity	Humans	Renal failure
Mercury toxicity	Humans, dogs	
Tungsten toxicity	Humans	Rare
Zirconium toxicity	Humans	

DD: causes of renal failure, diabetes hyperglycemia, hypertension, chronic glomerulonephritis, kidney infection, polycystic kidneys, toxin exposure, urinary tract obstruction.

Anxiety

Nervousness, unease, or worry about something.

Element	Species
Barium toxicity	Humans
Bismuth toxicity	Humans
Calcium toxicity	Humans
Lead toxicity	Humans (adults), dogs
Magnesium deficiency	Humans
Mercury toxicity	Humans, dogs
Phosphorus deficiency	Humans

DD: humans: panic disorder, phobias (agoraphobia, claustrophobia), posttraumatic stress disorder (PTSD), social anxiety disorder. Dogs: separation anxiety, loud noise phobias.

Aortic aneurysm/rupture

A bulge or swelling in the wall of the aorta due to weakness in its structure resulting in rupture.

Element	Species
Copper deficiency	Birds (?ratites), dogs, guinea pigs, pigs, poultry, sheep (lambs)

DD: cardiovascular disease, hypercholesterolemia, genetic factors, hypertension, chronic obstructive pulmonary disease (COPD), cigarette smoking.

Aortic calcification

Deposition of calcium in the aorta wall or in atherosclerotic plaques lining the aorta.

Element	Species
Magnesium deficiency	Cats, dogs, horses

DD: aortitis, arteritis, atherosclerosis, chronic kidney disease, syphilis.

Aortic wall thickening/plaques

Called atherosclerotic plaques are formed by a lipid-rich necrotic core and a fibrous cap that separates the core from the lumen.

Element	Species	Comments
Chromium deficiency	Mice, rabbits, rats	Plaques
Copper deficiency	Poultry	

DD: causes of hypercholesterolemia, trauma to aorta, hypertension.

Apathy

Lack of concern, enthusiasm, or interest.

Element	Species
Bismuth toxicity	Humans
Sodium deficiency	Humans
Zinc deficiency	Nonhuman primates

DD: neurological and psychiatric conditions: Alzheimer's disease, dementia (frontotemporal and vascular) depression, dysthymia, Huntington's disease, Parkinson's disease, progressive supranuclear palsy, schizophrenia, stroke.

Apnoea—Obstructive sleep apnoea

Cessation of breathing (temporary).

Element	Species
Antimony toxicity	Humans

DD: obesity, smoking, drinking alcohol, enlarged adenoids or tonsils, sleeping position.

Appetite increased

An increase in food intake.

Element	Species
Chromium deficiency	Humans, mice, rats

DD: anxiety, depression, diabetes, emotional disturbances, hyperthyroidism, Graves' disease (autoimmune causing hyperthyroidism), psychiatric conditions, stress.

Arched back

Element	Species
Calcium deficiency	Camels
Europium toxicity	Mice
Holmium toxicity	Mice
Lutetium toxicity	Mice
Potassium deficiency	Goats
Praseodymium toxicity	Mice

DD: malalignment of vertebrae, neuromuscular causes, pain.

Arginase activity increased

A manganese-containing enzyme and the final enzyme in the urea cycle responsible for converting arginine to urea. It is used as a marker for liver disease or inherited arginase deficiency.

Element	Species	Comments
Arsenic deficiency	Poultry (chicks)	In kidneys

DD: diabetes mellitus, liver disease (seen in cattle, dogs, goats, horses, and sheep).

Argyria

Blue-gray discoloration of skin, eyes, internal organs, nails, and gums due to silver accumulation.

Element	Species
Silver toxicity	Humans, mice, rats

DD: other causes of altered skin pigmentation.

Arrhythmia

Heart beats with an irregular or abnormal rhythm.

Element	Species	Comments
Antimony toxicity	Humans	Bradyarrhythmia, ventricular fibrillation
Arsenic toxicity	Various	Ventricular arrhythmias
Barium toxicity	Humans, dogs	
Calcium toxicity	Humans	
Magnesium deficiency	Humans	
Magnesium toxicity	Humans	
Potassium deficiency	Humans	
Thallium toxicity	Humans	

DD: cardiac disease (cardiac arrest), cardiac fibrosis, cardiomyopathy, coronary artery disease, diabetes, hypertension, hyperthyroidism, hypothyroidism, infections (including covid-19, sleep apnoea).

Arthralgia

Joint pain.

Element	Species
Antimony toxicity	Humans
Calcium deficiency	Camels
Molybdenum toxicity	Humans, cattle
Selenium toxicity	Humans

DD: cancer, chondromalacia patella, dislocations, fractures, gout (pseudogout), hemarthrosis, inflammatory joint disease, infective (septic) arthritis, intraarticular joint "mice" (fragments of cartilage/bone), necrosis (avascular), Osgood-Schlatter's disease, osteochondrosis, osteoarthritis, psoriatic arthritis, rheumatoid arthritis, trauma.

Arthropathy

Enlarged joints.

Element	Species	Comments
Cadmium toxicity	Various	
Calcium deficiency	Camels, dogs, various	Impaired endochondral ossification
Copper deficiency	Cattle, poultry	Thickened epiphyseal cartilage
Fluorine toxicity	Cattle, sheep	
Manganese deficiency	Goats, pigs, poultry (chickens and ducks)	Goats: Tarsus Pig: hocks Poultry: tibiometatarsal joint
Phosphorus deficiency	Various	Impaired endochondral ossification
Zinc deficiency	Poultry	Enlarged hock

DD: cancer, dislocations, fractures, gout (pseudogout), hemarthrosis, inflammatory joint disease, infective (septic) arthritis, intraarticular joint "mice" (fragments of cartilage/bone), necrosis (avascular), Osgood-Schlatter's disease, osteochondrosis, osteoarthritis, psoriatic arthritis, rheumatoid arthritis, trauma.

Ascites

Accumulation of fluid in the abdominal cavity causing distension.

Element	Species
Iron deficiency	Pigs
Iron toxicity	Birds
Selenium toxicity	Birds, dogs

DD: cancer (gastrointestinal tract, liver, spleen, ovary), heart failure, kidney failure, pancreatitis, peritonitis.

Ascorbic acid (vitamin C) levels—Low

A natural water-soluble vitamin, a potent reducing and antioxidant agent important for immune function, detoxification and formation of collagen in bones, capillaries, connective tissue, fibrous tissue, skin, and teeth.

Element	Species
Copper deficiency	Buffaloes

DD: alcoholism, anorexia, deficient diet, dialysis, liver failure (in species dependent on hepatic synthesis), severe mental illness, smoking.

Aspartate aminotransferase—AST—High

AST: enzyme found mostly in the heart and liver and some in skeletal muscle. Concentrations go up when these organs are damaged.

Element	Species	Comments
Aluminum toxicity	Rats	
Antimony toxicity	Rats	
Cerium toxicity	Rats	Following IV injection
Gadolinium toxicity	Rats	
Molybdenum toxicity	Rats	
Selenium deficiency	Horses	

DD: causes of cardiac, liver or muscle damage.

Astasia

Inability to stand upright unassisted.

Element	Species
Bismuth toxicity	Humans

DD: frontal, lenticular, midbrain, or thalamic lesions or psychogenic causes.

Asthma

Attacks of bronchospasm that cause difficulty breathing, usually associated with an allergic or hypersensitivity reaction.

Element	Species	Comments
Aluminum toxicity	Humans	
Chromium toxicity	Humans	Inhalation
Nickel toxicity	Humans	Inhalation

DD: other triggers including allergies [animal dander, dust mites, feathers, medications (NSAIDs), pollens] fumes, infections, pollutants, smoke, stress.

Ataxia

Neurological loss of voluntary coordination of muscle movements that can include gait abnormality.

Element	Species	Comments
Aluminum toxicity	Cats, dogs, rabbits	
Arsenic toxicity	Cattle	Acute toxicity
Bismuth toxicity	Humans	
Cerium toxicity	Rats	
Chlorine deficiency	Dogs	
Copper deficiency	Buffaloes, camels, cats, cattle, deer, goats, pigs, sheep	Sheep: swayback in lambs
Dysprosium toxicity	Mice	
Erbium toxicity	Mice	
Europium toxicity	Mice	
Holmium toxicity	Mice	
Iron toxicity	Pigs	
Lead toxicity	Humans (children), dogs	
Lithium toxicity	Humans, goats	
Lutetium toxicity	Mice	
Magnesium deficiency	Cats, cattle, dogs, guinea pigs, horses, pigs, poultry (rare)	
Magnesium toxicity		
Manganese deficiency	Humans, birds, mice, goats, guinea pigs, poultry (chicks), rats	Mice: neonatal ataxia
Manganese toxicity	Humans	
Mercury toxicity	Humans, cats, cattle, dogs, horses, mink, pigs	
Palladium toxicity	Rats	
Phosphorus deficiency	Cats	
Potassium deficiency	Cats, pigs	
Praseodymium toxicity	Mice	
Selenium toxicity	Humans, cattle	
Sodium deficiency	Cattle	

Strontium toxicity	Pigs	With low Ca intake
Thallium toxicity	Humans	Cerebellar ataxia
Vanadium toxicity	Cattle, sheep	
Zinc toxicity	Cattle, sheep	

DD: brain degeneration, cerebellar disease, cerebral palsy, medication, multiple sclerosis, stroke, tumors.

Atelectasis

Incomplete inflation or partial collapse of the lung.

Element	Species	Comments
Nickel toxicity	Hamsters	Inhalation

DD: pleural effusions, pneumonia, pneumothorax, pulmonary fibrosis, trauma, tumors.

Atherosclerosis

Fatty deposits in arterial walls called plaques.

Element	Species	Comments
Chromium deficiency	Mice, rabbits, rats	Aortic plaques
Copper deficiency	?Humans	Suggested not proved.
Silicon deficiency	?Humans	Suggested not proved. Si is not an essential nutrient for humans.

DD: factors involved are diabetes, hypercholesterolemia, hypertension.

ATPase activity—Reduced

Enzymes that catalyze the breakdown of the high energy compound adenosine triphosphate (ATP) into the diphosphate ADP releasing energy and a free phosphate ion or the inverse reaction.

Element	Species	Comments
Lanthanum toxicity	Poultry	Ca and Mg ATPases

DD: other causes of reduced ATPase activity.

Attention span loss

Inability to concentrate for a length of time.

Element	Species
Boron deficiency	Humans

DD: mental health disorders.

Autism (ASD)

Autism spectrum disorder (ASD) is a range of conditions causing inappropriate social skills, repetitive behaviors, speech, and nonverbal communications.

Element	Species	Comments
Aluminum toxicity	Humans	Association not necessarily a cause

DD: genetic inheritance, metabolic disturbances, virus infection.

Autoimmune disease—Increased prevalence

Abnormal immune response in which the immune system attacks the person/animals own body tissues.

Element	Species	Comments
Silicon toxicity	Humans	*Observed:* rheumatoid arthritis, systemic lupus erythematosus

DD: other causes of impaired immune function: malnutrition, infections, medications (e.g., corticosteroids).

Azotemia

An increase in blood urea nitrogen (BUN) and creatinine.

Element	Species
Gallium toxicity	Humans

Authors may not have mentioned azotemia specifically in their outcome measures, but azotemia will have been present in cases diagnosed with kidney failure—see also kidney failure and acute kidney injury (AKI).
DD: causes of renal disease, nonrenal causes: hypoperfusion of kidneys, adrenal insufficiency, congestive heart failure, hemorrhage, renal artery narrowing, shock.

Balance loss/vestibular syndrome

Feeling of giddiness and/or falling to the side of the vestibular apparatus in the inner ear affected often accompanied by nausea or vomiting.

Element	Species	Comments
Bismuth toxicity	Humans	
Gallium toxicity	Humans	Acute
Lead toxicity	Birds	
Manganese deficiency	Rats, sheep, guinea pigs	Otolith abnormality
Mercury toxicity	Humans, cats	

DD: aural infections, chemical imbalance in the brain, head trauma, hypertension, hypotension, the impaired vascular supply to the ear, medications.

Baritosis

Accumulation of barium in the lungs.

Element	Species
Barium toxicity	Humans

Basophil stippling

Basophilic granules consisting of aggregates of degenerating mitochondria, ribosomes; and siderosomes distributed throughout the cytoplasm of erythrocytes.

Element	Species	Comments
Arsenic toxicity	Humans	Acute
Lead toxicity	Humans, dogs	
Mercury toxicity	Humans	
Silver toxicity	Humans	
Zinc toxicity	Humans	

DD: anemia (congenital dyserythropoietic, hemolytic, megaloblastic, sickle cell, sideroblastic), CPD-choline phosphotransferase deficiency, erythroleukemia, gastrointestinal hemorrhage, impaired hemoglobin synthesis, leukemia, myelodysplastic syndrome, myelofibrosis, pyrimidine 5′-nucleotidase deficiency, thalassemia.

Beak abnormalities

Abnormal development or wear [quality of beak (soft, crumbling); abnormal shape; misalignment].

Element	Species	Comments
Calcium deficiency	Birds	Malformations
Manganese deficiency	Poultry	Chickens, duck: short lower beak ("parrot beak")
Phosphorus deficiency	Birds	Malformations

DD: other causes of abnormal beak formation genetic, nutritional abnormalities, parasites, toxins, trauma.

Behavioral changes

Variation from normal, expected behaviors.

Element	Species	Comments
Aluminum toxicity	Mice	Altered escape behaviors, impaired learning and memory
Arsenic toxicity	Humans	Cognitive impairment, memory loss
Cadmium toxicity	Rats	Reduced activity, reduced ability to learn avoidance behavior
Lanthanum toxicity	Rats	Reduced spatial learning, memory
Lead toxicity	Humans (adults, children)	*Children:* balance problems; cognitive function compromised; mental retardation; language deficits; poor school performance, less playful *Adults:* Hostility
Magnesium deficiency	Cats	
Mercury toxicity	Humans, mice	*Humans:* Anger, reduced mental capabilities

Sodium deficiency	Birds	Cannibalism, pecking
Uranium toxicity	Rats	Altered neurobehavior
Vanadium toxicity	Rats	Impaired open field and active avoidance tests
Zinc deficiency	Nonhuman primates	Infants play less, cognitive function impaired

DD: brain damage, genetic inheritance, malnutrition, mental disease, physical disability, physical illness.

Bilirubin—High blood levels

Bilirubin is a breakdown product of heme in red blood cells of vertebrates, and blood levels are used in the diagnosis of hemolytic anemia, jaundice, and liver disease.

Element	Species	Comments
Selenium toxicity	Rats	Inhaled

DD: causes of red cell hemolysis (autoimmune disorders, bone marrow failure, or infections bartonella in cats), inherited disease (sickle cell disease, thalassemia), liver disease.

Bipartite sternebrum

The right and left sternum ossification centers appear as two sites due to delayed ossification.

Element	Species	Comments
Uranium toxicity	Mice	Inhaled

DD: infections, toxins.

Birth rate—Low

Lower than expected number of offspring per pregnancy in a group of animas.

Element	Species
Boron toxicity	Rabbits
Manganese deficiency	Cattle, goats, rats
Rubidium deficiency	Goats

DD: other causes of low birth rate: malnutrition, infections, toxin exposure during pregnancy.

Birth weight—Low

Lower than expected birth weight of offspring.

Element	Species	Comments
Boron toxicity	Mice	
Lithium deficiency	Goats	Low Li intake during pregnancy
Rubidium deficiency	Goats	And low weaning weight. *Not confirmed essential nutrient*

DD: poor maternal nutrition, infections, toxin exposure.

Blackouts

Sudden temporary loss of consciousness.

Element	Species	Comments
Gallium toxicity	Humans	Acute

DD: cardiac disease, epileptiform seizure, hypotension, syncope.

Blanching of ears

Low blood circulation to the ears.

Element	Species
Magnesium deficiency	Rabbits

DD: anemia, poor circulation, heart disease, compromised vascular supply.

Blindness

Loss of vision—maybe partial or complete.

Element	Species
Barium toxicity	Humans
Lead toxicity	Buffaloes, cattle, dogs
Mercury toxicity	Humans, cats, dogs, mink, pigs, sheep
Sulfur toxicity	Cattle

DD: astigmatism, cataracts, diabetic retinopathy, glaucoma, hyperopia, hypertension, macular degeneration, myopia.

Blink reflex loss

Involuntary blink caused by contact stimulation of the cornea.

Element	Species
Magnesium deficiency	Cats

DD: central nervous system dysfunction, trigeminal nerve (V) damage, facial nerve (VII) damage.

Blisters—Skin

A bubble on the skin filled with serum.

Element	Species	Comments
Sulfur toxicity	Humans	Dermal contact

DD: friction, heat, infections [staphylococci bacteria; viral infections: herpes simplex virus (types 1 and 2); varicella zoster virus (chickenpox, shingles), foot and mouth disease].

Blood urea nitrogen (BUN)—High

The amount of urea nitrogen in the blood, it increases after high protein dietary intake, and also if it is not eliminated from the body by the kidneys in a timely fashion.

Element	Species
Iron toxicity	Sheep
Lanthanum toxicity	Mice, rats
Molybdenum toxicity	Rats
Selenium deficiency	Horses

DD: causes of impaired renal excretion of urea: severe burns, dehydration, congestive heart failure, gastrointestinal hemorrhage, heart attack, medication use, high protein diet, shock, urinary tract obstruction.

Blood volume expansion

An increase in the total amount of fluid circulating within the blood vessels (arteries, arterioles, capillaries, veins, venules) and chambers of the heart.

Element	Species
Cobalt toxicity	Rats

DD: renin-angiotensin-aldosterone (RAAS) system activation: cirrhosis, right ventricular heart failure, hepatitis, kidney failure, nephrotic syndrome.

Blood volume loss (severe)

A decrease in the total amount of fluid circulating within the blood vessels (arteries, arterioles, capillaries, veins, venules) and chambers of the heart.

Element	Species	Comments
Arsenic toxicity	Humans	Acute, due gastrointestinal losses

DD: severe burns, chronic diarrhea, hemorrhage.

Blurred vision

Loss of sharpness of vision and loss of clarity of fine detail.

Element	Species
Bismuth toxicity	Humans
Zirconium toxicity	Humans

DD: astigmatism, cataracts, cornea injury, far sightedness, infectious retinitis, macular degeneration, migraine, nearsightedness, optic neuritis.

Body fat content—Increased

Accumulation of adipose tissue in the body—subcutaneous, intraabdominal, or within organs (intercellular or intracellular).

Element	Species
Chromium deficiency	Humans, pigs

DD: excessive calorie intake, reduced exercise, hypothyroidism, stress.

Body righting reflexes—Delayed development

Early life development of righting reflexes delayed.

Element	Species
Manganese deficiency	Rats, guinea pigs

DD: central nervous system damage.

Bone damage

Element	Species	Comments
Palladium toxicity	Rabbits	
Selenium toxicity	Humans	Erosion of long bones

DD: trauma, causes of impaired endochondral ossification resulting in abnormal bone (hormonal abnormalities; malnutrition), infection, neoplasia.

Bone fragility

Undermineralised bone resulting in weak bones, thin bone cortices, reduced strength, and prone to spontaneous fractures.

Element	Species
Calcium deficiency	Humans, bears (polar bears), birds, buffaloes, camels, cats, cattle, deer, dogs, goats, horses, mice, nonhuman primates (lemurs) rabbits, rats, reptiles, sheep, squirrels
Phosphorus deficiency	Humans, birds, deer, goats, sheep
Phosphorus toxicity	Birds

DD: diabetes, chronic obstructive pulmonary disease (COPD), genetic defect affecting type 1 collagen synthesis, primary hyperparathyroidism, hypogonadism, chronic kidney disease, chronic liver disease, malabsorption, rheumatoid arthritis.

Bone loss

Removal of mineral content from bone, resulting in thin cortices and weakness.

Element	Species	Comments
Titanium toxicity	Humans	Due dental implants

DD: causes of bone loss: diabetes, chronic obstructive pulmonary disease (COPD), genetic defect affecting type 1 collagen synthesis, primary hyperparathyroidism, hypogonadism, chronic kidney disease, chronic liver disease, malabsorption, rheumatoid arthritis.

Bone marrow depression

Decreased number of hematopoietic cells (erythrocytes, leukocytes, thrombocytes) in bone marrow.

Element	Species
Arsenic toxicity	Humans
Gold toxicity	Humans
Zinc toxicity	Humans

DD: neoplasia, drugs (antiinflammatory drugs (NSAIDs, chemotherapeutic agents), viral infections (parvovirus)).

Bone marrow hyperplasia

An increase in the number of hematopoietic cells in the bone marrow.

Element	Species
Cobalt toxicity	Rats

DD: bone marrow disease (aplastic anemia, neoplasia), chronic anemia, hemolytic anemia, intrinsic hemorrhage.

Bone marrow hypoplasia

Bone marrow contains no or few hematopoietic cells (erythrocytes, leukocytes, thrombocytes).

Element	Species	Comments
Antimony toxicity	Humans	
Bismuth toxicity	Humans	
Cobalt deficiency	Cattle, sheep	Vitamin B12 deficiency
Gallium toxicity	Mice	

DD: autoimmune anemia, chemotherapy, ionizing radiation, toxin exposure (benzene, pesticides), medications (antibiotics, rheumatoid arthritis treatments), viral infections (cytomegalovirus, Epstein-Barr virus, HIV, parvovirus B19).

Bone pain

Pain when bone compressed, during weight bearing or exercise.

Element	Species	Comments
Aluminum toxicity	Humans	Osteomalacia
Calcium deficiency	Cats	
Calcium toxicity	Humans	
Phosphorus deficiency	Humans	

DD: blood supply compromise, cancer (primary, metastatic), demineralization (fragility or weakness), infection (osteomyelitis), leukemia, neoplasia, osteoporosis, trauma.

Bone tumors

A lump or mass of abnormal neoplastic tissue formed within bone.

Element	Species	Comments
Beryllium toxicity	Mice, rabbits	*Rabbits: Intravenous or intramedullary injection, or Be implants:* Mainly osteosarcoma

DD: genetic inheritance, ionizing radiation exposure, toxin exposure, trauma.

Bone undermineralization (metabolic bone disease, osteomalacia; osteoporosis; osteopenia, rickets)

Osteomalacia: softened, weakening of bone resulting in bowing and fractures.
Osteoporosis: an absolute decrease in bone mass.
Rickets (young growing).
Poor bone density and thin bone cortices on radiographs. Often due to hyperparathyroidism.

Element	Species	Comments
Aluminum toxicity	Humans, pigs, rats	Al deposits in mitochondria inhibit bone phosphate formation—osteomalacia
Bismuth toxicity	Humans	Osteoporosis
Cadmium toxicity	Humans, dogs, mice, rats	Humans: itai-itai disease—linked to renal disease
Calcium deficiency	Humans, amphibians, bears (*polar bears*), birds, buffaloes, camels, cats, cattle, deer, dogs, elephants, ferrets, foxes, goats, horses, marsupials, meercats, mice, nonhuman primates (*lemurs*) rabbits, rats, reptiles, sheep, squirrels	Metabolic bone disease Common in *Cats, Dogs and Elephants:* metabolic bone disease, nutritional secondary hyperparathyroidism, osteodystrophy *Ferrets:* metabolic bone disease, osteoporosis, hyperparathyroidism *Reptiles:* Due to nutritional secondary hyperparathyroidism due to all meat diet *Birds:* fractures, osteomalacia, osteoporosis, rickets *Goats, Sheep:* osteomalacia (adults); rickets (kids) *Marsupials: Koalas* Opossums, Sugar-gliders
Calcium toxicity	Cattle	
Copper deficiency	Humans (infants), camels, cattle, poultry, rabbits	

Fluorine toxicity	Amphibians (*Frogs*)	Metabolic bone disease
Lead toxicity	Sheep	Osteoporosis
Manganese deficiency	Humans (children), rabbits	
Molybdenum toxicity	Cattle	
Phosphorus deficiency	Humans, birds, deer, fish, goats, mink, sheep	
Phosphorus toxicity	Goats, sheep	
Potassium deficiency	Poultry (chicks)	
Selenium toxicity	Reptiles	
Sodium deficiency	Poultry, rats	
Uranium toxicity	Mice	
Zinc deficiency	Birds, fish, nonhuman primates, pigs	*Birds:* chondrodystrophy

DD: causes of bone loss: diabetes, chronic obstructive pulmonary disease (COPD), genetic defect affecting type 1 collagen synthesis, primary hyperparathyroidism, hypogonadism, chronic kidney disease, chronic liver disease, malabsorption, rheumatoid arthritis.

Bone—Increased mineralization (osteopetrosis)

Higher than expected bone density due to excessive mineralization.

Element	Species
Calcium toxicity	Cats, dogs

DD: genetic disorder.

Bowel obstruction

Blockage of the intestinal tract resulting in impaired movement of ingesta through the lumen.

Element	Species
Iron toxicity	Humans

DD: adhesions, foreign bodies, hernias, impaction, intussusception, scar tissue, tumors.

Bradycardia

Abnormally slow heart rate.

Element	Species
Potassium deficiency	Pigs
Potassium toxicity	Dogs, pigs
Praseodymium toxicity	Dogs
Samarium toxicity	Dogs

DD: hypothyroidism, sinoatrial (SA) node disease, disease of cardiac conduction network.

Brain degeneration

Decline and death of neurones.

Element	Species	Comments
Copper deficiency	Humans	Menkes disease

DD: toxin exposure.

Brain development impaired

Failure or delay in normal brain development.

Element	Species	Comments
Copper deficiency	Humans	Menkes disease
Iodine deficiency	Goats	
Uranium toxicity	Rats	

DD: possible causes of delayed brain development, genetic factors, infections, malnutrition.

Brain hemorrhage

Bleeding into the brain.

Element	Species	Comments
Ionizing Radiation	Various	Exposure

DD: hypertension, stroke, trauma, tumor.

Brain edema

Fluid accumulation around the brain increasing intracranial pressure.

Element	Species
Sodium toxicity	Ferrets

DD: hepatic encephalopathy, infections, ionizing radiation exposure, stroke, trauma, tumors.

Brain spongiosis

Vacuolation of cortical gray matter.

Element	Species
Tin toxicity	Rats

DD: prion disease.

Breathlessness

Sensation of shortness of breath or difficulty breathing.

Element	Species	Comments
Boron toxicity	Humans	Inhalation
Cadmium toxicity	Humans	
Mercury toxicity	Humans	

DD: allergic reaction, anemia, anaphylaxis, anxiety, arrhythmias, asthma, bronchitis, chronic obstructive pulmonary disease (COPD), congestive heart failure, inhalation of irritants, interstitial lung disease, lung cancer, pneumonia, pneumothorax, pulmonary embolism, subglottic stenosis.

Bronchoalveolar lavage (BAL)—Positive findings

A diagnostic method of the lower respiratory system by washing out lower airway lumen contents for examination.

Element	Species	Comments
Berylliosis toxicity	Humans	
Zinc toxicity	Humans, guinea pigs, mice, rabbits, rats	*Humans:* T-cells, T suppressor cells, natural killer cells. *Guinea pigs:* Macrophages *Mice:* lymphocytes and macrophages *Rabbits and Rats:* polymorphonuclear leukocytes *Rats:* Increased lactate dehydrogenase activity (LDH)

DD: hypersensitivity pneumonitis, infections, interstitial lung diseases, pulmonary alveolar proteinosis (PAP), eosinophilic pneumonia.

Bronchitis—Chronic

Inflammation of the bronchi.

Element	Species
Antimony toxicity	Humans
Arsenic toxicity	Humans
Sulfur toxicity	Humans

DD: infection (bacteria, viruses—influenza, flu), inhalation of irritant substance (aerosol, fumes, mist).

Bronchial obstruction

Obstruction of the bronchial airways.

Element	Species	Comments
Titanium toxicity	Humans	Part of "yellow nail syndrome"

DD: allergic reaction, foreign body inhalation, smoke inhalation, airway trauma, infections (bacterial, virus).

Bronchiectasis

Airways of the lungs become abnormally wide resulting in mucus build up.

Element	Species	Comments
Sulfur toxicity	Humans	Industrial exposure

DD: allergy, aspergillosis, cilia abnormalities, connective tissue diseases, cystic fibrosis, infections,

Bronchopneumonia

Infection and inflammation of the alveoli.

Element	Species
Zinc toxicity	Cats

DD: infection bacteria, fungi, viruses.

Bronchospasm

Sudden constriction of the muscles in the walls of the bronchiole making the airways narrower and breathing difficult.

Element	Species	Comments
Phosphorus toxicity	Humans	
Palladium toxicity	Cats	After IV injection
Selenium toxicity	Humans	Inhalation
Sulfur toxicity	Humans	

DD: allergen stimulation, chemical inhalation, inhalation of irritant fumes, gas, mists.

Cachexia

Weakness and severe loss of body condition due to severe chronic illness.

Element	Species
Selenium toxicity	Pigs

DD: AIDs, cancer, liver cirrhosis, congestive heart failure, chronic obstructive pulmonary disease (COPD), chronic kidney disease.

Cadmiuria—Increase

Cadmium in urine.

Element	Species	Comments
Cadmium toxicity	Humans	With metallothionein

Calcinosis: Soft tissue calcification

Blood calcium concentrations are usually maintained within a very narrow reference range however, chronic hypercalcemia resulting in soft tissue calcification can occur in certain circumstances.

Element	Species	Comments
Calcium deficiency	Nonhuman primates (Lemurs)	Due secondary hyperparathyroidism, high P
Calcium toxicity	Cattle, dogs, guinea pigs, rats, sheep	Dogs: often due to excess vitamin D intake Ruminants: due ingestion of solanaceous plants
Dysprosium toxicity	Mice	
Erbium toxicity	Mice	
Magnesium deficiency	Cats, dogs, fish, guinea pigs, horses, rats	Aortic calcification
Phosphorus toxicity	Humans	

DD: cancer (paraneoplastic syndrome), connective tissue disease, hypercalcemia, primary hyperparathyroidism, hyperphosphatemia, inflammation, hypervitaminosis D, infections, trauma, tumors, varicose veins.

Calcium absorption reduced from intestinal tract

Element	Species	Comments
Cadmium toxicity	Rats	P absorption also reduced

DD: small intestine disease (inflammatory bowel disease, neoplasia), interference from other nutrients (minerals), kidney failure (lack of conversion to active form vitamin D), vitamin D deficiency (dietary, lack of UV light exposure).

Calcium metabolism—Impaired

May result in hypocalcemia or hypercalcemia.

Element	Species
Iodine deficiency	Cats

DD: ATPase deficiency, malnutrition, nutrient interference.

Calculi (urinary)

Formation of crystals and stones in the urinary tract.

Element	Species	Comments
Calcium toxicity	Camels	
Molybdenum toxicity	Sheep	Xanthine uroliths
Phosphorus toxicity	Cattle	

DD: chronic cystitis, metabolic acidosis or alkalosis, urinary tract infection, urinary tract nidus.

Cannibalism

Element	Species	Comments
Copper deficiency	Cats	Newborn
Sodium deficiency	Birds	

DD: malnutrition, mental illness (humans), high population density.

Carcinogenic effect

Induces the development of cancer by damaging cells.

Element	Species	Comments
Aluminum toxicity	Humans	Al accumulation in breast cancer—not necessarily a cause
Antimony toxicity	Humans, rats	*Rats:* lung tumors (inhalation)
Arsenic toxicity	Humans	Bladder (adenocarcinoma and transition cell carcinoma TCC), bone, colon, kidney (TCC), larynx, liver, lung, lymphoma, nasal cavity, prostate, skin (Bowen's disease; basal cell carcinoma), ureter (TCC), urethra (all types).
Cadmium toxicity	Humans	*Humans:* liver, lungs, prostate
Chromium toxicity	Mice, rabbits, rats	*Mice: inhalation:* nasal papilloma, pulmonary adenoma, *intramuscular:* injection site sarcoma *Rats: Inhalation:* Lung tumors (adenoma) *injection site (subcutaneous or intramuscular*) rhabdomyosarcomas, spindle-cell sarcomas, fibrosarcomas
Nickel toxicity	Hamsters, mice, rats, salamander	*Hamsters: Intratracheal:* adenoma, adenocarcinoma, *intramuscular:* sarcomas *Mice: intramuscular:* fibrosarcoma, sarcoma *Rabbits: intramuscular implant:* rhabdomyosarcoma, *Rats: inhalation:* lung tumors: adenoma, *intrapleural:* mesothelioma, spindle cell carcinoma, rhabdomyosarcoma, *intramuscular:* fibrosarcoma, sarcoma, rhabdomyosarcoma, *subcutaneous* sarcoma *intratracheal:* adenocarcinoma, squamous cell carcinoma *intraocular:* glioma, melanoma, retinoblastoma *intratesticular:* testicular cancer *Salamander:* melanoma (after intraocular injection)
Palladium toxicity	Mice	
Rhodium toxicity	Mice	
Silicon toxicity	Humans	*Inhalation (silicosis):* Lung cancer
Silver toxicity	Rats	Subcutaneous injection site
Tungsten toxicity	Humans	Lung cancer
Vanadium toxicity	Mice	Lung cancer (carcinoma)
Yttrium toxicity	Humans, rats	
Zinc toxicity	Mice	Alveogenic carcinoma
Zirconium toxicity	Humans	Gastric carcinoma
Ionizing radiation	Various	Exposure—bladder, breast, leukemia, lung, multiple myeloma, esophagus, ovarian, stomach

DD: drugs, oncogenic infections (feline leukemia virus), toxins.

Cardiac arrest

Abrupt cessation of cardiac function.

Element	Species	Comments
Barium toxicity	Humans	Due hypokalemia
Magnesium deficiency	Humans	
Magnesium toxicity	Humans	
Silicon toxicity	Humans	Silicosis

DD: cardiac disease, toxemia.

Cardiac arrhythmia

Heart beats with an irregular or abnormal rhythm.

Element	Species	Comments
Aluminum toxicity	Humans	
Antimony toxicity	Humans	Ventricular tachycardia and fibrillation
Arsenic toxicity	Humans	Chronic
Fluorine toxicity	Humans	Due hypocalcemia
Magnesium deficiency	Cats	
Selenium deficiency	Goats	
Sodium deficiency	Cattle	

DD: cardiac disease (cardiac arrest), cardiac fibrosis, cardiomyopathy, coronary artery disease, diabetes, hypertension, hyperthyroidism, hypothyroidism, infections (including covid-19), sleep apnoea.

Cardiac atrophy

Catabolic remodeling of the heart due to reduced workload.

Element	Species
Selenium toxicity	Cattle

DD: cancer, lack of exercise, infection (virus), inflammatory disease (myocarditis).

Cardiac deformities (fetal)

Abnormal in utero development of the heart.

Element	Species	Comments
Aluminum toxicity	Mice, poultry, rats	*Mice:* also bone deformities *Poultry:* ventricular septal and myocardial defects *Rats:* conotruncal defects, right ventricular outflow defects, septal defects
Boron toxicity	Rabbits, rats	
Selenium deficiency	Cattle	Fetal heart lesions

DD: malnutrition during pregnancy, genetic factors, infection, toxin exposure.

Cardiac enlargement (cardiomegaly)

Increase in heart size beyond normal limits.

Element	Species
Copper deficiency	Pigs, poultry
Iron deficiency	Pigs, rats

DD: carnitine deficiency (dogs), congenital abnormality, congestive heart failure, coronary heart disease, hypertension, advanced valve regurgitation, taurine deficiency (cats and dogs).

Cardiac fibrosis

Increased collagen type I deposition and cardiac fibroblast activation with differentiation into myofibroblasts—resulting in scarring.

Element	Species	Comments
Cerium toxicity	Humans	Endomyocardial fibrosis
Magnesium deficiency	Rabbits	

DD: other heart disease aortic stenosis, coronary heart disease, hypertension.

Cardiac failure/heart failure

Cardiac output insufficient to meet the needs of the individual.

Element	Species	Comments
Antimony toxicity	Humans	After treatment: ventricular fibrillation
Chlorine deficiency	Cattle	Cardiovascular depression
Cobalt (cobalamin) deficiency	Humans (children)	Congestive heart failure
Copper deficiency	? Humans, cattle, rats	Humans—suggested link to atherosclerosis and ischemia heart disease
Lutetium toxicity	Cats	Acute toxicity study
Potassium deficiency	Cats, dogs, poultry	Dogs—reduced cardiac output, stroke volume, renal blood flow
Samarium toxicity	Dogs	Circulatory failure—acute toxicity
Selenium deficiency	Horses	Myocardial myopathy
Selenium toxicity	Humans	
Thallium toxicity	Humans	
Ytterbium toxicity	Guinea pigs	

DD: arrhythmia, cardiac arrest, cardiomyopathy, congenital heart defect, diabetes, heart valve disease, hypertension, pulmonary disease (emphysema).

Cardiac murmurs

An abnormal sound heard between heart beats.

Element	Species	Comments
Zinc toxicity	Dogs	Secondary to anemia

DD: innocent causes: anemia, exercise, fever, rapid growth phase, hyperthyroidism, pregnancy, fever other causes: carnitine deficiency (dogs), heart valve defects (acquired, congenital), endocarditis, taurine deficiency (cats and dogs), thiamine deficiency (pigs).

Cardiomyopathy

Disease of heart muscle which becomes thickened, thinned, or stretched—results in reduced cardiac output.

Element	Species	Comments
Aluminum toxicity	Humans	Myocardial dysfunction, hypokinesia, myocarditis (toxic)
Arsenic toxicity	Humans	
Cobalt toxicity	Humans, dogs	
Gallium toxicity	Humans	Acute—irreversible
Iron toxicity	Humans	Hemochromatosis
Selenium deficiency	Humans, birds, nonhuman primates (squirrel monkey), pigs	*Birds:* Ventricular muscle degeneration *Pigs:* Mulberry heart disease (Se vitamin E) *Squirrel monkeys:* if low protein intake as well
Selenium toxicity	Humans	

DD: alcoholism, amyloidosis, carnitine deficiency (dogs), postchemotherapy, connective tissue disorders, covid-19 infection, diabetes, drug abuse, heart valve disease, hypertension, infections (myocarditis), postmyocardial infarction, radiation exposure, sarcoidosis, chronic tachycardia, taurine deficiency (cats and dogs), thiamine deficiency, thyroid disease.

Cardiovascular abnormalities

Element	Species	Comments
Aluminum toxicity	Humans, poultry, rats	*Poultry, Rats:* congenital deformities *Humans:* myocardial disease, hypokinesia, myocarditis
Antimony toxicity	Humans	*Cardiotoxicity, ECG changes:* low T wave height, inverted T wave, prolonged QT, ventricular ectopics, tachycardia, torsade de ports, ventricular fibrillation
Arsenic toxicity	Humans	Myocardial injury
Boron toxicity	Rats	
Calcium toxicity	Reptiles	Cardiac disorders
Cesium toxicity	Humans	Cardiotoxicity
Cerium toxicity	Humans	Endomyocardial fibrosis, myocardial infarction

Chromium toxicity	Humans	Myocardial lesions
Copper deficiency	Humans, nonhuman primates	Myocardial and vasculature weakness
Iron toxicity	Humans	Impaired cardiac function
Palladium toxicity	Rabbits, rats	Heart damage
Selenium deficiency	Nonhuman primates, pigs	*Pigs:* Mulberry heart disease (cardiomyopathy) *Primates:* Cardiomyopathy—only if low protein intake as well
Selenium toxicity	Humans, cattle	Cardiomyopathy, ECG changes, heart failure *Cattle:* cardiac atrophy

Caries

Permanent cavities in the enamel of the tooth.

Element	Species	Comments
Fluorine deficiency	Humans	On low F intake—however F not considered essential
Molybdenum toxicity	Rats	

DD: bacteria, sugary foods/drinks, plaque accumulation.

Carpal hyperextension

Overextension of the carpal joint so leg sinks and carpus can touch the ground.

Element	Species
Copper deficiency	Dogs
Magnesium deficiency	Dogs

DD: third degree sprain, trauma (landing after jumping), connective tissue degeneration/weakness.

Cartilage development compromised

Abnormal cartilage formation.

Element	Species	Comments
Copper deficiency	Horses, rabbits	+ other species with skeletal signs
Molybdenum toxicity	Rabbits, rats	*Rabbits:* Epiphyseal and Metaphyseal cartilage calcification abnormalities *Rats:* Cartilage dysplasia

DD: arthropathies (inflammatory), genetic disorders, osteoarthritis, trauma.

Catalase activity (reduced)

An antioxidant enzyme that catalyzes the reduction of hydrogen peroxide, helping to mitigate against oxidative stress.

Element	Species	Comments
Copper deficiency	Buffaloes	
Iron deficiency	Cattle, pigs, rats	Young animals—activity in blood

DD: Alzheimer's disease, anemia, bipolar disorder, cancer, diabetes mellitus, hypertension, Parkinson's disease, bipolar disorder, schizophrenia, vitiligo.

Cataracts

Opacity in the lens of the eye.

Element	Species	Comments
Arsenic toxicity	Humans	Chronic toxicity
Calcium toxicity	Fish	
Magnesium deficiency	Fish	
Selenium toxicity	Rats	
Zinc deficiency	Fish	Rainbow trout; common carp
Ionizing radiation	Various	Exposure

DD: aging, diabetes mellitus, inherited disease, posttrauma.

Cecal enlargement

Distension of the cecum.

Element	Species
Iron deficiency	Rats

DD: atony of cecal wall, nutritional change, microbiome change, impaction.

Central nervous system lesions

Lesions of the brain or spinal cord.

Element	Species	Comments
Aluminum toxicity	Humans, dogs, mice	*Humans:* CNS accumulation associated with Alzheimer's and Parkinson's diseases *Dogs:* Cerebral and peripheral neuropathy, paresis (para- and tetra-), reflexes depressed (patellar; withdrawal) *Mice:* behavioral changes, reduced cell proliferation, impaired learning, impaired memory, impaired reflexes, motor disturbances, reduced number neural stem cells, reduced neuroblast cell differentiation, startle response increased
Arsenic toxicity	Humans	Cerebrovascular infarction, cognitive impairment, confusion, demyelination; encephalopathy, precapillary hemorrhages, polyneuritis
Boron toxicity	Humans	Irritability, seizures

Cadmium toxicity	Mice, rats	Encephalopathy, hydrocephaly
Chlorine deficiency	Humans	Cerebral hemorrhage, confusion, anxiousness, muscle twitches, coma, death
Copper deficiency	Humans (infants), guinea pigs, rats	*Humans*: neurological disturbances *Guinea pigs*: cerebellar agenesis, cerebral edema, *Rats*: behavioral changes
Iron toxicity	Humans, Horses	*Humans:* impaired central nervous system function
Lanthanum toxicity	Rats	Neural degeneration; changes in hippocampus (morphological and histological)
Lead toxicity	Birds	Central nervous system damage
Magnesium deficiency	Humans	Mental derangement
Manganese toxicity	Humans	
Mercury toxicity	Humans, ferrets, horses, mice, mink, rats	*Humans:* permanent brain damage *Ferrets:* convulsions, incoordination, tremors *Horses:* head bobbing, depression, trembling, neurological signs *Mice:* degeneration corpus striatum, cerebral cortex, hypothalamus *Mink:* chewing, circling, convulsions, dysphonia, falling to recumbency, head tremors, salivation *Rats:* neurotoxicity, degeneration of cerebellum, dorsal root ganglia
Molybdenum deficiency	Humans	Brain atrophy, coma, mental retardation, seizures.
Molybdenum toxicity	Rats, sheep	*Rats:* neurocyte degeneration (cerebral cortex, hippocampus) *Sheep:* CNS degeneration, motor neurone disease
Potassium deficiency	Humans, cattle	*Humans*—mental confusion *Cattle*—neurological disorders
Selenium toxicity	Humans, cattle, pigs	*Humans:* anesthesia, coma, confusion, delirium, hemiplegia, hyperreflexia, hyporeflexia, paresthesia *Cattle:* ataxia, coma, depression, incoordination, polioencephalomalacia, paralysis *Pig:* CNS depression, poliomyelomalacia
Sulfur toxicity	Rabbits	Neurological signs
Tin toxicity	Various	CNS edema (organic salts)
Thallium toxicity	Humans	Cerebellar ataxia, cranial nerve palsies, mental retardation, motor neuropathy, organic brain syndrome, psychoses, sensory polyneuropathy

DD: autoimmune disease, congenital (structural) defects, degenerative diseases, infection, stroke, trauma, tumors, degeneration.

Cerebellar agenesis

Failure of the cerebellum to form.

Element	Species	Comments
Copper deficiency	Guinea pigs	Cerebellar folia

DD: in humans an autosomal recessive inherited mutation in the PTF1A gene.

Cerebellar ataxia

Lack of fine control of voluntary movements due to a cerebellar lesion.

Element	Species
Mercury toxicity	Humans, horses, mink
Thallium toxicity	Humans

DD: autoimmune disease (coeliac disease, hypothyroidism, multiple sclerosis, sarcoidosis), brain abscess, cerebral palsy, infections (covid-19, Lyme disease, viral infections), paraneoplastic syndrome, stroke, thiamine deficiency, toxins (medications, paint thinner, vitamin B6), trauma,

Cerebellar coning (herniation)

The cerebellum moves down through the foramen magnum; may cause compression of the lower brainstem and upper cervical spinal cord.

Element	Species
Sodium toxicity	Ferrets

DD: other causes of increased brain pressure/swelling, brain hemorrhage, stroke, trauma, tumor.

Cerebral edema

Fluid accumulation in the brain causing an increase in intracranial pressure.

Element	Species
Antimony toxicity	Humans
Copper deficiency	Guinea pigs

DD: brain cancer, brain infection, hematoma (epidural, intracerebral or subdural), hemorrhage (subarachnoid), high altitude, hydrocephalus, hyponatremia, liver failure (acute), stroke (ischemia), trauma.

Cerebral palsy

A group of disorders causing motor disability (impaired balance, movement, and posture).

Element	Species	Comments
Mercury toxicity	Humans	congenital

DD: infection (e.g., chickenpox, cytomegalovirus, rubella, toxoplasmosis), hypoglycemia, hypoxia, meningitis, periventricular leukomalacia (PVL), stroke, trauma.

Cerebrovascular infarction

Necrosis of brain tissue due to reduced oxygen supply. Causes ischemic stroke.

Element	Species
Arsenic toxicity	Humans

DD: angiopathy, arterial dissection, atherosclerosis, atherothrombosis, cardioembolism, fibromuscular dysplasia, hypercoagulability (antiphospholipid antibody syndrome; factor V Leiden; prothrombin 20210A mutation), meningitis, inherited metabolic disorders (Fabry disease, homocystinuria, mitochondrial disorders), polycythemia, sickle cell disease, small vessel disease, systemic lupus erythematosus, vasculitis.

Ceruloplasmin (serum)—Reduced

A ferroxidase enzyme (encoded by CP gene in people). The major copper-carrying protein in blood and important in iron metabolism.

Element	Species
Copper deficiency	Buffaloes
Zinc toxicity	Humans

DD: kidney disease, Menkes disease, nephrotic syndrome.

Cheilitis

Inflamed lips.

Element	Species	Comments
Selenium toxicity	Nonhuman primates	*Macaque*
Thallium toxicity	Humans	
Zinc deficiency	Humans, rabbits	

DD: infection (bacteria, fungi, viruses), excess salivation.

Chest pain

Element	Species	Comments
Beryllium toxicity	Humans	
Chlorine toxicity	Humans	Inhalation
Gallium toxicity	Humans	
Mercury toxicity	Humans	Burning sensation
Selenium toxicity	Humans	
Yttrium toxicity	Humans	Industrial exposure
Zinc toxicity	Humans	Inhalation

DD: cancer, coronary artery disease, hiatal hernia, hypertrophic cardiomyopathy, gastro-esophageal reflux, lung abscess, mitral valve prolapse, muscle strain, myocardial infarction, myocarditis, pericarditis, pleuritis, pneumonia, pulmonary embolism, pulmonary hypertension, shingles, viral infections.

Chest tightness

Sensation of heaviness, pressure, or tightness in the chest.

Element	Species	Comments
Cadmium toxicity	Humans	
Mercury toxicity	Humans	
Sulfur toxicity	Humans	Inhalation
Zinc toxicity	Humans	Inhalation

DD: anxiety, gallbladder disease, intercostal muscle strain, heart disease, liver disease, lung disease, esophageal disease, pancreatic disease, stomach disease, stress.

Chewing constantly

Element	Species
Fluorine toxicity	Cattle
Mercury toxicity	Mink

Chewing uncoordinated

Element	Species
Sodium deficiency	Horses

DD: neurological disorders.

Cholangitis

Inflammation of the bile ducts.

Element	Species
Tin toxicity	Humans

DD: bacteremia, bile duct blockage, blood clot, gallstones, infection (ascending), pancreatic swelling, parasitism, bile duct stenosis postsurgery, tumor.

Cholesterol ratio change

Total cholesterol number divided by high density cholesterol HDL number. High ratios suggest increased risk of developing heart disease.

Element	Species	Comments
Zinc toxicity	Humans	Increased LDL:HDL cholesterol, due to Cu deficiency

DD: diabetes, liver disease, obesity, poor diet.

Cholinesterase—Reduced activity

An enzyme that hydrolyses choline esters, example acetylcholinesterase.

Element	Species
Lanthanum toxicity	Poultry

DD: exposure to organophosphorus toxins.

Chondrodystrophy

Maldevelopment of cartilage.

Element	Species	Comments
Manganese deficiency	Poultry	Chick embryos

Chronic obstructive pulmonary disease (COPD)

A group of lung diseases that cause respiratory difficulty.

Element	Species
Silicon toxicity	Humans

Chvostek sign

Twitching of muscles innervated by facial nerve.

Element	Species
Magnesium deficiency	Humans

DD: hypocalcemia.

Circling

Element	Species
Mercury toxicity	Mink

DD: brain tumor, cerebellar disorders, infection (e.g., listeriosis), otitis externa, vestibular syndrome.

Circulatory failure

Reduced arterial pressure poor capillary blood flow leading to impaired organ function.

Element	Species	Comments
Iodine deficiency	Birds	Circulatory collapse
Iron toxicity	Sheep	

DD: other causes of heart failure, vasoconstriction.

Cirrhosis

Liver cell degeneration, inflammation, and fibrous thickening (chronic).

Element	Species
Arsenic toxicity	Humans
Copper toxicity	Humans
Iron toxicity	Humans
Selenium toxicity	Dogs, rats

DD: alcoholism, alagille syndrome, alpha-1 antitrypsin, biliary atresia, cystic fibrosis, fatty liver, glycogen storage disease, hemochromatosis, hepatic copper accumulation (Wilson's) hepatitis (autoimmune, viral), infection (brucellosis), medications (e.g., methotrexate) sclerosing cholangitis, toxin exposure.

Choreoathetosis

A serious movement disorder causing involuntary twitching, jerky movements, or writhing.

Element	Species
Mercury toxicity	Humans

DD: cerebral palsy, copper accumulation in the liver, Huntington's disease, kernicterus (jaundice in newborn), medication, Tourette syndrome, trauma, tumors, Wilson's disease.

Cleft palate/lip

Element	Species
Antimony toxicity	Pigs
Cadmium toxicity	Hamsters, mice, rats
Chromium toxicity	Hamsters
Uranium toxicity	Mice
Vanadium toxicity	Mice, rats

DD: genetic inheritance (22q11 deletion syndrome), folate deficiency, hypervitaminosis A.

Cloacal prolapse

Element	Species
Calcium deficiency	Reptiles

DD: abnormal egg size, abdominal mass, behavioral problems, cloacal papilloma, cloacitis, constipation, diarrhea, obesity, peritonitis, poor nutrition.

Club foot

Deformed foot so that sole cannot touch floor normally.

Element	Species
Cadmium toxicity	Mice, rats

DD: genetic causes (possible).

Clumsiness worsens

Element	Species
Lead toxicity	Humans (children)

DD: anxiety, brain tumor, epilepsy, Parkinson's diseases, stress, stroke.

Coagulation/clotting time—Prolonged

The time required for a sample of blood to coagulate in vitro under standard conditions.

Element	Species	Comments
Cerium toxicity	Dogs	
Lutetium toxicity	Dogs, mice	And prothrombin time
Selenium toxicity	Humans	

Coagulopathy

Impaired hemostasis resulting in impaired clot formation or bleeding.

Element	Species
Iron toxicity	Humans, dogs, horses
Uranium toxicity	Humans
Zinc toxicity	Humans, dogs

Cognitive impairment/deficits

Trouble remembering, learning new things, concentrating, or making decisions.

Element	Species	Comments
Arsenic toxicity	Humans	
Bismuth toxicity	Humans	
Copper deficiency	Rats	Behavioral changes
Iron deficiency	Humans (children)	Attention span, intelligence, sensory perception functions emotional and behavior (Lozoff et al., 1982)

Molybdenum deficiency	Humans	Heritable condition
Zinc deficiency	Humans, nonhuman primates	Learning and hedonic

DD: delerium, medication side-effects, psychiatric disease, thyroid disease, vitamin deficiency.

Collagen synthesis reduced

Collagen synthesis occurs mainly in fibroblasts. It is a major component of connective tissue.

Element	Species
Sulfur deficiency	Rats

DD: vitamin A deficiency.

Collapse/fainting/syncope

Collapse: Sudden fall down.
Fainting/syncope: Temporary loss of consciousness usually due to lack of oxygen.

Element	Species	Comments
Arsenic toxicity	Dogs	
Barium toxicity	Dogs	
Cesium toxicity	Humans	Syncope
Calcium deficiency	Birds, dogs, cattle	*Cattle:* Milk fever
Copper toxicity	Various	
Fluorine toxicity	Cattle	
Gallium toxicity	Humans	
Germanium toxicity	Humans	
Magnesium deficiency	Cattle, horses pigs, sheep	
Sodium deficiency	Humans, marine mammals	*Marine mammals:* Addisonian crisis
Sodium toxicity	Cattle	
Vanadium toxicity	Cattle	After ingestion
Zirconium toxicity	Humans	

DD: blood flow reduction, body position change, cardiac arrhythmia (bradycardia, tachycardia), dehydration, exhaustion, hypotension, neurally mediated syncope (vasovagal), overheating, pooling of blood in legs, heavy sweating.

Colloid depletion

Loss of colloid, e.g., albumin.

Element	Species
Cobalt toxicity	Human (children)

DD: hepatic disease, gastrointestinal loss, renal disease.

Color vision loss (color blindness)

Inability to see colors normally.

Element	Species
Mercury toxicity	Humans
Thallium toxicity	Humans

DD: diabetes, genetic inheritance, glaucoma, macular degeneration, medication side-effect (digoxin, chloroquine, phenytoin, sildenafil), multiple sclerosis, toxin exposure (carbon disulfide, styrene).

Coma

Long period of deep unconsciousness.

Element	Species
Antimony toxicity	Humans
Arsenic toxicity	Humans
Bismuth toxicity	Humans
Bromine toxicity	Humans
Calcium deficiency	Various—cattle (milk fever)
Chlorine deficiency	Humans
Iron toxicity	Humans, horses (foals)
Lithium toxicity	Humans
Magnesium deficiency	Various
Mercury toxicity	Humans, pigs
Molybdenum deficiency	Humans
Selenium deficiency	Dogs
Selenium toxicity	Humans, cattle, pigs
Sulfur toxicity	Cattle
Thallium toxicity	Humans
Tin toxicity	Humans
Tungsten toxicity	Humans
Ionizing radiation	Various

DD: diabetes, drug overdose, hyperglycemia, hypoglycemia, hypoxia, infections, strokes, toxin exposure, seizures, brain trauma, tumors.

Concentration loss

Difficulty focusing on things.

Element	Species
Bismuth toxicity	Humans
Mercury toxicity	Humans

DD: ADHD, anxiety, brain trauma, stress.

Conception rate—Low

Reduced number of pregnancies to usual.

Element	Species	Comments
Bromine deficiency	Goats	
Cadmium deficiency	Goats	Impaired ability to conceive
Copper deficiency	Cats, deer	*Cats:* Delayed time to conception
Iodine deficiency	Goats	
Manganese deficiency	Cattle, goats	
Mercury toxicity	Monkeys	
Molybdenum deficiency	Goats	
Molybdenum toxicity	Cattle, rats	
Phosphorus deficiency	Goats, sheep	
Selenium deficiency	Deer	
Selenium toxicity	Pigs	
Zinc deficiency	Cattle, sheep	

DD: infections, poor nutrition, systemic disease, trichomoniasis, vibriosis.

Confusion

Uncertain about what is going on.

Element	Species
Aluminum toxicity	Humans
Arsenic toxicity	Humans
Bismuth toxicity	Humans
Bromine toxicity	Humans
Calcium toxicity	Humans
Chlorine deficiency	Humans
Lithium toxicity	Humans
Magnesium toxicity	Humans
Manganese toxicity	Humans
Mercury toxicity	Humans
Selenium toxicity	Humans
Thallium toxicity	Humans
Tin toxicity	Humans (*Organotin compounds*)
Tungsten toxicity	Humans
Ionizing radiation	Various

DD: dementia, hypoglycemia, infections [e.g., urinary tract infections (UTIs), stroke (or TIA), trauma (head)].

Congenital malformations

Element	Species	Comments
Antimony toxicity	Pigs	Cleft palate
Arsenic toxicity	Humans	Chronic
Cadmium toxicity	Hamsters, mice, rats	Cleft palate, lip
Chromium toxicity	Hamsters	Cleft palate
Selenium toxicity	Fish, poultry, sheep	*Chicks, Turkeys:* embryo deformities *Sheep:* eyes
Tellurium toxicity	Rats	Teratogenic: exophthalmos, hydrocephalus, ocular hemorrhage, edema, small kidneys, umbilical hernia, retained testicles
Uranium toxicity	Mice	Cleft palate, bipartite sternebrum, impaired bone ossification
Vanadium toxicity	Mice, rats	Skeletal and visceral anomalies and malformations (cleft palate, hematomas)
Ytterbium toxicity	Fish	Tail malformation
Zinc deficiency	Humans, birds, nonhuman primates	

DD: environmental factors, genetic factors, infections, poor nutrition.

Congestion (eyes)

Hyperemia of the conjunctiva/sclera.

Element	Species	Comments
Arsenic toxicity	Humans	Chronic toxicity

DD: exposure to irritant substance, increased intraocular pressure, glaucoma.

Congestion (internal organs)

Abnormal accumulation of fluid in organs.

Element	Species	Comments
Cerium toxicity	Mice	
Iron toxicity	Various	Acute toxicity
Selenium toxicity	Humans	

DD: blood vessel blockage, congestive heart failure, inflammatory disease, mitral stenosis, pneumonia, tachycardia, thrombosis, renal disease.

Congestion (pulmonary)

Accumulation of fluid in lungs, causing impaired gas exchange and arterial hypoxemia.

Element	Species	Comments
Zinc toxicity	Cats, rats	Inhaled

DD: cardiac insufficiency, hypertension, infection.

Conjunctivitis

Inflamed conjunctiva.

Element	Species	Comments
Aluminum toxicity	Humans	
Boron toxicity	Humans, rabbits	Exposure
Beryllium toxicity	Humans	
Chlorine deficiency	Cattle	Infections
Chlorine toxicity	Humans	
Dysprosium toxicity	Rabbits	
Erbium toxicity	Rabbits	
Europium toxicity	Rabbits	
Fluorine toxicity	Humans and others	Gas exposure or HF acid
Gadolinium toxicity	Humans	
Holmium toxicity	Rabbits	
Lutetium toxicity	Rabbit	
Molybdenum toxicity	Rabbits	
Palladium toxicity	Humans, rabbits, rats	
Praseodymium toxicity	Humans, rabbits	
Sulfur toxicity	Humans	
Thulium toxicity	Rats	Conjunctival ulcers
Tin toxicity	Humans	
Vanadium toxicity	Cattle	After ingestion
Ytterbium toxicity	Humans, rats	*Rats:* conjunctival ulcers
Zinc deficiency	Dogs	
Zirconium toxicity	Humans	Industrial exposure

DD: irritant exposure (dusts, chemicals, fumes, mists), infections (bacteria, e.g., chlamydia, viruses, e.g., adenovirus).

Conjunctival edema (chemosis)

Fluid swelling of the conjunctiva.

Element	Species	Comments
Selenium toxicity	Humans	Airborne exposure

DD: allergy, conjunctivitis, following eye surgery, eye rubbing, hyperthyroidism, lymphatic obstruction, trauma, venous obstruction.

Consciousness changes

Element	Species
Lead toxicity	Humans
Lithium toxicity	Humans
Mercury toxicity	Humans

Consolidation (lungs)

Air in the alveoli are replaced with fluids (e.g., blood, pus, water) or solids (e.g., stomach contents, cells).

Element	Species	Comments
Zinc toxicity	Guinea pigs	Inhaled

DD: aspiration, congestive heart failure, gastrointestinal disease, hemorrhage, infection (pneumonia), lung cancer, lung torsion, pulmonary hemorrhage, pulmonary edema.

Constipation

Accumulation of feces in the large intestine due to difficulty in evacuating bowels.

Element	Species
Aluminum toxicity	Humans
Bismuth toxicity	Humans
Chlorine deficiency	Cattle
Lead toxicity	Humans
Lithium toxicity	Humans
Selenium toxicity	Cattle

DD: reduced activity, dehydration, lack of fiber, physical intestinal obstruction.

Convulsions (fits)

Sudden onset, violent, irregular movement of the body, caused by involuntary contraction of muscles. Associated especially with brain disorders (e.g., epilepsy); toxins or other agents in the blood: or fever (see also seizures).

Element	Species	Comments
Aluminum toxicity	Humans, cats, dogs, rabbits	*Cats and Rabbits*: associated with neurofilament formation
Antimony toxicity	Humans	
Arsenic toxicity	Humans	Acute and chronic toxicities
Barium toxicity	Humans	
Beryllium toxicity	Rabbits	
Bismuth toxicity	Humans	Rare
Boron toxicity	Humans	
Calcium deficiency	All—humans (child), dogs	
Chlorine deficiency	Humans	
Copper deficiency	Birds	
Copper toxicity	Various	
Iron toxicity	Humans, pigs	
Lead toxicity	Humans (children), cattle, mink	Convulsions
Lithium toxicity	Humans	
Magnesium deficiency	Birds, cats, fish	
Manganese deficiency	Rats	Epileptiform
Mercury toxicity	Humans, cats, cattle, dogs, ferrets, mink, sheep	Seizures *Ferrets, Mink:* Convulsions
Palladium toxicity	Rat	Convulsions
Potassium deficiency	Fish, poultry (chicks)	
Sodium deficiency	Marine mammals	
Sodium toxicity	Birds	
Sulfur toxicity	Horses	Accidental poisoning
Thallium toxicity	Humans	
Tin toxicity	Humans	
Tungsten toxicity	Humans	Rare
Zirconium toxicity	Humans	

Copper accumulation in body tissues

Element	Species	Comments
Copper toxicity	Humans, dogs, marine mammals, marsupials	*Humans:* Menkes syndrome; cornea, kidney, liver *Dogs:* Hepatic accumulation in some breeds leading to liver disease *Marine mammals Dolphins:* Liver *Marsupials (wombats):* Liver
Sulfur toxicity	Rabbits	Thyroid, pancreas, spinal cord

DD: genetic disorder (Wilson's disease).

Copper—Low plasma concentrations

Element	Species
Copper deficiency	Buffaloes, poultry
Iron toxicity	Sheep
Molybdenum toxicity	Deer

DD: malabsorption, postgastrointestinal surgery, gastrointestinal disease, Menkes disease.

Copper—Low liver concentrations

Element	Species
Molybdenum toxicity	Deer
Zinc toxicity	Fish

DD: malabsorption, postgastrointestinal surgery, gastrointestinal disease, Menkes disease.

Corneal lesions

Element	Species	Comments
Antimony toxicity	Humans	Corneal burns
Chromium deficiency	Humans, mice, monkeys, rats	*Squirrel monkeys:* irreversible, haziness, superficial macule, vascularization
Selenium toxicity	Humans	Corneal ulcers
Sodium deficiency	Poultry, rats	Keratinization (Poultry)
Sulfur toxicity	Humans	Corneal burns

DD: infection, exposure to irritant substances (fumes, gases, liquids, mist) trauma.

Coronary band inflammation

Inflammation of the tissue at the top of the hoof capsule.

Element	Species	Comments
Selenium toxicity	Cattle, horses	Inhaled

DD: abscess, trauma.

Cough

Sudden expulsion of air from the airways/lungs.

Element	Species	Comments
Antimony toxicity	Humans	Inhalation
Beryllium toxicity	Humans	Dry cough
Boron toxicity	Humans	Inhalation
Bromine toxicity	Humans	
Cadmium toxicity	Humans	
Chlorine toxicity	Humans	Inhalation
Iodine toxicity	Cattle	Dry, hacking cough
Mercury toxicity	Humans	Dry
Molybdenum toxicity	Humans	
Nickel toxicity	Humans	Inhalation
Selenium toxicity	Birds	
Silicon toxicity	Humans	Inhalation
Sulfur toxicity	Humans	
Tellurium toxicity	Humans	Inhalation
Titanium toxicity	Humans	Part of "yellow nail syndrome"
Vanadium toxicity	Humans,	
Yttrium toxicity	Humans	Industrial exposure
Zinc toxicity	Humans	Inhalation—dry throat

DD: antihypertensive drugs (ACE inhibitors), aspiration, asthma, bronchiectasis, bronchiolitis, chronic obstructive pulmonary disease (COPD), pulmonary fibrosis, cystic fibrosis, eosinophilic bronchitis, gastrointestinal reflux, infection, laryngopharyngeal reflux, lung cancer, sarcoidosis, upper airway cough syndrome (nasal drip).

Cracks on palms of hands and soles of feet

Element	Species	Comments
Arsenic toxicity	Humans	Hyperkeratosis

DD: athletes foot, diabetic neuropathy, dry skin, eczema, keratolysis, psoriasis, xerosis (dry skin).

Cranial nerve palsies

Loss of function of one or more cranial nerves.

Element	Species
Thallium toxicity	Humans

DD: aneurysm, congenital, diabetes, hypertension, infection, migraines, high intracranial pressure stroke, traumatic, tumors.

Creatine kinase (CK)

An enzyme expressed by various tissues.

Element	Species	Comments
Selenium deficiency	Horses	Myopathy and donkeys
Selenium toxicity	Humans	High

DD: CK can go up after a heart attack, skeletal muscle injury, or strenuous exercise.

Creatinine—High blood concentrations

Creatinine is a breakdown product of protein catabolism. It is released constantly from skeletal muscle. High levels occur in impaired kidney function or kidney disease due to reduced excretion rate.

Element	Species
Aluminum toxicity	Rats
Antimony toxicity	Rats
Gallium toxicity	Humans
Iron toxicity	Sheep
Lanthanum toxicity	Mice
Selenium toxicity	Humans
Yttrium toxicity	Rats

DD: causes of kidney disease and impaired function or skeletal muscle damage.

Cretinism

A syndrome consisting of mental deficiency, deaf mutism, spastic diplegia, neurological deficits.

Element	Species
Iodine deficiency	Humans, dogs

DD: causes of hypothyroidism.

Crystalluria

Crystals in the urine.

Element	Species	Comments
Mercury toxicity	Pigs	White crystals

DD: toxin exposure (e.g., antifreeze ingestion, brassica vegetables), urine pH change (acidification or alkalinization), urinary tract infection (UTI).

Curled toe paralysis

Element	Species	Comments
Boron toxicity	Poultry (Chicks)	Due riboflavin deficiency

DD: riboflavin deficiency.

Cyanosis

A bluish discoloration of the skin or mucous membranes because of poor circulation or inadequate oxygenation of blood.

Element	Species	Comments
Cadmium toxicity	Humans	
Cobalt toxicity	Humans	
Nickel toxicity	Humans	Inhalation
Samarium toxicity	Humans	Inhalation
Sulfur toxicity	Horses	Accidental poisoning—due respiratory depression
Selenium toxicity	Humans	Extremities
Yttrium toxicity	Humans	Industrial exposure
Zinc toxicity	Birds	

DD: airway obstruction, asthma, blood clots, low cardiac output, hemoglobin abnormalities, heart failure, pneumonia, acute pulmonary edema, Raynaud's disease, vasoconstriction [cold, medications (beta-blockers, ergot poisoning), venous stasis].

Cytochrome C in tissues (decreased amounts)

A transmembrane enzyme involved in the respiratory electron transfer chain.

Element	Species	Comments
Iron deficiency	Rats	Heart, intestine, kidneys, liver, skeletal muscles

DD: bone and skeletal diseases, cancer, diabetes, inherited mitochondrial cytochrome C oxidase deficiency, myocardial ischemia/reperfusion, neurodegenerative diseases.

Cytochrome CYT P450 change

Terminal oxidase enzymes in electron transfer chains and oxidize steroids, fatty acids, and xenobiotics.

Element	Species
Cerium toxicity	Mice

Cytokine—Increased activity

Cytokines are involved in autocrine, endocrine, and paracrine signaling as immunomodulating agents.

Element	Species	Comments
Cerium toxicity	Rats	Proinflammatory cytokines

Cytokine suppression

Cytokines are involved in autocrine, endocrine, and paracrine signaling as immunomodulating agents.

Element	Species	Comments
Zinc deficiency	Humans	IL-1Beta, IL-2, IL-6, TNFalpha

Cytotoxic effects

Element	Species	Comments
Cerium toxicity	Rats	Reduced cell viability
Lanthanum toxicity	Human cells	*In vitro: Lung macrophages, chromosome abnormalities*

Deafness/hearing loss

Element	Species	Comments
Arsenic toxicity	Humans (children)	Chronic toxicity
Mercury toxicity	Humans	

DD: diabetes, infections (viruses-measles, mumps) injury, loud noise exposure, meningitis, ototoxic drugs, shingles.

Death

Element	Species	Comments
Aluminum toxicity	Humans, rats	*Humans:* dialysis patients with Al encephalopathy *Rats:* neonatal deaths
Antimony toxicity	Humans	
Arsenic deficiency	Goats (lactating), rats	
Arsenic toxicity	Humans, dogs, fish, rabbits, toads, turkeys (young)	
Barium toxicity	Humans, dogs, mice, rats	Due cardiac arrest
Beryllium toxicity	Humans, hamsters, monkeys, rabbits	

Boron toxicity	Humans	
Cadmium deficiency	Goats	
Cadmium toxicity	Humans, fish, mice, rats	
Cesium toxicity	Humans, mice, rats	
Calcium deficiency	Camels, cattle, reptiles, sheep	*Cattle, Sheep:* milk fever
Calcium toxicity	Dogs, poultry	Due soft tissue calcification due chronic hypercalcemia, e.g., myocardium
Cerium toxicity	Mice, rats	
Chlorine deficiency	Humans	
Chlorine toxicity	Humans	Ingested, inhaled
Chromium toxicity	Humans, hamsters, mice (*inhalation*)	
Cobalt deficiency	Cattle, deer, sheep	
Cobalt toxicity	Cattle, rats, sheep	
Copper deficiency	*Dogs*, deer, guinea pigs, poultry, sheep (lambs)	Major blood vessel aneurysm and rupture, e.g., aorta *Dogs:* unpublished study on sudden death in racing Grayhounds
Copper toxicity	Dogs, fish, goats, marine mammals, marsupials, sheep	Wombats
Europium toxicity	Mice	
Fluorine deficiency	Goats (*kids*), mice	Reduced life expectancy
Fluorine toxicity	Humans, cattle (neonates), goats, honeybees, pigs, sheep (neonates)	*Goats:* intrauterine and neonatal *Honeybees:* F in insecticides
Indium toxicity	Humans	
Iodine deficiency	Cats, fish, horses (neonates), pigs	
Iodine toxicity	Rabbits	
Iron deficiency	Mink, pigs, poultry (embryos)	
Iron toxicity	Humans, bats, cats, cattle, dogs, fish, horses (foals), pigs, sheep	
Lead toxicity	Humans, amphibians (*frogs*), birds, buffaloes, cattle, fish, marine mammals (*dolphins*), poultry, sheep, reptiles	*Waterfowl:* lead fishing weights
Lithium toxicity	Humans, goats	
Lutetium toxicity	Cats, dogs	Acute toxicity studies
Magnesium deficiency	Birds (*chicks*), buffaloes, cattle, gerbils, goats, horse, mice, pigs, poultry (rare), rats, sheep	
Magnesium toxicity	Humans, goats, poultry	*Poultry:* increased mortality
Manganese deficiency	Birds (*embryos*), guinea pigs, poultry (chickens, ducks), rats, sheep	Rats, Guinea pigs: neonatal
Mercury toxicity	Humans, cats, dogs, ferrets, horses, marine mammals (*seals*) mink, pigs, sheep	
Molybdenum deficiency	Humans (postnatal), goats (kids), poultry (chicks)	*Humans:* genetic disease
Molybdenum toxicity	Cattle, fish, guinea pigs, rabbits, rats	*Fish:* death of eggs postfertilization in rainbow trout
Nickel toxicity	Humans, fish, hamsters	
Palladium toxicity	Mice, rabbits, rats	
Phosphorus deficiency	Buffaloes, horses, poultry	

Potassium deficiency	Cattle, fish, poultry, rats, sheep	
Potassium toxicity	Cattle (calves)	
Praseodymium toxicity	Dogs	
Samarium toxicity	Dogs	
Selenium deficiency	Goats, pigs, sheep	*Sheep:* early embryonic death *Pigs:* cardiomyopathy
Selenium toxicity	Humans, birds, buffaloes, cattle, deer, fish, pigs, nonhuman primates, rabbits, sheep	*Macaque*
Silicon toxicity	Humans	Silicosis
Sodium deficiency	Humans, cattle, marine mammals, rats	
Sodium toxicity	Humans, birds, dogs, ferrets, marsupials (wombats)	Corrosive chemicals, e.g., bleach (sodium hypochlorite)
Sulfur deficiency	Goats, sheep	
Sulfur toxicity	Dogs, goats, horses, rabbits, rats, sheep	Dogs—experimental Horses—accidental poisoning
Thallium toxicity	Humans, mice, rats	
Thyroid toxicity	Rabbits	
Titanium toxicity	Fish	
Tungsten toxicity	Humans	Increased mortality rate due lung cancer
Uranium toxicity	Humans	
Vanadium deficiency	Goats	Kids
Vanadium toxicity	Cattle, poultry (chicks), rats, sheep	
Ytterbium toxicity	Guinea pigs	Cardiovascular collapse
Zinc deficiency	Birds (*embryos*), nonhuman primates (fetal), pigs, poultry, sheep, fish	
Zinc toxicity	Humans, cattle, dogs, fish, pigs, poultry (chicks), rats, sheep (postnatal)	
Zirconium toxicity	Birds, fish, guinea pigs, mice, rats	
Ionizing radiation	Various	Exposure

DD: cancer, infections, major organ system failures, toxicity, trauma.

Decubital ulcers

Element	Species
Selenium toxicity	Cattle

DD: infection, contact irritation, trauma.

Defecation frequency reduced

Element	Species
Palladium toxicity	Rabbits

Dehydration

Inadequate fluid status in the blood and tissues.

Element	Species	Comments
Arsenic toxicity	Humans	Acute
Chlorine deficiency	Humans, poultry	
Chlorine toxicity	Humans	
Iron toxicity	Horses (foals)	
Lead toxicity	Birds, mink	
Mercury toxicity	Humans	
Phosphorus toxicity	Cats	Only one report
Sodium deficiency	Cattle, dogs	*Dogs:* dry, tacky mucous membranes
Potassium deficiency	Dogs	
Zinc toxicity	Dogs	

DD: inadequate water intake, excessive losses (diabetes, diarrhea, urination, vomiting).

Delayed gastric emptying

Food retained in the stomach for longer than usual.

Element	Species
Aluminum toxicity	Humans

DD: gastric atony, impaired gastric outflow (foreign body, tumor, pyloric stenosis).

Delirium

Disturbed state of mind with illusions, incoherence and restlessness.

Element	Species
Bismuth toxicity	Humans
Lithium toxicity	Humans
Selenium toxicity	Humans
Thallium toxicity	Humans

DD: alcohol or drug intoxication or withdrawal, heart attack, hypocalcemia, hyponatremia, liver diseases, lung disease, medications, metabolic imbalances, terminal illness, trauma.

Delusions

Idiosyncratic beliefs or impressions contradicted by reality or rational argument.

Element	Species
Bismuth toxicity	Humans

DD: alcoholism, brain disease, drug abuse, poor sight, poor hearing, psychotic disease, schizophrenia, stress.

Dementia

A group of signs associated with declining brain function.

Element	Species	Comments
Aluminum toxicity	Humans	Lapresle et al. (1975) and McLaughlin et al. (1962)
Bismuth toxicity	Humans	

DD: Alzheimer's disease, amyloidosis, corticobasal degeneration, frontotemporal dementia, Huntingdon's disease, hydrocephalus, Lewy body dementia, progressive supranuclear palsy, vascular dementia.

Demyelination nervous tissue

Loss of the protective myelin sheath around nerve fibers in brain, optic nerves, and spinal cord.

Element	Species	Comments
Arsenic toxicity	Humans	
Tellurium toxicity	Humans	On CT and MRI scans

DD: autoimmune disease, genetic inheritance, brain hypoxia, infection (viruses), vascular damage to brain.

Dendriform pulmonary ossification

Mature bone is in peripheral lung interstitium and typically occurs in interstitial pneumonia (UIP).

Element	Species
Cerium toxicity	Humans

DD: infections, other toxin inhalation.

Dental abnormalities

Element	Species	Comments
Boron toxicity	Rats	Incisors lack pigmentation
Calcium deficiency	Humans, dogs, ferrets, guinea pigs, marsupials (sugar-gliders), sheep, squirrels	*Dogs, ferrets*: Tooth loss *Humans*: Poor mineralization enamel and dentin *Guinea pigs:* enamel hypoplasia *Squirrels:* Malocclusion *Sugar-gliders:* periodontal disease
Cadmium toxicity	Humans	Discolored teeth
Fluorine toxicity	Humans, cattle, horses, pigs, sheep	*Humans, Cattle, Pigs:* mottled, pitted, white chalky enamel, discolored teeth (yellow, red, green, black) *Horses:* abscesses, pulp exposed, jaw swelling, cut cheek and gums, excess wear, teeth discoloration
Iron deficiency	Rats	White incisor teeth
Magnesium deficiency	Guinea pigs	Enamel discoloration and erosion

Molybdenum toxicity	Rats	
Phosphorus deficiency	Humans	Poor tooth development
Uranium toxicity	Rats	Teeth malformation

Depression

Feeling despondent and dejected.

Element	Species	Comments
Arsenic toxicity (acute)	Various	
Calcium deficiency	Cattle	Milk fever
Calcium toxicity	Humans	
Cobalt deficiency	Sheep	
Iodine deficiency	Cattle, sheep	
Iron toxicity	Cattle, dogs, horses, sheep, reptiles	
Lead toxicity	Humans (adults), birds, cattle, reptiles	
Lithium toxicity	Goats	
Magnesium deficiency	Cats, buffaloes	
Magnesium toxicity	Humans	
Mercury toxicity	Humans, dogs, horses, sheep	
Phosphorus toxicity	Cats	Only one report
Selenium deficiency	Dogs	
Selenium toxicity	Birds, cattle	
Sodium toxicity	Ferrets	
Zinc deficiency	Humans	Anhedonic
Zinc toxicity	Humans, dogs, reptiles	

DD: genetics, major illness, medications (corticosteroids, isotretinoin, interferon-alpha), mental illness, social pressures/ stress.

Depth perception—Loss

Element	Species
Lead toxicity	Birds

DD: optic nerve hypoplasia, optic nerve edema, papilledema.

Dermatitis

Inflammation of the skin.

Element	Species	Comments
Aluminum toxicity	Humans	Contact dermatitis
Antimony toxicity	Humans	Antimony spots: Papules, pustules around sweat and sebaceous glands with eczema and lichenification Palmar keratitis
Arsenic deficiency	Goats	
Arsenic toxicity	Humans, various	*Humans:* exfoliative
Beryllium toxicity	Humans	Contact dermatitis, skin ulcers, hypersensitivity reaction
Bismuth toxicity	Humans	Erythematous rash
Boron toxicity	Humans	Exfoliative
Cadmium toxicity	Various	Scaly skin
Calcium deficiency	Birds	
Chlorine toxicity	Humans	Dermal contact: irritation, inflammation
Chromium toxicity	Humans	Industrial contact exposure. Irritation, ulcers
Copper deficiency	Rabbits	
Dysprosium toxicity	Rabbits	Severe irritation to abraded skin
Erbium toxicity	Rabbits	Severe irritation to abraded skin
Gallium toxicity	Humans	Acute
Magnesium toxicity	Birds	
Manganese deficiency	Birds	
Mercury toxicity	Humans, horses	Rare except contact dermatitis *Horses:* ulcerative
Molybdenum toxicity	Rabbits	And alopecia
Nickel deficiency	Goats, minipigs	
Palladium toxicity	Humans, rabbits	Contact dermatitis
Samarium toxicity	Humans, rabbits	Contact abraded skin: severe irritation and (*humans*) ulcers
Selenium deficiency	Birds	
Selenium toxicity	Humans, nonhuman primates	*Humans: acute:* chemical burns, erythema, heat, pain, rash, swelling; *chronic:* chronic dermatitis *Macaques*
Sulfur toxicity	Humans	Erythema, pruritus
Tellurium toxicity	Humans	Skin dry, erythema
Tin toxicity	Humans	
Tellurium toxicity	Rats	*Experimental:* Erythema and edema feet
Tungsten toxicity	Humans	Skin exposure—suspected but may be due to cobalt exposure
Ytterbium toxicity	Humans, guinea pigs	*Humans:* Skin irritation (contact) *Guinea pigs:* Ulcers on abraded skin

Zinc deficiency	Humans, birds, guinea pigs, nonhuman primates (*squirrel monkey*) poultry, rabbits, rats	*Humans:* skin rash *Birds:* exfoliative dermatitis on face and legs *Guinea pigs, Poultry:* scales *Rabbits:* sores around mouth, wet matted hair lower jaw and ruff *Rats:* epidermal thickening
Zirconium toxicity	Humans	Contact dermatitis after exposure

DD: allergies, ectoparasites, infections (bacteria, fungal, viral), contact irritation, nutritional deficiencies (vitamins).

Dexterity loss

Inability to perform manual functions.

Element	Species
Boron deficiency	Humans

DD: multiple sclerosis, neurological trauma, osteoarthritis.

Diarrhea/scours

The passage of liquid feces (in humans three times or more a day).

Element	Species	Comments
Antimony toxicity	Humans	
Arsenic toxicity	Humans, cattle, dogs	Acute and chronic may be hemorrhagic
Barium toxicity	Humans, rats	Watery diarrhea
Boron toxicity	Humans	
Bromine toxicity	Humans	
Cadmium toxicity	Humans	
Cesium toxicity	Humans	
Chlorine deficiency	Humans	
Chlorine toxicity	Humans	
Chromium toxicity	Humans	May be hemorrhagic
Copper deficiency	Cattle, deer, goats, sheep	Severe if Mo high in pasture
Fluorine toxicity	Cattle	
Gadolinium toxicity	Rats	
Gallium toxicity	Humans	
Iron deficiency	Cats, dogs, horses	
Iron toxicity	Humans, cats, dogs, fish, horses (foals), pigs, sheep	
Lead toxicity	Birds	Bright green feces
Lithium toxicity	Humans, goats	

Magnesium toxicity	Humans, birds, cattle, goats, poultry, sheep	
Mercury toxicity	Humans, dogs, pigs, sheep	Bloody
Molybdenum toxicity	Cattle, rats	
Nickel toxicity	Humans, pigs	
Phosphorus toxicity	Cattle	
Samarium toxicity	Mice	
Selenium toxicity	Humans, cattle	
Silver toxicity	Humans	
Sodium toxicity	Birds	
Sulfur toxicity	Goats, sheep	
Tin toxicity	Humans	
Vanadium toxicity	Humans, rats, sheep	
Zinc deficiency	Humans	
Zinc toxicity	Humans, birds, cattle, dogs, rats, sheep	
Zirconium toxicity	Humans	Watery or bloody
Ionizing radiation	Various	Exposure

DD: dietary allergy, dietary intolerance, dietary indiscretion, gastrointestinal disease, infection, liver disease, malabsorption, pancreatic insufficiency, parasitism, toxins.

Dilated pupils/miosis

Element	Species	Comments
Calcium deficiency	Cattle	Milk fever

DD: aphakic, brain stem stroke, headaches, Horner's syndrome, intracranial hemorrhage, iridocyclitis, Lyme disease, multiple sclerosis (MS), neurosyphilis, uveitis.

Diplopia

Double vision.

Element	Species
Bismuth toxicity	Humans

DD: brain aneurysm, brain swelling, brain tumor, cataracts, corneal abnormalities, cranial nerve palsies, dry eye, stroke, trauma to head.

Disorientation (directional)

Element	Species	Comments
Aluminum toxicity	Humans	Especially dialysis patients with Al encephalopathy
Bismuth toxicity	Humans	
Europium toxicity	Mice	
Lithium toxicity	Humans	
Selenium toxicity	Humans	
Ionizing radiation	Various	Exposure

DD: delirium, dementia infections, medications, trauma.

Disseminated intravascular coagulation (DIC)

Abnormal blood clotting throughout the body's blood vessels.

Element	Species
Zinc toxicity	Dogs

DD: blood transfusion reaction, cancer (especially leukemia), pancreatitis, sepsis, fungal infection, liver disease, retained placenta, postsurgery/general anesthesia.

Dizziness

Feeling faint, unsteady, weak, woozy, unsteady, vertigo.

Element	Species	Comments
Antimony toxicity	Humans	
Barium toxicity	Humans	
Bromine toxicity	Humans	
Gadolinium toxicity	Humans	After injection for MRI scan
Germanium toxicity	Humans	
Iron deficiency	Humans	Due anemia
Selenium toxicity	Humans	
Sodium deficiency	Humans	
Tellurium toxicity	Humans	
Tin toxicity	Humans	Organotin compounds
Zinc toxicity	Humans	
Ionizing radiation	Various	Exposure

DD: anemia, anxiety, dehydration, diabetes, ear infection, heat exhaustion, hypoglycemia, migraine, motion sickness, stress dehydration, or heat exhaustion, stress.

DNA (deoxyribonucleic acid) damage

Change in the basic structure of DNA which can lead to mutations and epimutations that progress to cancer.

Element	Species	Comments
Indium toxicity	Humans	As indium tin oxide

DD: chemotherapy, ionizing radiation exposure.

Dropped wings

Element	Species
Lead toxicity	Birds

DD: botulism, muscle weakness, toxin exposure, trauma, severe weakness.

Drowsiness

Feeling abnormally sleepy during the day.

Element	Species	Comments
Arsenic toxicity	Humans	Acute and chronic toxicity
Iodine deficiency	Dogs	
Lithium toxicity	Humans	
Magnesium toxicity	Various	
Tellurium toxicity	Humans	
Thallium toxicity	Humans	Can be extreme

DD: anxiety, depression, emotional state, insomnia, medications, mental state, psychological state, sleep apnoea, sleep deprivation, stress.

Dry mouth

Element	Species
Lithium toxicity	Humans
Tellurium toxicity	Humans

Dullness

Element	Species
Iodine deficiency	Dogs
Iodine toxicity	Cattle

Duodenal ulcers

Element	Species
Zinc toxicity	Dogs

Dwarfism

Element	Species	Comments
Iodine deficiency	Humans, dogs, pigs	
Manganese deficiency	Fish, guinea pigs, pigs, poultry (chickens, ducks), rabbits, rats	Shortened limbs *Fish:* short body dwarfism
Zinc deficiency	Humans, various	

Dysarthria

Slurred speech.

Element	Species
Bismuth toxicity	Humans
Lead toxicity	Humans (children)
Lithium toxicity	Humans
Mercury toxicity	Humans

Dysgeusia

Distorted sense of taste.

Element	Species
Gadolinium toxicity	Mink

Dysmenorrhoea

Element	Species	Comments
Mercury toxicity	Humans	
Selenium toxicity	Nonhuman primates	Macaques

Dysphagia

Element	Species	Comments
Iodine toxicity	Cattle	Difficulty swallowing
Iron deficiency	Humans	

Magnesium deficiency	Humans	
Mercury toxicity	Humans	
Selenium deficiency	Horses	Myopathy

Dysphonia

Element	Species
Mercury toxicity	Mink

Dyspnoea

Element	Species	Comments
Aluminum toxicity	Humans	
Barium toxicity	Humans	
Beryllium toxicity	Humans	
Cadmium toxicity	Humans	
Calcium deficiency	Squirrels	Labored breathing
Cerium toxicity	Rats	
Chlorine deficiency	Humans	
Dysprosium toxicity	Mice	Labored breathing
Gallium toxicity	Humans	
Germanium toxicity	Humans	
Holmium toxicity	Mice	
Iodine deficiency	Birds	
Iron toxicity	Cattle	
Lutetium toxicity	Mice	
Mercury toxicity	Humans, cattle	
Molybdenum toxicity	Humans	
Nickel toxicity	Humans	Inhalation
Praseodymium toxicity	Mice	
Samarium toxicity	Humans, mice	
Selenium deficiency	Dogs	
Selenium toxicity	Birds, cattle	
Sulfur toxicity	Humans, goats, sheep	
Vanadium toxicity	Rats	Inhalation
Yttrium toxicity	Rats	
Zinc toxicity	Humans, cats	Inhalation
Zirconium toxicity	Humans	

Dyspraxia (speech impediment)

Element	Species
Aluminum toxicity	Humans

Dystocia

Difficult birth.

Element	Species
Aluminum toxicity	Humans
Calcium deficiency	Reptiles
Zinc deficiency	Humans

Ear drooping

Element	Species
Iron deficiency	Pigs

ECG changes

Element	Species	Comments
Antimony toxicity	Humans	Prolonged QT, small T wave, inverted T wave, torsade de pointes, ventricular ectopics, ventricular tachycardia, ventricular fibrillation
Arsenic toxicity	Humans	Arrhythmias
Cesium toxicity	Humans	Prolonged QT interval
Cobalt toxicity	Humans	Low voltage ECG
Lithium toxicity	Humans	Flat T waves, long QT, U waves
Magnesium deficiency	Humans	Arrhythmias
Magnesium toxicity	Humans	Arrhythmias, prolonged PR, wide QRS, peaked T waves
Nickel toxicity	Humans	Myocarditis
Selenium deficiency	Sheep (*lambs*)	Tachycardia, increased sinus rhythm, tall P wave, shorter PR, QT and ST intervals, narrow QRS interval, short T wave duration.
Selenium toxicity	Humans	Flat inverted T wave; long ST; prolonged QT interval

Eczema

Element	Species
Zinc deficiency	Humans

Edema

Fluid collecting in the cavities or tissues of the body.

Element	Species	Comments
Copper deficiency	Humans, guinea pigs	*Guinea pigs:* cerebral edema
Iron deficiency	Horses, pigs	*Pigs*—subcutaneous Horses—pulmonary
Lead toxicity	Birds	Cephalic edema
Mercury toxicity	Horses	Ventral edema
Palladium toxicity	Humans, rabbits	Topical
Selenium deficiency	Birds, horses	*Birds:* subcutaneous edema *Horses:* ventral edema
Selenium toxicity	Humans, dogs	*Humans (inhalation):* pulmonary edema *Dogs:* ventral edema
Sodium toxicity	Cattle, dogs	*Cattle:* udder edema *Dogs:* pharyngeal edema, e.g., after swallowing corrosives like bleach (sodium hypochlorite)
Tellurium toxicity	Rats	*Teratogenic effect*
Zinc deficiency	Cattle	Limb edema

Egg production—Poor quality/reduced

Element	Species	Comments
Calcium deficiency	Birds, Reptiles	*Birds:* Abnormal shape; soft, thin shells, egg binding *Reptiles:* Poor mineralization, deformities
Copper deficiency	Poultry	Reduced number, reduced hatchability, shell-less, abnormal shape, abnormal texture
Iodine deficiency	Poultry (experimental)	Reduced number and size
Iodine toxicity	Poultry	Number, size, hatchability
Magnesium deficiency	Poultry	
Magnesium toxicity	Birds, poultry	*Birds, Poultry:* reduced number, thin shells
Manganese deficiency	Birds, fish	*Birds:* reduced number and hatchability *Fish:* reduced Mn in eggs, reduced hatchability
Phosphorus deficiency	Birds	Abnormal shape; egg binding
Phosphorus toxicity	Poultry	
Potassium deficiency	Poultry	Number, weight, shell thickness, albumin content

Selenium toxicity	Poultry	Poor production, low weight
Sodium deficiency	Poultry	Number, size and hatchability all reduced
Ytterbium toxicity	Fish	Reduced hatchability and survival
Zinc deficiency	Birds, fish, poultry	*Birds, Fish:* reduced production, reduced hatchability *Poultry:* delayed production, Reduced number.

Electrolyte imbalance

Element	Species
Antimony toxicity	Humans
Chromium toxicity	Humans
Gallium toxicity	Humans
Ionizing radiation	Various

Emaciation

Element	Species
Calcium deficiency	Camels
Chlorine deficiency	Cattle
Cobalt deficiency	Cattle, deer, sheep
Copper deficiency	Buffaloes, deer
Iron deficiency	Mink
Lead toxicity	Birds
Palladium toxicity	Rats
Phosphorus deficiency	Buffaloes, dogs
Potassium deficiency	Cattle, goats, mice, sheep
Selenium toxicity	Buffaloes, cattle, horses, pigs
Sodium deficiency	Goats
Sulfur deficiency	Goats, sheep
Zinc deficiency	Dogs, mice

Embolism

Element	Species
Yttrium toxicity	Humans

Embryotoxicity

Element	Species	Comments
Ytterbium toxicity	Fish	Developmental abnormalities

Emotional lability

Element	Species
Mercury toxicity	Humans

Emphysema

Element	Species	Comments
Antimony toxicity	Humans	chronic
Cerium toxicity	Humans	
Cadmium toxicity	Humans	
Nickel toxicity	Hamsters	Inhalation
Sulfur toxicity	Humans	
Zinc toxicity	Guinea pigs	Inhalation

Encephalopathy

Element	Species	Comments
Aluminum toxicity	Humans	Especially in patients on dialysis with renal failure
Arsenic toxicity	Humans	
Bismuth toxicity	Humans	Confusion, disorientation, seizures (latter rare)
Cadmium toxicity	Mice, rats	Fetal development
Chromium deficiency	Humans	
Lead toxicity	Humans, dogs	
Selenium deficiency	Birds	Encephalomalacia
Tungsten toxicity	Humans	Rare

Endomyocarditis

Element	Species
Cerium toxicity	Humans

Endurance activity reduced

Element	Species
Lead toxicity	Humans

Enteropathy

Element	Species	Comments
Iron deficiency	Humans (children)	Associated with anemia
Sulfur toxicity	Cattle	Enteritis

Eosinophilia

Element	Species
Barium toxicity	Rats

Epigastric pain

Element	Species
Zinc toxicity	Humans

Epiphyseal dysplasia

Element	Species
Manganese deficiency	Rats

Epiphyseal plate widened

Element	Species
Calcium deficiency	Camels, dogs

Epistaxis

Element	Species	Comments
Bismuth toxicity	Humans	Due thrombocytopenia
Bromine toxicity	Humans	

Erectile dysfunction

Element	Species
Arsenic toxicity	Humans

Erythema

Reddening of the skin.

Element	Species	Comments
Bismuth toxicity	Humans	
Boron toxicity	Various	
Mercury toxicity	Humans	Contact dermatitis
Palladium toxicity	Humans, rabbits	Contact dermatitis
Selenium toxicity	Humans	Contact dermatitis
Sulfur toxicity	Humans	Contact dermatitis

Erythrocyte abnormalities

Red blood cell abnormalities.

Element	Species	Comments
Copper deficiency	Pigs	Reduced mean corpuscular volume
Iron deficiency	Poultry (rare)	
Lead toxicity	Horses, poultry	*Horses:* stippling

Erythrocytopenia

Element	Species	Comments
Barium toxicity	Rats	
Lead toxicity	Amphibians	Frogs

Erythropoiesis impaired

Element	Species
Aluminum toxicity	Rats

Eschar

Dead skin that sheds off healthy skin.

Element	Species
Palladium toxicity	Rabbits

Esophageal impaction

Element	Species
Lead toxicity	Birds

Esophagitis

Element	Species	Comments
Arsenic toxicity	Humans	Acute
Chlorine toxicity	Humans	Caustic effect—hypochlorite

Esophageal perforation

Element	Species	Comments
Chorine toxicity	Humans	Caustic effect—hypochlorite
Zirconium toxicity	Humans	

Estradiol hormone—Reduced synthesis

Element	Species
Aluminum toxicity	Rats
Chromium toxicity	Amphibians

Estrus abnormalities

Element	Species	Comments
Molybdenum toxicity	Rat	
Phosphorus deficiency	Goats, sheep	Estrus suppression
Zinc deficiency	Cattle, hamster, monkeys (*squirrel monkeys*), sheep	*Cattle, Hamster, Sheep:* abnormal cycle *Monkeys:* cessation of estrus cycle

Excitability

Element	Species	Comments
Lead toxicity	Cats, amphibians	Frogs
Mercury toxicity	Humans	
Potassium toxicity	Cattle	Experimental
Selenium toxicity	Cattle	

Exhaustion

Element	Species
Sodium deficiency	Humans, dogs

Exophthalmia

Bulging of the eye out of the orbit.

Element	Species	Comments
Tellurium toxicity	Rats	Teratogenic effect

Exostosis

New bone formed on the surface of a bone.

Element	Species	Comments
Fluorine toxicity	Humans, cattle, horses, pigs, sheep	
Molybdenum toxicity	Cattle, rats	Especially mandible/maxilla

Expectoration

Ejection of phlegm or mucus from the throat or lungs by coughing or spitting.

Element	Species	Comments
Sulfur toxicity	Humans	With blood streaks

Extensor rigidity

Rigid contractions of extensor muscles: sometimes indicate the site of the motor neuron lesion associated with the disorder. Injury to the cerebellum produces an increased tone of extensor muscles.

Element	Species
Mercury toxicity	Dogs

Exudate formation

When fluid leaks out of blood vessels into tissues.

Element	Species	Comments
Palladium toxicity	Rats	Response to contact
Selenium deficiency	Birds	With vitamin E deficiency

Eye irritation

Element	Species	Comments
Antimony toxicity	Humans	Occupational exposure
Bismuth toxicity	Humans	
Chlorine toxicity	Humans	Gas exposure
Europium toxicity	Humans	
Mercury toxicity	Humans	Occupational exposure
Samarium toxicity	Rabbits	Blinking, not conjunctivitis
Selenium toxicity	Humans	
Ytterbium toxicity	Humans	

Eyelid drooping

Element	Species	Comments
Iron deficiency	Pigs	
Phosphorus toxicity	Humans	Ptosis

Facial deformity

Element	Species
Calcium deficiency	Camels

Facial expression changes

Element	Species
Mercury toxicity	Humans

Falling off perch

Element	Species
Calcium deficiency	Birds

Falling to recumbency

Element	Species
Mercury toxicity	Mink

Falls

Element	Species
Bismuth toxicity	Humans

Fanconi syndrome

A renal tubule disorder resulting in excess amino acids, bicarbonate, glucose, phosphates, potassium, and uric acid being excreted in urine.

Element	Species
Uranium toxicity	Humans

Fasciculations

Element	Species
Selenium toxicity	Humans

Fatigue

Element	Species
Beryllium toxicity	Humans
Calcium toxicity	Humans, reptiles
Cadmium toxicity	Humans
Chlorine deficiency	Humans
Iron deficiency	Humans, horses
Lead toxicity	Humans
Lithium toxicity	Humans
Magnesium deficiency	Humans
Manganese toxicity	Humans

Mercury toxicity	Humans
Selenium toxicity	Humans
Sodium deficiency	Human, dog, horses
Tellurium toxicity	Humans
Titanium allergy	Humans
Zinc toxicity	Humans
Ionizing radiation	Various

Fatty liver

Element	Species	Comments
Antimony toxicity	Guinea pigs	
Arsenic toxicity	Humans	Hepatic steatosis—acute
Bromine deficiency	Goats	
Cerium toxicity	Rats	
Cobalt deficiency	Cattle, sheep	
Cobalt toxicity	Goats, sheep	*Fatty degeneration*
Germanium toxicity	Humans, mice, rats	*Experimental*
Iron toxicity	Marine mammals	Dolphins
Manganese deficiency	Dogs, mice, pigs	

Fat infiltration of muscle

Element	Species
Bromine deficiency	Goats

Feather abnormalities

Element	Species	Comments
Calcium deficiency	Birds	Brittle; frayed feathers
Copper deficiency	Birds	Pale colored feathers
Iodine deficiency	Poultry	*Experimental*
Iron deficiency	Poultry	
Magnesium deficiency	Poultry	
Magnesium toxicity	Birds	Brittle; frayed feathers
Manganese	Various	
Molybdenum deficiency	Poultry	Feather defects
Selenium deficiency	Birds	Brittle; frayed feathers
Sodium deficiency	Birds	Feather plucking
Zinc deficiency	Birds, poultry	*Birds:* brittle; frayed feathers *Poultry:* poor and broken feathers
Zinc toxicity	Birds	Feather plucking

Fetal abnormalities

Element	Species	Comments
Aluminum toxicity	Mice, poultry, rats	*Mice:* bone, micronucleated erythrocytes *Poultry, Rats:* cardiac defects
Boron toxicity	Mice, rabbits	Cardiac and skeletal deformities
Cadmium toxicity	Hamsters, mice, rats	Encephalopathy, hydrocephaly, cleft lip, cleft palate, microphthalmia, micrognathia, club foot, dysplastic tail
Chromium toxicity	Guinea pigs, hamsters, mice	*Hamsters:* cleft palate, hydrocephalus, skeletal abnormalities *Mice:* anencephaly, exencephaly
Copper deficiency	Cats, cattle	Cats: kinked tail, deformed carpi Cattle: rickets
Iodine deficiency	Humans, cattle, sheep	Impaired brain development
Manganese deficiency	Guinea pigs, mice, mink, rabbits, rats	Otolith abnormalities
Mercury toxicity	Cats	
Selenium toxicity	Poultry, sheep	*Poultry:* deformities *Sheep:* eyes
Uranium toxicity	Mice	Cleft palate, bipartite sternebrum
Zinc deficiency	Nonhuman primates, poultry, rats	*Nonhuman primates:* malformations, small offspring *Poultry:* embryo deformities, short thick limbs, spinal curvature *Rats:* agenesis skeleton, limbs, toes

Fetal death rate increased

Element	Species	Comments
Molybdenum toxicity	Rats	
Selenium toxicity	Pigs, poultry	*Poultry:* fail to hatch
Zinc deficiency	Nonhuman primates	

Fetal resorption

Element	Species	Comments
Aluminum toxicity	Rats	
Boron toxicity	Rabbits	
Cobalt (cobalamin) deficiency	Poultry	Embryonic death
Copper deficiency	Guinea pigs, rats	
Iodine deficiency	Camels, cats, cattle, sheep	
Molybdenum toxicity	Rats	

Fetal size reduced

Element	Species
Molybdenum toxicity	Rats
Zinc deficiency	Nonhuman primates

Fever

Element	Species
Arsenic toxicity	Humans
Beryllium toxicity	Humans
Cadmium toxicity	Humans
Lead toxicity	Humans
Nickel toxicity	Humans
Palladium toxicity	Humans
Selenium deficiency	Horses
Selenium toxicity	Humans (*inhalation*)
Zinc toxicity	Humans
Ionizing radiation	Various

Fibrosis—Tissue

Element	Species	Comments
Arsenic toxicity	Humans	Hepatic fibrosis
Dysprosium toxicity	Mice	
Erbium toxicosis	Mice	
Mercury toxicity	Horses	Interstitial renal fibrosis

Fibrous osteodystrophy (replacement bone with fibrous tissue)

Element	Species	Comments
Calcium deficiency	Various—camels, pigs	Ca:P less than 1:1
Phosphorus toxicity	Various—camels, pigs	Ca:P less than 1:1

Fin and skin erosions

Element	Species	Comments
Zinc deficiency	Fish	Rainbow trout; Common carp

Fingernail changes

Element	Species	Comments
Arsenic toxicity	Humans	Chronic toxicity
Selenium toxicity	Humans	Brittle, deformed, discolored, loss, white spots, longitudinal striations

Flatus

Element	Species
Lithium toxicity	Humans

Fluid losses

Element	Species	Comments
Arsenic toxicity	Humans	Gastrointestinal losses
Antimony toxicity	Humans	

Flushes

Element	Species	Comments
Cobalt toxicity	Humans	Due vasodilation
Gadolinium toxicity	Humans	
Magnesium toxicity	Humans	Facial
Selenium toxicity	Humans	Facial

Folic acid—Low in serum

Element	Species	Comments
Iron deficiency	Rats	Due increased folic acid use in heme catabolism

Follicle stimulating hormone (FSH)—Increased production

Element	Species
Chromium toxicity	Amphibians

Food conversion efficacy reduced

Element	Species	Comments
Calcium deficiency	Dogs	
Phosphorus deficiency	Fish	
Potassium deficiency	Cattle	
Selenium toxicity	Fish	
Sodium deficiency	Goats, pigs, poultry	
Sodium toxicity	Fish	
Sulfur deficiency	Goats, sheep	*Sheep:* reduced digestibility
Tin toxicity	Rats	
Zinc deficiency	Cattle, poultry	
Zinc toxicity	Cattle, pigs, sheep	

Food intake decreased (anorexia; inappetence)

Element	Species
Arsenic deficiency	Goats
Beryllium toxicity	Humans
Boron toxicity	Humans, cattle
Bromine deficiency	Goats
Cadmium toxicity	Various
Calcium deficiency	Fish
Calcium toxicity	Cats, cattle, pigs, poultry
Chlorine deficiency	Humans, cattle, rats
Cobalt (cobalamin) deficiency	Cattle, sheep
Cobalt (cobalamin) toxicity	Cattle, sheep
Iodine deficiency	Cattle
Iodine toxicity	Cattle, sheep
Iron deficiency	Humans (children), cattle (calves), pigs
Iron toxicity	Various
Magnesium deficiency	Cats, cattle, dogs, nonhuman primates, fish
Nickel toxicity	Cattle, dogs, ducks, mice, monkeys, pigs, poultry (chicks), rabbits, rats
Phosphorus deficiency	Various—cattle, dogs, horses, poultry, sheep
Potassium deficiency	Cats, cattle, horses, pigs, sheep, rats
Selenium toxicity	Cattle, dogs, poultry

Sodium deficiency	Cats, cattle, goats, horses, pigs, poultry, rats
Sodium toxicity	Cattle
Sulfur deficiency	Cattle, sheep
Sulfur toxicity	Cattle
Tin toxicity	Various
Uranium toxicity	Mice
Vanadium deficiency	Goats, sheep
Zinc deficiency	Humans, cattle, nonhuman primates, pigs, poultry, rabbits, rats, fish

Food intake increased

Element	Species
Chromium deficiency	Humans, mice, rats
Fluorine deficiency	Goats

Food utilization—Decreased

Element	Species
Magnesium toxicity	Poultry

Foot pad changes

Element	Species	Comments
Boron toxicity	Rats	Desquamation pads on feet

Forgetfulness

Element	Species
Mercury toxicity	Humans

Fractures

Element	Species	Comments
Aluminum toxicity	Humans	Postdialysis osteomalacia
Bismuth toxicity	Humans	
Calcium deficiency	Humans, amphibians, bears (polar), birds, buffaloes, camels, cats, cattle, dogs, horses, nonhuman primates, pigs, poultry, rabbits, reptiles, sheep, squirrels	Cats and dogs: common when fed an all meat ration due to nutritional secondary hyperparathyroidism

Copper deficiency	Cattle, goats, poultry	Spontaneous due poor bone mineralization
Molybdenum toxicity	Rabbits, sheep	
Phosphorus deficiency	Humans, birds, buffaloes	

Free radical formation

Element	Species
Iron toxicity	Humans

Fur chewing

Element	Species
Magnesium deficiency	Rabbits

Gait abnormalities

Element	Species	Comments
Bismuth toxicity	Humans	
Selenium toxicity	Cattle	Ataxia
Sodium deficiency	Goats, horses	Unsteady gait (wobbly)

Gamma-glutamyl transferase (GGT) increased activity

Element	Species
Bromine deficiency	Goats
Copper toxicity	Goats, sheep
Selenium deficiency	Horses

Gamma-glutamyl transpeptidase

Element	Species	Comments
Lanthanum toxicity	Humans	Dialysis patients

Gangrene

Element	Species	Comments
Arsenic toxicity	Humans	Due peripheral vascular disease
Selenium toxicity	Buffaloes, cattle	Distal extremities (India)

Garlic smell on breath/alliaceous breath

Breath smells of garlic or onions.

Element	Species
Antimony toxicity	Humans
Arsenic toxicity	Humans
Selenium toxicity	Humans, cattle
Tellurium toxicity	Humans, rats

Gastric achlorhydria

Element	Species	Comments
Iron deficiency	Humans (children)	Associated with anemia

Gastric atrophy

Element	Species	Comments
Iron deficiency	Humans (children)	Associated with anemia

Gastric carcinoma

Element	Species
Zirconium toxicity	Humans

Gastric hemorrhage

Element	Species	Comments
Arsenic toxicity	Humans	Acute
Barium toxicity	Humans	And dysplasia
Iron toxicity	Humans	
Palladium toxicity	Rats	

Gastric hyperkeratosis

Element	Species	Comments
Lanthanum toxicity	Rats	Also: eosinocyte infiltration

Gastric hyperplasia

Element	Species	Comments
Boron toxicity	Mice	And dysplasia

Gastrointestinal obstruction

Element	Species
Iron toxicity	Humans, dogs

Gastric pain

Element	Species	Comments
Antimony toxicity	Humans	Cramps
Arsenic toxicity	Humans	
Barium toxicity	Humans	And dysplasia
Calcium toxicity	Humans	

Gastric perforation

Element	Species
Iron toxicity	Humans
Zirconium toxicity	Humans

Gastric ulcers

Element	Species
Antimony toxicity	Humans
Iron toxicity	Humans
Palladium toxicity	Rats
Zinc toxicity	Dogs
Zirconium toxicity	Humans

Gastritis

Element	Species	Comments
Arsenic toxicity	Humans	Maybe bloody
Iron deficiency	Humans (children)	Associated with anemia
Sodium toxicity	Humans, dogs	Corrosive chemicals, e.g., bleach (sodium hypochlorite)

Gastrointestinal dysfunction

Element	Species	Comments
Selenium deficiency	Birds	Malabsorption, maldigestion

Gastrointestinal fibrosis

Element	Species
Iron toxicity	Humans

Gastrointestinal inflammation and ulceration

Element	Species	Comments
Aluminum toxicity	Humans	Granulomatous enteritis
Antimony toxicity	Humans	Check
Chromium toxicity	Cattle	Inflammation and ulceration of stomach, abomasum, and rumen
Iron toxicity	Humans, cats, dogs, horses (foals)	Jejunal erosions
Selenium toxicity	Humans (*severe*), Cattle	
Zinc toxicity	Birds	Ulceration; perforation

Gastrointestinal tract desquamation/necrosis

Element	Species	Comments
Cadmium toxicity	Rats	
Cerium toxicity	Mice	
Iron toxicity	Humans	Hemorrhagic necrosis
Mercury toxicity	Humans, pigs	Necrosis

Giant cell infiltration

Element	Species	Comments
Erbium toxicity	Mice	At injection site

Giddiness

Element	Species
Nickel toxicity	Humans

Gill filament degeneration

Element	Species	Comments
Iodine toxicity	Amphibians	Gill sloughing in axolotls
Magnesium deficiency	Fish	

Gingivitis

Inflammation of the gums.

Element	Species	Comments
Bismuth toxicity	Humans	
Mercury toxicity	Humans, horses	*Horses:* gingival swelling
Thallium toxicity	Humans	

Gliosis

Reactive change of glial cells (e.g., astrocytes, microglia, and oligodendrocytes) in response to damage to the central nervous system (CNS).

Element	Species	Comments
Mercury toxicity	Cats	Fibrillary

Glomerular filtration rate (GFR)—Reduced

Element	Species	Comments
Bismuth toxicity	Humans	
Niobium toxicity	Dogs	Single IV dose

Glomerulonephritis

Element	Species
Bismuth toxicity	Humans
Mercury toxicity	Humans
Palladium toxicity	Rats

Glossitis—Inflamed tongue

Element	Species
Iron deficiency	Human

Glucose-6-phosphatase—Decreased

Element	Species	Comments
Molybdenum toxicity	Rats	Liver

Glucose tolerance—Impaired

Element	Species
Aluminum toxicity	Rats
Chromium deficiency	Humans, guinea pigs, mice, monkeys, rats

Manganese deficiency	Humans
Manganese toxicity	Guinea pigs, rats

Glutamic oxaloacetic transaminase in serum (SGOT)

An enzyme in liver, heart, and other tissues. High SGOT in blood indicates liver or heart damage, cancer, or other diseases.

Element	Species
Arsenic toxicity	Fish
Cadmium toxicity	Fish
Copper toxicity	Fish, goats, sheep
Lead toxicity	Fish
Nickel toxicity	Fish
Selenium toxicity	Pigs
Titanium toxicity	Fish
Zinc toxicity	Fish
Zirconium toxicity	Fish

Glutamic pyruvic transaminase in serum (SGPT)—Increased

An enzyme present in liver and heart cells.

Element	Species
Arsenic toxicity	Fish
Cadmium toxicity	Fish
Copper toxicity	Fish, goats, sheep
Lead toxicity	Fish
Nickel toxicity	Fish
Selenium toxicity	Pigs
Titanium toxicity	Fish
Zinc toxicity	Fish
Zirconium toxicity	Fish

Glutathione peroxidase—In serum decreased activity

Element	Species
Selenium deficiency	Humans

Glutathione peroxidase—In serum increased activity

Element	Species	Comments
Selenium deficiency	Fish	In liver and plasma CHECK
Selenium toxicity	Pigs, rats	

Glycosuria

The presence of glucose in urine.

Element	Species
Bismuth toxicity	Humans
Cadmium toxicity	Humans
Chromium deficiency	Humans, rats
Chromium toxicity	Rats
Mercury toxicity	Dogs
Selenium toxicity	Humans
Zirconium toxicity	Rats

Goiter

Enlarged thyroid.

Element	Species	Comments
Cobalt (cobalamin) deficiency	Humans	
Cobalt (cobalamin) toxicity	Humans	
Iodine deficiency	Humans, birds, cats, cattle, deer, dogs, fish, goats, hamsters, horses, mice, pigs, poultry, sheep	
Iodine toxicity	Humans, horses (foals)	

Gout

Sudden severe erythema, heat swelling and pain in any joint (big toe, elbows, fingers, knees, wrists) due to deposition of uric acid crystals.

Element	Species	Comments
Calcium toxicity	Birds, poultry (growing)	Causes visceral gout
Molybdenum toxicity	Humans	
Sodium toxicity	Birds	Visceral gout

Granulocytopenia

Element	Species	Comments
Zinc toxicity	Humans	Due to Cu deficiency

Granuloma formation

Element	Species	Comments
Antimony toxicity	Rats	Pulmonary
Beryllium toxicity	Humans, dogs	Pulmonary and elsewhere
Lanthanum toxicity	Rats	Pulmonary
Zirconium toxicity	Humans	Pulmonary, skin

Growth impairment/retardation/poor weight gain

Element	Species	Comments
Aluminum deficiency	Goats, poultry (*chicks*)	Anke et al. (2005a)
Aluminum toxicity	Humans, pigs	*Humans:* Al deposits in mitochondria inhibit bone phosphate formation
Arsenic deficiency	Goats	
Arsenic toxicity	Fish, poultry, rats	
Boron deficiency	Poultry (chicks)	Vit D deficient chicks
Boron toxicity	Cattle	
Bromine deficiency	Goats, mice, poultry	
Cadmium toxicity	Fish, various	
Calcium deficiency	Birds, buffaloes, cattle, mice, nonhuman primates, fish, sheep, squirrels	*Buffalo:* delayed maturity, osteopenia, stunted growth, rickets
Calcium toxicity	Cats, dogs, fish, pigs, poultry	*Fish:* dwarfism
Chlorine deficiency	Dogs, poultry, rats	
Chromium deficiency	Humans, mice, rats	
Chromium toxicity	Hamsters	
Cobalt deficiency	Cattle, goats, sheep	
Cobalt toxicity	Cattle, goats, sheep	
Copper deficiency	Humans, camels, cattle, deer, fish, guinea pigs, poultry	
Copper toxicity	Fish, various	
Fluorine deficiency	Goats, mice	*Goats:* low F during pregnancy and neonates
Fluorine toxicity	Dogs, cattle, goats, pigs, poultry	
Gallium toxicity	Mice, rats	
Iodine deficiency	Cattle, poultry (*experimental*)	
Iodine toxicity	Birds, pigs, sheep	
Iron deficiency	Humans, dogs, horses, mink	
Iron toxicity	Fish	
Lead deficiency	Rats	
Lead toxicity	Fish, cattle	
Lithium deficiency	Goats	
Magnesium deficiency	Birds, fish, guinea pigs	
Magnesium toxicity	Pigs, poultry	

Manganese deficiency	Humans (children), birds, dogs, fish, goats, guinea pigs, rabbits, rats, sheep	
Manganese toxicity	Sheep	
Molybdenum deficiency	Goats, poultry (chicks)	
Molybdenum toxicity	Cattle, deer, guinea pigs, poultry (chicks), rabbits, rats	
Nickel deficiency	Cattle, goats, minipigs, sheep, rats	
Nickel toxicity	Cattle, dogs, fish, rats	
Phosphorus deficiency	Humans, birds, buffaloes, cattle, dogs, fish, horses, pigs	
Phosphorus toxicity	Fish	
Potassium deficiency	Cats, cattle, dogs, goats, horses, pigs, poultry, rats, sheep	
Samarium toxicity	Mice, rats	
Selenium deficiency	Fish, sheep	
Selenium toxicity	Dogs, fish, pigs, rats	
Silicon deficiency	Rats	Schwarz and Milne (1972)
Sodium deficiency	Cats, dogs, goats, pigs, poultry, rabbits, rats, sheep	
Sodium toxicity	Fish	
Sulfur deficiency	Pigs, poultry, rats, sheep	
Thulium toxicity	Rats	
Tin toxicity	Rats	Due reduced food intake
Titanium toxicity	Fish	
Uranium toxicity	Mice	Low fetal weight, body length and weight gain
Vanadium toxicity	Cattle, mice, poultry (chicks), rats, sheep	
Yttrium toxicity	Mice, rats	
Zinc deficiency	Humans, cats, cattle, dogs, fish, goats, hamster, horses (*foals*), mice, nonhuman primates (*squirrel monkeys*) pigs, poultry, rats, sheep, fish	
Zinc toxicity	Fish, pigs, poultry (*chicks*) sheep	
Zirconium toxicity	Fish	

GSH-px blood concentration—Decreased

Element	Species	Comments
Selenium deficiency	Cats	

GSH-R blood concentration—Decreased

Element	Species	Comments
Copper deficiency	Buffaloes	

Gum disease

Element	Species	Comments
Bismuth toxicity	Humans	Poor oral hygiene—blue-black gum line
Zinc deficiency	Cattle	Mucosal overgrowth dental pad and lips

Hair coat changes

Element	Species	Comments
Arsenic toxicity (chronic)	Cattle	Rare
Boron toxicity	Rats	Coarse hair, delayed maturation prepubescent hair
Cobalt (cobalamin) deficiency	Cattle	Rough haircoat, thickened skin, loss of color
Copper deficiency	Humans, camels, cats, cattle, deer, dogs, goats, guinea pigs, poultry, rabbits, rats, sheep	*Camels:* Hypotrichia, rough coat, depigmentation Poor keratinization *Humans:* Menkes steely hair syndrome (inherited) *Goats:* depigmentation skin and hair
Dysprosium toxicity	Guinea pigs	Epilation
Fluorine toxicity	Horses	Dry rough coat
Iodine deficiency	Goats, pigs, sheep	*Goats:* hairless neonates *Pigs:* dry, rough, thickened skin *Sheep:* reduced wool
Iron deficiency	Horses mink, pigs	Rough coat
Manganese deficiency	Humans	Depigmentation
Molybdenum toxicity	Cattle, sheep	*Cattle:* dull staring coat *Sheep:* depigmentation, wool loses crimp
Nickel deficiency	Rats	Rough uneven coat
Nickel toxicity	Pigs	Shaggy coarse hair
Phosphorus deficiency	Various—cattle	Dull dry coat
Potassium deficiency	Cattle, mice, pigs, rats, sheep	Rough hair coat—*sheep, pigs, rats* Dull coat—*mice* Dry scaly tail—*mice* Reduced hide pliability—*cattle* Loss of wool—*sheep*
Sodium deficiency	Cattle, dog, goats	*Cattle:* rough haircoat *Dog:* dry skin *Goats:* shaggy, dull haircoat
Sulfur deficiency	Sheep	Reduced wool production
Thyroid deficiency	Dogs	Dull, dry, sparse hair coat
Zinc deficiency	Cattle, guinea pigs, nonhuman primates (*squirrel monkey*), rats	*Cattle, Guinea pigs:* Rough haircoat *Nonhuman primates:* unkempt hair *Rats:* loss hair follicles

Halitosis

Element	Species	Comments
Sulfur toxicity	Cattle, goats, sheep	Hydrogen sulfide

Hallucinations

Element	Species	Comments
Aluminum toxicity	Humans	Especially dialysis patients
Bismuth toxicity	Humans	
Bromine toxicity	Humans	
Lead toxicity	Humans	
Manganese toxicity	Humans	
Molybdenum toxicity	Humans	
Tin toxicity	Humans	

Headaches

Element	Species	Comments
Arsenic toxicity	Human	Acute and chronic toxicity
Bromine toxicity	Human	
Cadmium toxicity	Humans	
Cesium toxicity	Humans	
Gadolinium toxicity	Humans	After injection for MRI scan
Germanium toxicity	Humans	
Gold toxicity	Humans	
Lanthanum toxicity	Humans	
Lead toxicity	Humans	
Mercury toxicity	Humans	
Molybdenum toxicity	Humans	
Nickel toxicity	Humans	
Phosphorus toxicity	Humans	
Sodium deficiency	Human	
Tellurium toxicity	Humans	
Thallium toxicity	Humans	
Tin toxicity	Humans	Organotin compounds
Zinc toxicity	Humans	
Ionizing radiation	Humans	Exposure

Head pressing

Element	Species
Cobalt deficiency	Sheep
Lead toxicity	Buffaloes

Head retention

Element	Species
Manganese deficiency	Guinea pigs, rats

Head shaking

Element	Species
Magnesium deficiency	Buffaloes

Head tilt

Element	Species
Mercury toxicity	Dogs

Head tremors

Element	Species	Comments
Manganese deficiency	Guinea pigs, rats	
Mercury toxicity	Horses, mink	*Horses:* head bobbing

Hearing loss

Element	Species
Lead toxicity	Sheep
Mercury toxicity	Humans

Heart failure (congestive)/cardiac failure

Element	Species	Comments
Antimony toxicity	Humans	Myocardial depression, ventricular fibrillation
Cobalt deficiency	Humans (children)	Congestive heart failure Vitamin B12 deficiency
Cobalt toxicity	Humans	
Copper deficiency	? Humans, cattle, rats	Humans—suggested link to atherosclerosis and ischemia heart disease

Lithium toxicity	Humans	
Potassium deficiency	Cats, dogs, poultry	Dogs—reduced cardiac output, stroke volume, renal blood flow
Selenium toxicity	Humans	

Heat sensitivity

Element	Species	Comments
Cerium toxicity	Humans	Industrial exposure

Heme oxygenase activity increased

An enzyme responsible for catalyzing the degradation of heme to form biliverdin/bilirubin, ferrous iron, and carbon monoxide.

Element	Species
Nickel toxicity	Hamsters, mice, rats

Heme synthesis impaired

Element	Species
Lead toxicity	Birds

Hematemesis

Element	Species	Comments
Iron toxicity	Humans, dogs	
Sodium toxicity	Dogs	Corrosive chemicals, e.g., bleach (sodium hypochlorite)
Thallium toxicity	Humans	
Zinc toxicity	Humans	
Zirconium toxicity	Humans	
Ionizing radiation	Various	Exposure

Hematochezia

Fresh blood in feces.

Element	Species
Iron toxicity	Humans, dogs
Ionizing radiation	Various

Hematocrit (PCV)—High

Element	Species	Comments
Arsenic toxicity	Poultry (chicks)	Increases erythrocyte formation
Cobalt toxicity	Cattle	
Sodium deficiency	Cats, dogs, poultry	

Hematocrit (PCV)—Low

Element	Species	Comments
Arsenic deficiency	Poultry (chicks)	
Boron toxicity	Cattle, rats	Renal damage
Cobalt (cobalamin) deficiency	Cattle, sheep	
Iron deficiency	Cats, dogs, fish, poultry (rare)	
Nickel deficiency	Poultry (chickens), rats, sheep	
Phosphorus deficiency	Fish	
Potassium deficiency	Cattle	
Selenium deficiency	Horses	
Vanadium toxicity	Rats	
Zinc deficiency	Rabbits	
Zinc toxicity	Fish	*Catfish*
Zirconium toxicity	Dogs	Marginal

Hematoma

Element	Species
Palladium toxicity	Rats

Hematopoiesis—Impaired

Element	Species	Comments
Tin toxicity	Various	Due altered Cu concentrations

Hematuria

Element	Species
Barium toxicity	Humans
Bismuth toxicity	Humans *Due thrombocytopenia*
Iron deficiency	Cats
Vanadium toxicity	Cattle
Zinc toxicity	Humans, birds
Zirconium toxicity	Humans

Hemochromatosis

In humans, an inherited condition in which excess Fe is absorbed and deposited in organs eventually causing toxic effects including abdominal pain, cardiac arrhythmia, chest pain, dysmenorrhoea, erectile dysfunction, fatigue, jaundice, joint pain, loss of libido, polydipsia, polyuria, shortness of breath, swollen hands or feet, skin pigmentation, reduced testicular size, tiredness, weakness, weight loss.

Element	Species	Comments
Iron toxicity	Humans, bats, birds, coati, horses, marine mammals, nonhuman primates (*lemurs*), racoons, rhinoceros, tapir	*Birds:* frugivores and insectivores *Marine mammals:* Dolphins

Hemoglobin—High

Element	Species
Cobalt toxicity	Cattle
Sodium deficiency	Dogs

Hemoglobin—Low

Element	Species
Aluminum toxicity	Dogs
Arsenic deficiency	Poultry (chicks)
Boron toxicity	Cattle, rats
Bromine deficiency	Goats
Cobalt deficiency	Cattle, sheep
Copper deficiency	Cattle, pigs
Gallium toxicity	Humans
Iodine toxicity	Pigs
Iron deficiency	Cats, cattle (calves), dogs, fish, pigs, poultry (rare)
Lead toxicity	Cattle
Selenium toxicity	Humans, dogs, rats
Tin toxicity	Rats
Vanadium toxicity	Rats
Zinc toxicity	Fish (*catfish*)
Zirconium toxicity	Dogs (marginal)

Hemoglobinuria

Element	Species	Comments
Arsenic toxicity	Humans	Acute
Copper toxicity	Goats, marsupial, sheep	*Wombat*
Lead toxicity	Birds	*Amazon parrots*
Phosphorus deficiency	Buffaloes	
Zinc toxicity	Birds	

Hemolysis

Element	Species
Arsenic toxicity	Humans
Copper toxicosis	Goats, marsupial (*Wombat*), sheep
Germanium toxicity	Humans
Palladium toxicity	Humans
Zinc toxicity	Humans, dogs
Zirconium toxicity	Humans

Hemorrhage

Element	Species	Comments
Bismuth toxicity	Humans	Buccal, epistaxis, hematuria, petechie, rectal
Boron toxicity	Rats	Hemorrhage from eyes
Cesium toxicity	Rats	Gastrointestinal
Calcium toxicity	Pigs	Due to interference with Vitamin K
Chlorine deficiency	Humans	Cerebral hemorrhage
Chlorine toxicity	Humans	*Caustic effect (hypochlorite):* Gastrointestinal
Copper deficiency	*Dogs*, guinea pigs, horses, poultry, sheep	*Horses:* at parturition—ruptured uterine artery *Others:* due to ruptured of main blood vessel aneurysm, e.g., aorta *Dogs Unpublished study on sudden death in Grayhounds*
Gold toxicity	Humans	Gastrointestinal hemorrhage
Mercury toxicity	Sheep	Hemorrhagic gastroenteritis
Palladium toxicity	Rats	Gastrointestinal
Tellurium toxicity	Rats	*Teratogenic effect:* ocular hemorrhage
Vanadium toxicity	Cattle, sheep	*Cattle:* abomasal, pulmonary, tracheal (after ingestion) *Sheep:* in kidney, *and* small intestine mucosa
Zinc deficiency	Humans, cattle, pigs	*Cattle:* submucosal *Pigs:* ?linked vitamin K
Zirconium toxicity	Humans	Hematuria

Hemosiderosis—Siderosis

Deposits of hemosiderin in tissues.

Element	Species	Comments
Cobalt (cobalamin) deficiency	Cattle, sheep	In spleen
Copper deficiency	Deer	In liver

Gallium toxicity	Mice	Liver/spleen
Iron toxicity	Humans, bats, horses, nonhuman primates (*lemurs*), sheep	"Iron overload" *Bats:* captive fruit bats
Selenium toxicity	Birds	Hepatic

Hemianopsis

Blindness over half the field of vision.

Element	Species
Nickel toxicity	Humans

Hemiplegia

Element	Species
Selenium toxicity	Humans

Hepatic atrophy

Element	Species	Comments
Selenium toxicity	Dogs, rats	Inhaled

Hepatic fibrosis

Element	Species
Arsenic toxicity	Humans

Hepatic (hepatocellular) necrosis

Element	Species	Comments
Antimony toxicity	Humans	
Beryllium toxicity	Rabbits, rats	
Boron toxicity	Mice	
Cadmium toxicity	Rats	
Cerium toxicity	Mice, rats	
Chromium toxicity	Humans	
Iron toxicity	Humans, cattle, fish, sheep	Hepatic degeneration
Lanthanum toxicity	Mice	Hepatic degeneration
Selenium deficiency	Pigs, nonhuman primates (squirrel monkeys)	Hepatic degeneration
Selenium toxicity	Humans, dogs	*Humans:* hepatic degeneration
Thulium toxicity	Mice	Hepatic degeneration
Zinc toxicity	Humans	

Hepatitis

Element	Species	Comments
Mercury toxicity	Marine mammals	Seals

Hepatomegaly

Element	Species
Arsenic toxicity	Humans
Bismuth toxicity	Humans

Hepatorenal disease

Element	Species
Aluminum toxicity	Humans

Hepatotoxicity

Element	Species
Aluminum toxicity	Rats
Selenium toxicity	Humans

Histamine (serum) increased

Element	Species	Comments
Palladium toxicity	Cats	After intravenous injection, caused bronchospasm

Hives with angioedema

Hypersensitivity reaction.

Element	Species
Zinc toxicity	Humans

Hoarseness

Element	Species
Zirconium toxicity	Humans

Hoof abnormalities

Element	Species	Comments
Selenium toxicity	Buffaloes, cattle, horses, pigs	*Buffaloes:* deformities, detached hooves *Cattle, Horses:* coronary band inflammation All: cracks, deformities

Horn abnormalities

Element	Species	Comments
Calcium deficiency	Deer	delayed antler growth and velvet shedding
Phosphorus deficiency	Deer	delayed antler growth and velvet shedding
Selenium toxicity	Buffaloes, cattle	Deformities
Zinc deficiency	Goats, sheep	Soft, deformed, lack normal striations

Hydrocephalus

Element	Species	Comments
Cadmium toxicity	Mice, rats	Fetal developmental abnormality
Chromium toxicity	Hamsters	
Tellurium toxicity	Rats	*Teratogenic effect*

Hydronephrosis

Element	Species
Lead toxicity	Sheep

Hyperaldosteronism

High levels of aldosterone (a hormone produced in the adrenal glands) in the blood.

Element	Species
Sodium deficiency	Cats

DD: A benign adrenal adenoma (primary hyperaldosteronism).

Hyperammonemia

High amounts of ammonia in the blood.

Element	Species
Tellurium toxicity	Humans

Hyperbilirubinemia

Element	Species
Aluminum toxicity	Rats

Hypercalcemia

Element	Species
Calcium toxicity	Humans, cats, poultry
Manganese deficiency	Humans

Hypercalciuria

Element	Species
Cadmium toxicity	Humans
Gallium toxicity	Humans, rats

Hyperchloridemia

Element	Species
Chlorine toxicity	Humans

Hypercholesterolemia

High blood cholesterol concentrations.

Element	Species
Aluminum toxicity	Rats
Chromium deficiency	Humans, cattle, mice, pigs, rats
Copper deficiency	Humans, nonhuman primates, rats
Gadolinium toxicity	Rats
Zirconium toxicity	Rats

Hyperesthesia

Element	Species
Fluorine toxicity	Cattle
Mercury toxicity	Humans

Hyperexcitability

Element	Species	Comments
Magnesium deficiency	Pigs, rabbit	
Manganese deficiency	Guinea pigs, rats	Hyperreactive to stimuli

Hyperextension of carpi

Element	Species
Copper deficiency	Dogs
Magnesium deficiency	Cats, dogs

Hyperflexion—Hock

Element	Species	Comments
Manganese deficiency	Poultry	Chicken and ducks

Hyperflexion—Phalanges

Movement of a flexor muscles of the digits beyond normal limits.

Element	Species
Copper deficiency	Dogs

Hyperglobulinemia

Element	Species
Iron toxicity	Marine mammals (dolphins)

Hyperglycemia

Element	Species
Aluminum toxicity	Rats
Chromium deficiency	Humans, dogs (?)
Iron toxicity	Marine mammals (dolphins)
Manganese toxicity	Rats

Hyperinsulinemia

Excess insulin in blood.

Element	Species
Chromium deficiency	Humans, dogs (?), pigs, rats
Iron toxicity	Marine mammals (dolphins) (postprandial)

Hyperkalemia

Excess potassium in blood.

Element	Species	Comments
Antimony toxicity	Rats	
Potassium toxicity	Humans, cattle	Cattle—experimental
Selenium deficiency	Horses	
Magnesium toxicity	Humans	Rare

Hyperkaluria

Excess potassium salts in the urine.

Element	Species	Comments
Potassium toxicity	Cattle	Cattle—experimental

Hyperkeratinization

Element	Species	Comments
Nickel toxicity	Rats	Dermal contact

Hyperkeratosis

Element	Species	Comments
Arsenic toxicity	Humans	
Calcium toxicity	Elephants	Due Zn deficiency
Europium toxicity	Rats	Gastric hyperkeratosis
Lanthanum toxicity	Rats	Gastric hyperkeratosis
Zinc deficiency	Dogs, elephants, poultry	Due Ca excess

Hyperlipidemia

Element	Species
Aluminum toxicity	Rats

Hypermagnesemia

High magnesium in the blood.

Element	Species
Magnesium toxicity	Humans

Hypermineralization of bone

Element	Species
Fluorine toxicity	Cattle, pigs

Hypernatremia

High sodium in blood.

Element	Species
Antimony toxicity	Rats

Hyperphosphatemia

Element	Species	Comments
Barium toxicity	Humans	Barium carbonate toxicity
Calcium deficiency	Nonhuman primates	Lemurs
Iron toxicity	Sheep	
Magnesium deficiency	Humans, guinea pigs	
Phosphorus toxicity	Cats, cattle	

Hyperpigmentation

Element	Species
Arsenic toxicity	Humans

Hyperpnoea/tachypnoea

Element	Species
Europium toxicity	Mice
Iron toxicity	Pigs
Magnesium deficiency	Horses
Molybdenum deficiency	Humans
Praseodymium toxicity	Dogs
Zinc deficiency	Poultry (tachypnoea)

Hyperproteinemia

Element	Species
Lead toxicity	Humans (children)

Hyperreactive to external stimuli

Element	Species
Magnesium deficiency	Buffaloes

Hyperreflexia

Overactive or overresponsive reflexes, due to increased upper motor neurone excitability or disinhibition of lower motor neurones by the higher brain centers.

Element	Species
Bismuth toxicity	Humans
Lithium toxicity	Humans
Magnesium deficiency	Humans, cats
Selenium toxicity	Humans

Hypersalivation

Element	Species
Arsenic toxicity	Dogs
Cadmium toxicity	Humans
Copper toxicity	Various
Fluorine toxicity	Horses
Iodine toxicity	Humans, cats
Lead toxicity	Buffaloes, cattle, frogs
Lithium toxicity	Dogs
Magnesium deficiency	Cattle, sheep
Mercury toxicity	Humans, mink
Phosphorus toxicity	Humans
Potassium toxicity	Cattle
Selenium toxicity	Humans, cattle
Sodium toxicity	Dogs *Corrosive chemicals*, e.g., *bleach (sodium hypochlorite)*
Sulfur deficiency	Goats, sheep
Thallium toxicity	Humans
Zinc deficiency	Cattle

Hypersomnia

Excessive sleeping.

Element	Species	Comments
Tin toxicity	Humans	Organotin compounds

Hypertension

Element	Species	Comments
Arsenic toxicity	Humans	Chronic
Barium toxicity	Humans	
Cadmium toxicity	Various	
Calcium toxicity	Humans	
Chlorine deficiency	Humans	
Gadolinium toxicity	Humans	
Lead toxicity	Humans	
Lutetium toxicity	Dogs	
Mercury toxicity	Humans	
Silicon deficiency	Humans	Proposed not confirmed. Si nonessential in humans
Thallium toxicity	Humans	

Hyperthermia

Element	Species	Comments
Iron toxicity		Acute toxicity
Lithium toxicity	Humans	

Hypertonia

Element	Species
Lithium toxicity	Humans

Hypertriglyceridemia

Element	Species	Comments
Aluminum toxicity	Rats	
Bromine deficiency	Goats	
Chromium deficiency	Humans, mice, rats	
Copper deficiency	Rats	
Gadolinium toxicity	Rats	
Iron deficiency	Poultry (chicks), rats	Rats (pups) maybe associated with low tissue carnitine

Hyperthyroidism

Element	Species
Lithium toxicity	Humans

Hyperuricemia

Element	Species
Molybdenum toxicity	Humans

Hyperuricuria

Element	Species
Molybdenum toxicity	Humans

Hypoalbuminemia

Element	Species	Comments
Cerium toxicity	Rats	
Gadolinium toxicity	Rats	
Lanthanum toxicity	Humans	Dialysis patients

Hypocalcemia

Low calcium in blood.

Element	Species	Comments
Arsenic toxicity	Cattle	
Calcium deficiency	Birds, elephants, foxes, nonhuman primates, reptiles	Lemurs
Fluorine toxicity	Humans	
Gallium toxicity	Humans, rats	
Lanthanum toxicity	Mice	
Magnesium deficiency	Humans, cattle (calves) dogs, nonhuman primates (*rhesus monkeys*), pigs, poultry, sheep	
Mercury toxicity	Dogs	
Phosphorus deficiency	Cattle	
Phosphorus toxicity	Humans	
Potassium deficiency	Sheep	And high hepatic glycogen

Potassium toxicity	Sheep (milk fever)	
Tungsten deficiency	Humans	Rare

Hypochloridemia

Low chloride in blood.

Element	Species
Chlorine deficiency	Cats, cattle, dogs, poultry
Sodium deficiency	Marine mammals

Hypocholesterolemia

Element	Species
Lanthanum toxicity	Humans
Manganese deficiency	Humans

Hypocupremia

Element	Species
Molybdenum toxicity	Deer

Hypodynamia

Abnormally reduced power.

Element	Species	Comments
Tin toxicity	Humans	Organotin compounds

Hypoglycemia

Element	Species	Comments
Arsenic toxicity	Cattle	
Beryllium toxicity	Rabbits, rats	Due liver necrosis
Gadolinium toxicity	Rats	
Iron toxicity	Humans	

Hypogonadism

Element	Species
Zinc deficiency	Humans

Hypoinsulinemia

Element	Species
Manganese toxicity	Rats
Zinc deficiency	Rats

Hypokalemia

Low potassium in the blood.

Element	Species
Barium toxicity	Humans, dogs (*severe*)
Cesium toxicity	Humans
Chlorine deficiency	Cattle, dogs
Gallium toxicity	Humans
Mercury toxicity	Dogs
Potassium deficiency	Cattle, horses
Tellurium toxicity	Humans

Hypomagnesemia

Low magnesium in the blood.

Element	Species
Magnesium deficiency	Cats, cattle, dogs, horses, poultry, sheep
Potassium toxicity	Sheep (grass tetany)

Hyponatremia

Element	Species
Chlorine deficiency	Cattle

Hypophosphatemia

Low levels of phosphate in the blood.

Element	Species	Comments
Calcium toxicity	Some: poultry	
Lanthanum toxicity	Mice	
Phosphorus deficiency	Buffaloes, cattle	
Phosphorus toxicity	Cats	Only one report

DD: most commonly caused by hyperparathyroidism and vitamin D deficiency.

Hypoproteinemia

Low levels of protein in the blood.

Element	Species
Aluminum toxicity	Rats
Mercury toxicity	Dogs

DD: inadequate protein intake; liver disease (hepatitis, cirrhosis); kidney disease and proteinuria, gastrointestinal protein loss (gluten sensitivity—celiac disease, inflammatory bowel disease, lymphangiectasis, neoplasia).

Hyporeflexia

Muscles are less responsive to stimuli.

Element	Species
Magnesium toxicity	Humans
Selenium toxicity	Humans

DD: drug induced (e.g., benzodiazepinse), electrolyte imbalance (e.g., excess magnesium), hypothyroidism, drug induced, amyotrophic lateral sclerosis (ALS) or Lou Gehrig's disease, Guillain-Barré syndrome (GBS), chronic inflammatory demyelinating polyneuropathy (CIDP), spinal cord injuries, strokes.

Hypotension

Low blood pressure.

Element	Species
Antimony toxicity	Humans
Arsenic toxicity	Humans
Barium toxicity	Humans (*due hypokalemia*)
Cobalt toxicity	Humans
Iron toxicity	Humans, dogs
Lanthanum toxicity	Humans
Lithium toxicity	Humans
Magnesium toxicity	Humans
Potassium deficiency	Dogs
Praseodymium toxicity	Dogs
Samarium toxicity	Dogs
Selenium toxicity	Humans
Zirconium toxicity	Humans
Ionizing radiation	Various

DD: adrenal insufficiency (hypoadrenocorticism), anaphylaxis, bradycardia, cardiac failure, dehydration, diabetes, endocrine disorders, hemorrhage, hypoadrenocorticism (Addison's disease), hypoglycemia, pregnancy, sepsis,

Hypothermia

Element	Species
Arsenic toxicity	Dogs
Cobalt toxicity	Mice
Iodine toxicity	Sheep
Iron deficiency	Rats
Selenium toxicity	Nonhuman primates (*macaques*)

Hypothyroidism

Element	Species
Cesium toxicity	Humans
Cobalt (cobalamin) deficiency	Rats
Iodine deficiency	Humans, birds, dogs
Lithium toxicity	Humans

Hypotonia

Poor muscle tone.

Element	Species
Copper deficiency	Humans

Hypouricemia

Low uric acid in blood.

Element	Species	Comments
Molybdenum deficiency	Humans	Acute toxicity

Hypouricuria

Low uric acid in urine.

Element	Species	Comments
Molybdenum deficiency	Humans	Acute toxicity

Hypovolemia

Low blood volume.

Element	Species
Iron toxicity	Humans

Icterus/Jaundice

Element	Species	Comments
Iron toxicity	Humans, cattle, horses (foals), pigs	
Selenium toxicity	Rats	High bilirubin
Zinc toxicity	Dogs	

Ileus

Element	Species	Comments
Magnesium toxicity	Humans	
Sulfur toxicity	Cattle	Reduced rumen activity

Immune humoral (antibody) response reduced

Element	Species
Chromium deficiency	Cattle
Cobalt deficiency	Rats
Copper deficiency	Mice

Immune response impaired

Element	Species	Comments
Calcium toxicity	Elephants	
Cobalt deficiency	Sheep	Poor response to vaccination
Copper deficiency	Goats	
Mercury toxicity	Humans	
Nickel toxicity	Mice, rats	*Rats:* natural killer cell activity reduced
Zinc deficiency	Humans, elephants, nonhuman primates	Due Cu deficiency
Ionizing radiation	Various	Exposure

Immunoglobulin—Impairment

Element	Species	Comments
Cerium toxicity	Mice	IgM lower

Impotence

Element	Species
Molybdenum toxicity	Cattle, rats
Zinc deficiency	Humans

Incoordination—Neurological

Element	Species
Aluminum deficiency	Goats (Anke et al., 2005a)
Bismuth toxicity	Humans
Calcium deficiency	Birds
Lead toxicity	Humans, dogs
Magnesium deficiency	Humans
Mercury toxicity	Humans, cats, cattle, dog, ferrets, mink, nonhuman primates, sheep
Selenium toxicity	Cattle
Sodium deficiency	Marine mammals
Ionizing radiation exposure	Various

Infarction

Element	Species	Comments
Selenium toxicity	Humans	Mesenteric, myocardial

Infection risk increased

Element	Species	Comments
Copper deficiency	Goats, mice, sheep	*Mice:* reduced antibody response
Iodine toxicity	Cattle	Reduced humeral and cell mediated immunity

Iron deficiency	Humans (children), cattle (calves), pigs, rats	
Zinc deficiency	Humans, cattle	
Ionizing radiation	Various	Exposure

Infertility

Element	Species
Copper deficiency	Camels, goats
Iodine deficiency	Goats
Selenium toxicity	Cattle

Inflammation

Element	Species	Comments
Gadolinium toxicity	Humans	After injection for MRI scan
Zinc toxicity	Humans	Chemical burns to oral cavity, pharynx, trachea, esophagus and stomach

Inflammatory bowel disease (IBD)

Element	Species	Comments
Aluminum toxicity	Humans	Association not necessarily a cause

Inflammatory nodule formation

Element	Species	Comments
Beryllium toxicity	Humans, dogs	Pulmonary and elsewhere

Injection site necrosis

Element	Species	Comments
Palladium toxicity	Humans	Subcutaneous

Injection site pain

Element	Species	Comments
Gadolinium toxicity	Humans	After injection for MRI scan

Insomnia

Trouble falling or staying asleep.

Element	Species
Antimony toxicity	Humans
Lead toxicity	Humans
Mercury toxicity	Humans

Insulin-binding reduced

Element	Species	Comments
Chromium deficiency	Humans	Binding to red blood cells

Insulin receptor numbers reduced

Element	Species
Chromium deficiency	Humans

Insulin resistance—Increased

Element	Species	Comments
Chromium deficiency	Humans	

Insulin synthesis reduced

Element	Species	Comments
Manganese deficiency	Rats	

Intellectual disability

Element	Species	Comments
Copper deficiency	Humans	Menkes disease
Molybdenum deficiency	Humans	Genetic defect

Internal hemorrhage

Element	Species	Comments
Arsenic toxicity	Humans	Gastrointestinal hemorrhage
Cerium toxicity	Mice	Pulmonary and tracheal hemorrhage
Chlorine deficiency	Cattle	Melena
Copper deficiency	Guinea pigs, poultry, sheep (lambs)	Major blood vessel aneurysm and rupture, e.g., aorta

Internal organ congestion/hemorrhages

Element	Species	Comments
Magnesium deficiency	Buffaloes	Congestion and hemorrhages

Internal organ discoloration

Element	Species	Comments
Silver toxicity	Humans	Gray—and skin

Interstitial nephritis

Element	Species
Mercury toxicity	Humans
Zinc toxicity	Humans

Intestinal distension

Element	Species	Comments
Potassium deficiency	Humans	Due loss of muscle tone

Intestinal inflammation

Element	Species
Aluminum toxicity	Mice

Intracranial pressure—Increased

Element	Species	Comments
Ionizing radiation	Various	Exposure

Intraocular pressure—Increased

Element	Species
Chromium deficiency	Humans

Intravascular coagulation

Element	Species	Comments
Arsenic toxicity	Humans	Acute

Iron in serum—Low

Element	Species
Copper deficiency	Buffaloes

Iron in serum—High

Element	Species
Iron toxicity	Sheep

Iron-binding capacity (total) (TIBC)—High result

Used as a measure of Fe status. In humans normal is 250–450 micrograms per deciliter (mcg/d).

Element	Species	Comments
Iron toxicity	Sheep	High

Irregular menses

Element	Species	Comments
Selenium toxicity	Nonhuman primates	*Macaques*
Zinc deficiency	Humans	

Irritability

Element	Species
Bismuth toxicity	Humans
Boron toxicity	Humans
Iron deficiency	Humans
Lead toxicity	Humans, cattle
Magnesium deficiency	Humans, cats, cattle, nonhuman primates (*rhesus monkeys*) pigs, rats
Magnesium toxicity	Birds

Manganese toxicity	Humans
Molybdenum deficiency	Humans
Mercury toxicity	Humans
Phosphorus deficiency	Humans
Selenium toxicity	Humans
Tellurium toxicity	Humans
Zinc deficiency	Humans, rats

Ischemic heart disease

Element	Species	Comments
Arsenic toxicity	Humans	Suspected

Itchiness

Element	Species	Comments
Cerium toxicity	Humans	Industrial exposure
Gadolinium toxicity	Humans	After injection for MRI scan
Mercury toxicity	Humans	Contact dermatitis

Jaundice/icterus

Element	Species	Comments
Arsenic toxicity	Humans	
Copper toxicity	Goats, marsupials, sheep	*Wombats*
Gold toxicity	Humans	
Iron toxicity	Humans	
Selenium toxicity	Rats	High bilirubinemia
Zinc toxicity	Dogs	
Zirconium toxicity	Humans	Hemolysis, liver damage

Jaw swelling

Element	Species	Comments
Fluorine toxicity	Horses, sheep	
Molybdenum toxicity	Cattle, rats	Due exostoses

Jejunal stricture

Narrowing of the jejunal lumen section of the small intestine.

Element	Species
Zirconium toxicity	Humans

Joint enlargement/swelling

In rickets affects joints in limbs and ribs.

Element	Species	Comments
Calcium deficiency	Humans, bears (polar), cats, cattle, dogs, guinea pigs, rats, squirrels	Wide area uncalcified cartilage at growth plate *Polar Bears:* swollen metaphyses
Copper deficiency	Cattle, poultry	Poultry: hock
Fluorine deficiency	Goats	
Fluorine toxicity	Cattle, sheep	
Manganese deficiency	Birds, goats, pigs, poultry	*Pigs:* hocks *Goat:* tarsus *Birds, Chickens, Ducks:* Tibiometatarsal joint
Phosphorus deficiency	Dogs, mink, various	Rickets
Zinc deficiency	Birds, horses	*Birds:* swollen hocks *Horses:* Epiphyseal enlargement

Joint pain/arthralgia

Element	Species
Antimony toxicity	Humans
Beryllium toxicity	Humans
Calcium deficiency	Camels
Lead toxicity	Humans
Molybdenum toxicity	Humans
Selenium toxicity	Humans

Junctionopathy

Abnormal function of the neuromuscular junction.

Element	Species
Aluminum toxicity	Dogs

Kallikrein—Urine concentration increased

Element	Species
Molybdenum toxicity	Rats

Keratinization—Poor

Element	Species	Comments
Copper deficiency	Humans, dogs, cats, sheep	*Humans* Menkes steely hair syndrome (inherited) *Sheep*: wool lacks crimp, steely or stringy (straight) wool

Keratoconjunctivitis

Element	Species
Palladium toxicity	Rats

Ketonuria

Element	Species
Palladium toxicity	Rats

Kidney disease

Element	Species	Comments
Aluminum toxicity	Koalas	Al accumulation in kidneys of Koalas with renal failure—but cause and effect not proven
Antimony toxicity	Rats	
Arsenic toxicity	Humans	Renal tubular acidosis, nephritis, kidney failure
Barium toxicity	Humans, mice	Renal degeneration, renal insufficiency, acute renal failure *Mice:* renal tubular dilation, hyaline casts, interstitial fibrosis, glomerulosclerosis
Boron toxicity	Humans	Tubular degeneration
Cadmium toxicity	Various: humans, mice, rats, rhesus monkeys	*Humans*: atrophy, interstitial fibrosis, glomerular sclerosis, tubular degeneration *Mice, Rats, Rhesus monkeys*: tubular necrosis
Calcium toxicity	Poultry	
Cerium toxicity	Mice	Renal damage
Chlorine deficiency	Rats	Renal damage
Chromium toxicity	Humans, laboratory animals	Renal damage
Copper toxicity	Dogs	

Dysprosium toxicity	Rats	Loss of urine concentrating ability, increased renal vascular resistance
Europium toxicity	Rats	*Kidney disease:* reduced creatinine clearance, low glomerular filtration rate (GFR) (Ohnishi et al., 2011)
Fluorine toxicity	Humans, cattle, pigs	*Cattle:* tubular fibrosis, degeneration and mineralization
Gallium toxicity	Humans, rats	Renal toxicity; tubular precipitates, azotemia
Iron toxicity	Humans, dogs, sheep	*Humans:* renal failure *Dogs:* renal failure (acute) *Sheep:* renal degeneration
Lead toxicity	Humans, birds, poultry	*Poultry:* renal intranuclear inclusions
Lithium toxicity	Humans	Sodium-losing nephritis
Mercury toxicity	Humans, dogs, marine mammals (*seals*), pigs, sheep	Acute renal failure
Molybdenum toxicity	Rats	Kidney damage
Nickel toxicity	Humans	Nephrotoxicity
Niobium toxicity	Dogs, rats	*Dogs, Rats:* damage to proximal and distal tubules, granular pigment deposits, hyaline casts, necrosis *Dogs:* reduced glomerular filtration rate, reduced maximum glucose resorption, *p*-aminohippurate secretion reduced, flattened tubule cells *Rats:* polydipsia and polyuria, low specific gravity, inability to concentrate urine
Palladium toxicity	Rabbits, rats	Kidney damage
Potassium deficiency	Dog	Reduced renal blood flow (GFR)
Selenium toxicity	Humans, sheep	Kidney failure
Silicon toxicity	Humans	Kidney damage
Sodium toxicity		
Thallium toxicity	Humans	Kidney damage
Tin toxicity	Humans, rats	*Humans:* renal damage *Rats:* nephrocalcinosis
Tungsten toxicity	Human, rats	*Humans:* renal failure—acute tubular necrosis
Yttrium toxicity	Rats	Reduced kidney function, reduced creatinine clearance, reduced urine production
Zinc toxicity	Humans	Interstitial nephritis, acute tubular necrosis

Kinked tail (congenital effect)

Element	Species
Copper deficiency	Cats

Koilonychia—Spoon nail

Element	Species
Iron deficiency	Humans

Lachrymation

Increased tear production.

Element	Species
Bromine toxicity	Humans
Cobalt deficiency	Sheep
Dysprosium toxicity	Rabbits
Erbium toxicity	Mice
Europium toxicity	Mice
Holmium toxicity	Mice
Iodine toxicity	Cats, cattle
Lutetium toxicity	Mice
Selenium toxicity	Humans (aerosol contact)
Sulfur deficiency	Goats, sheep
Zirconium toxicity	Humans

Lactate dehydrogenase (LDH)—Activity increased

An intracellular enzyme that catalyzes the conversion of lactate to pyruvate and vice versa.

Element	Species
Cerium toxicity	Rats
Gadolinium toxicity	Rats
Selenium deficiency	Horses (and donkeys)

Lactate dehydrogenase (LDH)—Activity decreased

An intracellular enzyme that catalyzes the conversion of lactate to pyruvate and vice versa.

Element	Species	Comments
Lithium deficiency	Goats	
Samarium toxicity	Rats	In testes

Lameness

Element	Species	Comments
Calcium deficiency	Humans, cats, cattle, dogs, foxes, horses, pigs, poultry	Impaired endochondral ossification and fractures
Calcium toxicity	Dogs	Developmental skeletal diseases
Copper deficiency	Camels, cattle, poultry (chicks)	
Fluorine toxicity	Cattle, horses	*Horses*—shifting body weight
Magnesium deficiency	Dogs	
Manganese deficiency	Pigs, sheep	
Molybdenum toxicity	Cattle, sheep	
Phosphorus deficiency	Cattle, horses, mink, pigs	
Selenium toxicity	Deer, cattle	
Zinc deficiency	Horses	
Zinc toxicity	Dogs	

Laryngeal edema

Element	Species
Zirconium toxicity	Humans

Laryngeal spasm

Sudden contraction of the laryngeal muscles narrowing the opening from the pharynx into the trachea making breathing difficult.

Element	Species
Sulfur toxicity	Humans

Lassitude

Mental or physical weariness.

Element	Species
Nickel toxicity	Humans
Selenium toxicity	Humans

Learning difficulties

Element	Species
Lead toxicity	Humans (children)

Lens luxation

Element	Species
Molybdenum deficiency	Humans

Lethargy/listlessness/apathy/tiredness

Element	Species	Comments
Aluminum toxicity	Dogs, mice	
Arsenic toxicity	Dogs	
Bismuth toxicity	Humans	
Boron toxicity	Humans	
Cadmium deficiency	Goats	
Calcium toxicity	Humans	
Cerium toxicity	Rats	
Chlorine deficiency	Cattle	
Cobalt deficiency	Cattle, sheep	
Cobalt toxicity	Humans	
Gadolinium toxicity	Rats	
Iodine deficiency	Dogs, cattle, pigs	
Iodine toxicity	Cattle	
Iron deficiency	Humans, cats, cattle (calves), dogs, horses, pigs, reptiles	
Iron toxicity	Dogs	
Lead toxicity	Humans, birds, dogs, reptiles	
Lutetium toxicity	Humans	Tiredness
Magnesium deficiency	Humans, birds, cattle, guinea pigs, fish	
Magnesium toxicity	Humans, goats	
Manganese deficiency	Sheep	Impaired locomotion
Mercury toxicity	Humans, marine mammals	Seals
Potassium deficiency	Sheep	
Samarium toxicity	Mice	
Selenium deficiency	Horses, nonhuman primates	*Squirrel monkeys*
Selenium toxicity	Cattle, pigs	Reluctant to move
Sodium deficiency	Marine mammals	

Sulfur toxicity	Horses	Accidental poisoning
Zinc deficiency	Humans, cattle, ferrets	
Zinc toxicity	Humans, birds, dogs, reptiles	

Leukocyte activity reduced

Element	Species	Comments
Arsenic toxicity	Poultry (chick)	Reduces leukocyte formation
Cobalamin deficiency	Cattle	
Copper deficiency	Sheep	Reduced phagocytosis

Leukocytosis

Element	Species
Barium toxicity	Rats
Calcium deficiency	Birds
Cerium toxicity	Rats
Gadolinium toxicity	Rats
Iron toxicity	Marine mammals (dolphins)
Nickel toxicity	Humans
Zinc toxicity	Humans

Leukoderma/achromoderma

White patches on the skin.

Element	Species
Copper deficiency	Buffaloes

Leukopenia

Element	Species	Comments
Copper deficiency	Humans	
Lead toxicity	Frogs	
Zinc toxicity	Humans	Cu deficiency

Libido loss

Element	Species
Manganese deficiency	Rabbits, rats
Molybdenum toxicity	Cattle

Licking objects (similar to pica)

Element	Species	Comments
Chlorine deficiency	Cattle	
Potassium deficiency	Cattle	
Sodium deficiency	Horses, pigs	For example, licking cages, rocks

Life expectancy (longevity; lifespan)—Extended

Element	Species	Comments
Palladium toxicity	Mice	Males only—not females

Life expectancy (longevity; lifespan)—Reduced

Element	Species	Comments
Aluminum deficiency	Goats	Anke et al. (2005a) Kids
Chromium deficiency	Mice, rats	
Fluorine deficiency	Mice	*Controversial study*
Gallium toxicity	Mice, rats	*Experimental* Females
Germanium toxicity	Mice, rats	*Experimental*
Lithium deficiency	Goats	
Tungsten toxicity	Mice, rats	
Zirconium toxicity	Guinea pigs, mice, rats	

Ligament mineralization

Element	Species	Comments
Fluorine toxicity	Cattle	And tendons

Light-headedness

Element	Species
Selenium toxicity	Humans

Limb movements (uncontrolled jerks, cycling, paddling)

Element	Species
Magnesium deficiency	Cattle, horses, sheep

Limb edema

Element	Species	Comments
Arsenic toxicity	Humans	Nonpitting edema
Boron toxicity	Cattle	
Potassium toxicity	Various	

Limb stretching while walking

Element	Species
Dysprosium toxicity	Mice
Erbium toxicity	Mice
Europium toxicity	Mice
Holmium toxicity	Mice
Lutetium toxicity	Mice
Praseodymium toxicity	Mice

Lip cracks/fissures

Element	Species
Zinc deficiency	Humans

Lipid peroxidase activity increased

Enzyme responsible for the oxidation of lipids resulting in cell membrane damage due to free radicles take electrons from cell membrane lipids.

Element	Species
Nickel toxicity	Rats

Lipofuscinosis

Element	Species	Comments
Mercury toxicity	Marine mammals (*dolphins*)	Accumulation in liver
Selenium deficiency	Dogs	In intestinal muscle

Listlessness

Having or showing little or no interest in anything; languid; spiritless; indifferent.

Element	Species	Comments
Lead toxicity	Humans	
Selenium deficiency	Nonhuman primates	*Macaques*
Vanadium toxicity	Cattle	

Litter size—Low

Element	Species
Manganese deficiency	Rabbits
Vanadium deficiency	Goats

Liver copper content increased

Element	Species
Molybdenum toxicity	Deer

Liver disease/lesions

Element	Species	Comments
Antimony toxicity	Humans	Liver enzymes elevated
Arsenic toxicity	Humans	Cirrhosis, fibrosis, hepatomegaly, hepatic steatosis
Beryllium toxicity	Rabbits	*Intravenous injection:* Hepatic necrosis
Bismuth toxicity	Humans	Elevated liver enzymes (AST, ALT), jaundice
Boron toxicity	Humans, mice	
Cadmium toxicity	Humans, rats	*Humans:* impaired liver function *Rats:* liver necrosis
Cerium toxicity	Mice, rats	*Mice, Rats:* hepatocellular necrosis, fatty degeneration, high liver glutathione and metallothionine *Rats:* fatty liver, degeneration, mitochondrial damage, invagination of nuclear membrane, low liver weight
Chromium toxicity	Humans, laboratory animals	Liver damage, necrosis
Cobalt deficiency	Goats	Fatty liver degeneration
Copper toxicity	Humans, dogs, goats, marsupials (wombats), sheep	*Humans:* Wilsons disease *Dogs:* inherited defect in specific breeds *Wombats:* liver congestion *Sheep:* liver damage, elevated liver enzymes
Fluorine toxicity	Humans	
Gadolinium toxicity	Rats	Liver necrosis
Iodine toxicity	Pigs	Low liver Fe stores
Iron deficiency	Pigs	Hepatomegaly, fatty liver
Iron toxicity	Humans, bats, dogs, horses	*Humans:* liver damage, acute failure *Bats:* hemosiderosis *Dogs:* liver failure *Horses:* liver degeneration, failure
Lead toxicity	Birds, marine mammals, poultry	Liver damage

Manganese deficiency	Dogs, pigs, mice	Fatty liver
Mercury toxicity	Marine mammals	Dolphins, lipofuscinosis
Molybdenum toxicity	Humans	
Neodymium toxicity	Humans	Accumulation over time: industrial exposure
Nickel deficiency	Poultry (chickens), rats	Rough endoplasmic reticulum and ribosomal changes
Nickel toxicity	Rats	Hepatic necrosis
Palladium toxicity	Rabbits, rats	Liver damage, *Rats:* reduced liver enzyme activity
Praseodymium toxicity	Humans, rats	Liver damage *Rats:* fatty degeneration, altered microsomal lipid, reduced RNA polymerase, reduced gluconeogenesis, reduced drug-metabolizing enzymes
Samarium toxicity	Rats	Liver damage
Selenium toxicity	Humans, dogs, rats	*Humans:* hepatic degeneration *Dogs and Rats:* atrophy, cirrhosis *Dogs:* necrosis
Thulium toxicity	Mice	Liver degeneration
Tin toxicity	Human	Liver damage (inorganic)
Uranium toxicity	Humans	Liver dysfunction
Yttrium toxicity	Rats	Liver transaminase enzyme elevation
Zinc toxicity	Dogs	Hepatic dysfunction
Zirconium toxicity	Humans	

Liver iron content decreased

Element	Species
Lead deficiency	Rats

Liver iron content increased

Element	Species
Iron toxicity	Horses

Liver zinc content decreased

Element	Species
Iron toxicity	Horses

Lung compliance—Decreased

Element	Species
Zinc toxicity	Guinea pigs

Lung disease

Element	Species	Comments
Aluminum toxicity	Humans Other species	Following inhalation, pneumoconiosis, pulmonary fibrosis-restrictive, obstructive airway disease Granulomatous lesions in bronchioles following inhalation of aerosols
Antimony toxicity	Humans, guinea pigs, pigs, rabbits, rats	*Humans:* cough, pneumoconiosis, chronic emphysema, pleural adhesions, respiratory irritation, wheezing *Guinea pigs, Rats:* interstitial pneumonitis, macrophage degeneration, pulmonary fibrosis *Pigs:* adenomatous hyperplasia, multinucleated giant cells, pigmented macrophages, pneumocyte hyperplasia, pulmonary fibrosis, *Rabbits:* interstitial pneumonia *Rats:* alveolitis, cellular metaplasia, multinucleated giant cells, pulmonary tumors
Arsenic toxicity	Humans	Bronchitis, restrictive and obstructive lung disease, respiratory failure
Beryllium toxicity	Humans, dogs, guinea pigs, hamsters, monkeys, rats	*Humans, Guinea pigs, Hamsters, Rats:* Lung cancer, *Humans:* pneumonia, *Dogs, Guinea pigs, Hamsters:* Granulomatous lung disease *Guinea pigs:* pulmonary fibrosis *Monkeys:* macrophages, lymphocyte and plasma cell infiltration
Cadmium toxicity	Hamsters, rats	*Mice, Hamsters:* Alveolar hyperplasia *Hamsters:* interstitial fibrosis *Rats:* lung cancer, pulmonary edema
Cerium toxicity	Humans	Interstitial lung disease, pneumoconiosis, pneumonia, pulmonary fibrosis
Chlorine toxicity	Humans	*Inhalation:* lung damage, toxic pneumonitis
Chromium toxicity	Humans, mice, rats	*Humans:* reduced respiratory function *Mice:* lung granulomas, lung tumors *Rats:* bronchial squamous metaplasia, lung tumors
Dysprosium toxicity	Mice	Labored breathing
Fluorine toxicity	Humans and others	Gas exposure
Gadolinium toxicity	Mice	Pneumonia, lung calcification
Gallium toxicity	Rats	Inflammation, neutrophil infiltration, necrosis, fibrosis
Lanthanum toxicity	Humans, rats	*Humans:* pneumoconiosis *Rats:* granulomas, giant cell infiltration
Molybdenum toxicity	Humans, rats	*Humans:* reduced lung function *Rats:* Alveolar cellular metaplasia, histiocyte infiltration, pulmonary inflammation, hyaline degeneration nasal epithelium, metaplasia nasal planum
Neodymium toxicity	Humans	Pulmonary embolism following inhalation

Nickel toxicity	Humans, hamsters, mice, rats	*Humans:* Pneumonitis (interstitial), pulmonary edema, shortness of breath, upper airway irritation *Hamsters:* alveolar septal fibrosis, bronchial hyperplasia, granulomatous pneumonia, interstitial pneumonia, lung tumors (adenoma, adenocarcinoma) reduced tracheal ciliary activity, squamous metaplasia, tracheal mucosal degeneration, *Mice:* atrophy olfactory epithelium, pulmonary hemorrhage *Mice, Rats:* bronchial epithelial degeneration, hypoplasia macrophages, hyperplasia lymph nodes, necrotizing pneumonia, pulmonary inflammation, fibrosis, lung tumors (adenocarcinomas,) *Rats:* alveolitis, lung tumors (adenoma, mesotheliomas, rhabdomyocarcinomas, spindle cell carcinomas), necrotizing pneumonia, squamous metaplasia
Selenium toxicity	Humans, cattle	*Humans:* acute respiratory distress syndrome, pneumomediastinum, *Humans, Cattle:* pulmonary edema
Silicon toxicity	Humans	Silicosis: reduced lung function
Sulfur toxicity	Humans	Pulmonary fibrosis and airway disease, reduced pulmonary function
Uranium toxicity	Humans	Pulmonary edema, pulmonary fibrosis
Vanadium toxicity	Nonhuman primates	*Nonhuman primates:* reduced lung function (reduced airflow, increased resistance) *Rats:* dyspnoea
Yttrium toxicity	Rats (*inhalation*)	Pleural effusion, pulmonary hyperemia, pulmonary edema
Zinc toxicity	Humans, cats, guinea pigs, mice, rabbits, rats	*Humans:* cough, dyspnoea, chest pain, pneumonitis, pneumothorax, acute respiratory distress syndrome, decreased vital capacity *Cats:* bronchopneumonia, leukocyte infiltration alveoli, congestion, dyspnoea *Guinea pigs:* alveolitis, consolidation, emphysema, fibrosis, reduced compliance, reduced functional reserve, reduced lung function. Inflammation *Mice:* macrophage and lymphocyte infiltration, increased alveogenic carcinoma *Rabbits, Rats:* gray areas congestion, peribronchial leukocyte infiltration, polymorphic leukocytes in bronchial lavage exudate *Rats:* high Lactate dehydrogenase in bronchial lavage exudate
Zirconium toxicity	Humans, hamsters, guinea pigs, rabbits, rats	*Humans:* lung granulomas, pulmonary fibrosis *Rabbits:* Pneumonia, pulmonary fibrosis *Hamsters, Guinea pigs, Rats:* interstitial pneumonitis

Lung tumors

Element	Species	Comments
Arsenic toxicity	Humans	
Beryllium toxicity	Humans, guinea pigs, hamsters, rats	*Bronchoalveolar tumors:* mainly adenocarcinomas, adenomas
Cadmium toxicity	Mice, rats	
Chromium toxicity	Mice	Adenomas
Nickel toxicity	Hamsters, mice, rats	*Hamsters:* adenoma, adenocarcinoma *Mice, Rats:* adenocarcinomas *Rats:* adenoma, mesotheliomas, rhabdomyocarcinomas, spindle cell carcinomas
Zinc toxicity	Mice	

Luteinizing hormone—Reduced synthesis

Element	Species
Aluminum toxicity	Rats

Luteinizing hormone (LH)—Increased production

Element	Species
Chromium toxicity	Amphibians

Lymphadenopathy

Abnormal size or consistency of lymph nodes.

Element	Species	Comments
Beryllium toxicity	Humans	
Titanium toxicity	Humans	Part of "yellow nail syndrome"

DD: autoimmune disease, benign reactive, infection, neoplasia.

Lymphocytic aneuploidy

The lymphocytes contain an abnormal number of chromosomes in a cell.

Element	Species
Mercury toxicity	Humans

Lymphocytopenia

Element	Species	Comments
Cerium toxicity	Mice	Reduced CD3 and CD8 lymphocytes
Mercury toxicity	Dogs	

Lymphoid hyperplasia

Element	Species
Cerium toxicity	Rats

Lysozymuria

The presence of lysozyme, a bacteriolytic enzyme, in urine—occurs in renal disease.

Element	Species
Chromium toxicity	Rats

Macrophagic myofasciitis

Inflammatory macrophages with aluminum-containing crystal inclusions and muscle necrosis. Injection site associated with Al-containing adjuvant in vaccines.

Element	Species
Aluminum toxicity	Humans

Magnesium in tissues (low)

Element	Species
Magnesium deficiency	Fish

Malabsorption

Element	Species
Iron deficiency	Humans

Malondialdehyde (MDA)—High

An organic compound that is a marker for stress.

Element	Species	Comments
Copper deficiency	Buffaloes	
Gallium toxicity	Humans	In blood and urine
Samarium toxicity	Rats	

Male infertility

Element	Species	Comments
Chromium deficiency	Rats	Rats: reduced sperm count
Iodine deficiency	Cattle, horses poultry, sheep	*Cattle*: lack libido; poor sperm *Poultry (experimental)*: small testes, low sperm count, small comb
Manganese deficiency	Rabbits, rats	Lack of libido, reduced sperm count, testicular degeneration
Molybdenum toxicity	Humans, cattle, rats	*Humans:* reduced testosterone production *Cattle, Rats:* testes damage, reduced spermatogenesis

Manganese accumulation in body organs

Element	Species	Comments
Arsenic deficiency	Goats (kids), pigs (piglets) rats	*Goats and Pigs*: in body organs and milk of mother *Rats*: in liver

Mean corpuscular volume—Decreased

Element	Species
Aluminum toxicity	Dogs
Copper deficiency	Pigs
Vanadium toxicity	Rats

Meat production reduced

Element	Species
Phosphorus deficiency	Buffaloes

Melena

Element	Species
Chlorine deficiency	Cattle
Iron toxicity	Cats, dogs
Nickel toxicity	Pigs
Zinc toxicity	

Melanin synthesis reduced

Element	Species
Copper deficiency	Various
Manganese deficiency	Mice (Pallid)

Memory loss

Element	Species	Comments
Arsenic toxicity	Humans	
Bismuth toxicity	Humans	
Boron deficiency	Humans	Short term memory loss
Magnesium deficiency	Humans	
Mercury toxicity	Humans	
Tungsten toxicity	Humans	

Meningitis

Element	Species	Comments
Sodium toxicity	Ferrets	Nonsupportive meningitis

Menkes syndrome

Caused by mutations in the ATP7A gene which provides instructions for the synthesis of a protein important for copper levels in the body. This disease results in Cu accumulation in some tissues (small intestine; kidneys) but low levels in others (brain and other tissues).
Characteristic signs include delayed brain development, sparse, kinky, or coarse hair; intellectual disability, poor growth rate, nervous system deterioration, hypotonia, joint laxity, sagging of facial skin, occipital horn syndrome (calcium deposits at the base of the skull), seizures, weakness, and poor weight gain.

Element	Species	Comments
Copper deficiency	Humans	Genetic defect results in low Cu
Copper toxicity	Humans	Genetic defect results in high Cu

Mental derangement

Element	Species
Magnesium deficiency	Humans

Mental retardation

Element	Species
Molybdenum deficiency	Humans
Thallium toxicity	Humans

Metabolic acidosis

Element	Species	Comments
Barium toxicity	Dogs	
Chlorine deficiency	Dogs	
Chromium toxicity	Humans	
Iron toxicity	Humans, dogs	Acute
Metabolic acidosis	Dogs	
Phosphorus deficiency	Cats	
Phosphorus toxicity	Cats	Only one report
Selenium toxicity	Humans	
Tellurium toxicity	Humans	

Metabolic alkalosis

Element	Species
Chlorine deficiency	Humans, cattle

Metal fume fever

Influenza type symptoms—a raised temperature, chills, aches and pains, nausea and dizziness. It is caused by exposure to the fume of certain metals.

Element	Species
Antimony toxicity	Humans
Zinc toxicity	Humans

Metallic taste in mouth

Element	Species	Comments
Antimony toxicity	Humans	
Bismuth toxicity	Humans	
Gallium toxicity	Humans	
Iodine toxicity	Humans	Described as "brassy" taste
Mercury toxicity	Humans	
Selenium toxicity	Humans	
Tellurium toxicity	Humans	
Zinc toxicity	Humans	

Metamorphosis delayed

Element	Species	Comments
Lead toxicity	Frogs	Tadpoles

Metritis

Element	Species	Comments
Selenium deficiency	Cattle	Se supplementation reduced occurrence rate

Micrognathia

Element	Species
Cadmium toxicity	Mice, rats

Micromelia

Short limb bones.

Element	Species
Zinc deficiency	Humans, poultry, and others

Microcytosis

Small red blood cells (erythrocytes).

Element	Species
Aluminum toxicity	Dogs

Microphthalmia

Element	Species
Cadmium toxicity	Mice, rats

Milk fever/parturient paresis

Element	Species	Comments
Calcium deficiency	Cats, cattle, dogs, sheep, rabbits	Usually NOT due to dietary deficiency, but due to excess calcium during pregnancy inhibiting parathyroid hormone and osteoclast activity when needed during milk production

Milk production—Reduced

Element	Species
Arsenic deficiency	Goats
Calcium deficiency	Buffaloes
Chlorine deficiency	Cattle
Cobalt (cobalamin) deficiency	Cattle, sheep
Copper deficiency	Camels, cattle
Fluorine toxicity	Cattle, pigs
Iodine deficiency	Cattle, goats
Lithium deficiency	Goats
Molybdenum toxicity	Rats
Nickel deficiency	Goats
Phosphorus deficiency	Buffaloes, cattle, goats, sheep
Potassium deficiency	Cattle
Potassium toxicity	Cattle
Sodium deficiency	Cattle, horses
Sodium toxicity	Cattle
Sulfur deficiency	Cattle
Sulfur toxicity	Cattle
Vanadium deficiency	Goats
Zinc deficiency	Cattle
Zinc toxicity	Cattle

Mineralization of soft tissue (not necessarily calcium)

Element	Species	Comments
Gadolinium toxicity	Rats	Lung and kidney
Lutetium toxicity	Guinea pigs	At injection site

Mitochondrial damage

Element	Species	Comments
Arsenic deficiency	Goats (kids) Rats	Goats: in cardiac muscle, skeletal muscle, and liver (less) Rats: in liver
Cadmium deficiency	Goats	Especially renal and hepatic mitochondria
Cerium toxicity	Rats	In liver

Mitral valve prolapse

Element	Species	Comments
Magnesium deficiency	? Dogs	? In Cavalier King Charles Spaniels

Mood swings

Element	Species
Mercury toxicity	Humans

Monoamine oxidase activity reduced

Element	Species
Copper deficiency	Poultry

Monocytopenia

Element	Species
Lead toxicity	Frogs

Morbidity increased

Element	Species	Comments
Arsenic toxicity	Humans	*Chronic:* specifically due hypertensive disease
Chromium deficiency	Cattle	

Motor coordination lost

Element	Species
Europium toxicity	Mice
Mercury toxicity	Cats, dogs

Motor neurone disease

Element	Species	Comments
Calcium deficiency	Nonhuman primates	Cynomolgus monkeys
Molybdenum toxicity	Sheep	

Motor neuropathy

Element	Species
Thallium toxicity	Humans

Motor speed reduced

Element	Species
Boron deficiency	Humans

Movement resisted (reluctance to move)

Element	Species
Phosphorus deficiency	Buffaloes

Mucosal lesions

Element	Species	Comments
Arsenic toxicity	Humans	Chronic toxicity
Boron toxicity	Humans	Congestion, exfoliation, inflammation
Palladium toxicity	Humans	Mucositis
Selenium toxicity	Humans	*Caustic*: mouth, esophagus, stomach
Sodium toxicity	Humans, dogs	Corrosive burns oral cavity to small intestine, e.g., sodium hypochlorite
Zinc deficiency	Cattle	Submucosal hemorrhage
Zirconium toxicity	Humans	Inflammation

Mucous membranes dry and tacky

A sign of dehydration.

Element	Species
Sodium deficiency	Dogs

Multiple organ system failure

Element	Species	Comments
Iron toxicity	Humans	
Ionizing radiation	Humans, various	Exposure

Multiple sclerosis

Chronic, progressive damage to nerve cell sheaths in brain and spinal cord, signs include fatigue, numbness, speech impairment, muscular incoordination, blurred vision.

Element	Species	Comments
Aluminum toxicity	Humans	
Molybdenum toxicity	Humans	Acute toxicity

Muscle disease

See also weakness.

Element	Species	Comments
Aluminum toxicity	Pigs, rats	*Pigs:* muscle atrophy *Rats:* muscle necrosis at injection site
Arsenic deficiency	Goats (kids)	Atrophy of cardiac and skeletal muscle fibers—see mitochondrial damage
Arsenic toxicity	Humans	Muscle aches in chronic toxicity
Barium toxicity	Humans	Rhabdomyolysis, tremors, weakness
Calcium deficiency	Reptiles	Soft, flabby muscles
Copper toxicity	Humans	Muscular atrophy
Lead toxicity	Humans, frogs	*Humans:* loss muscle strength *Frogs:* loss of muscle tone
Lithium toxicity	Humans	Muscle weakness
Magnesium deficiency	Humans	Myographic changes, muscle weakness
Magnesium toxicity	Humans	Muscle weakness
Mercury toxicity	Horses	Masseter muscle atrophy
Molybdenum toxicity	Rabbits	Muscle degeneration
Phosphorus toxicity	Humans	Muscle weakness

Potassium deficiency	Humans Rabbits	Soft flabby muscles Muscular dystrophy
Selenium deficiency	Dogs, fish, goats	Degeneration, discolored (white lines), myopathy, muscle pallor *Cardiac muscle:* necrosis *Intestinal muscle:* discolored brown-yellow due lipofuscinosis *Fish:* muscular dystrophy (salmon) *Goats:* white muscle disease
Selenium toxicity	Humans	Muscle spasms, myoclonus, tenderness,
Sulfur toxicity	Rabbits	Loss of muscle control
Tellurium toxicity	Rats	*Experimental:* necrosis of skeletal and cardiac muscle—similar to vitamin E/Selenium deficiency

Muscle cramps/spasms (dystonia)

Element	Species
Arsenic toxicity	Humans
Bismuth toxicity	Humans
Calcium deficiency	Squirrels
Magnesium deficiency	Humans
Manganese toxicity	Humans
Samarium toxicity	Mice
Selenium toxicity	Humans
Sodium deficiency	Humans, horses
Sodium toxicity	Ferrets (*Periodic choreiform spasm*)
Zinc toxicity	Humans

Muscle fasciculations

Element	Species
Magnesium deficiency	Humans, buffaloes

Muscle movements (uncontrolled)

Element	Species	Comments
Lithium toxicity	Humans	
Selenium toxicity	Humans	Fasciculations

Muscle necrosis

Element	Species
Magnesium deficiency	Dogs

Muscle pain

Element	Species	Comments
Cadmium toxicity	Humans	Aches
Selenium toxicity	Humans	Muscle tenderness
Thallium toxicity	Humans	Aches
Titanium allergy	Humans	

Muscle strength loss/weakness

Element	Species
Calcium toxicity	Humans, reptiles
Copper deficiency	Birds
Lead toxicity	Humans
Lithium toxicity	Humans
Magnesium toxicity	Humans
Sodium toxicity	Birds

Muscle tone—Loss (hypotonia)

Element	Species	Comments
Barium toxicity	Dogs	
Copper deficiency	Humans	
Potassium deficiency	Humans	Intestinal muscle—results in intestinal distension

Muscle tremors/twitches

Element	Species	Comments
Aluminum toxicity	Dogs	
Barium toxicity	Humans, dogs	
Calcium deficiency	Cats, dogs, rabbits, reptiles	*Dogs, Cats:* especially during milk fever/parturient paresis *Reptiles:* twitches in toes, legs
Chlorine deficiency	Humans	Twitches
Europium toxicity	Mice	Infrequent tremors
Fluorine toxicity	Cattle	
Gallium toxicity	Humans	Acute
Lead toxicity	Cattle, frogs	
Magnesium deficiency	Humans, cats, cattle, horses, pigs	*Humans:* tremors and fasciculations
Sulfur deficiency	Goats	
Sulfur toxicity	Cattle, goats, sheep	

Muscle wastage/sarcopenia

Loss of lean muscle mass.

Element	Species
Chromium deficiency	Humans, pigs, rats
Cobalt (cobalamin) deficiency	Cattle, sheep

Muscular dystrophy

Element	Species
Selenium deficiency	Birds, cattle, sheep

Myalgia

Muscle aches and pain.

Element	Species
Antimony toxicity	Humans
Selenium deficiency	Humans

Myasthenia

Element	Species
Cadmium deficiency	Goats

Mydriasis

Element	Species	Comments
Barium toxicity	Humans	And dysplasia

Myelination delayed

Element	Species
Copper deficiency	Guinea pigs

Myelodysplastic syndromes

Cancers in which immature blood cells in the bone marrow do not mature.

Element	Species	Comments
Zinc toxicity	Humans	Due Cu deficiency

Myocardial degeneration

Element	Species
Iron toxicity	Sheep

Myocardial depression

Element	Species	Comments
Aluminum toxicity	Humans	Hypokinesia
Phosphorus deficiency	Dogs	
Selenium toxicity	Humans	

Myocardial fibrosis

Element	Species
Copper deficiency	Cattle
Selenium toxicity	Deer

Myocardial hypokinesia

Decrease in amplitude of left ventricular wall motion.

Element	Species
Aluminum toxicity	Humans

Myocardial hypoxic contracture—Reduced

Element	Species
Cobalt toxicity	Rats

Myocardial infarction

Element	Species
Cerium toxicity	Humans

Myocardial mineralization

Element	Species
Selenium toxicity	Deer

Myocardial necrosis

Element	Species
Selenium toxicity	Deer

Myocarditis

Element	Species	Comments
Aluminum toxicity	Humans	Toxic
Uranium toxicity	Humans	

Myoclonus

Rapid involuntary muscle jerks.

Element	Species	Comments
Aluminum toxicity	Humans	Especially dialysis patients with Al encephalopathy
Bismuth toxicity	Humans	
Lithium toxicity	Humans	
Selenium toxicity	Humans	

Myofasciitis

Macrophagic myofasciitis (MMF) specific muscle lesion following aluminum hydroxide (a common vaccine adjuvant) injection.

Element	Species
Aluminum toxicity	Humans

Myoglobin concentrations: Decreased

Element	Species	Comments
Iron deficiency	*In muscle*: dogs, pigs, poultry, and rats *In serum:* Humans	In young animals In children

Myokymia

Involuntary, spontaneous, localized quivering of a few muscles.

Element	Species
Gold toxicity	Humans

Myopathy

Element	Species	Comments
Aluminum toxicity	Humans	Proximal myopathy
Selenium deficiency	Cattle (*calve*), deer, dogs, goats (*kids*), horses (*foals and donkey foals*), nonhuman primates (*squirrel monkeys*), pigs, sheep (*lambs*), reptiles	White muscle disease

Myxedema

Swelling of the skin and underlying tissue.

Element	Species
Cobalt toxicity	Humans (children)
Iodine deficiency	Humans, dogs

N-Acetyl-β-D-glucosaminidase (NAG)—Activity reduced

Element	Species
Europium toxicity	Rats

Nail/hoof abnormalities

Element	Species	Comments
Arsenic toxicity	Humans	Nail striations (Mees' lines)
Boron toxicity	Rats	Abnormally long nails
Selenium toxicity	Humans, cattle, horses, pigs, reptiles	*Humans:* nails brittle, deformed, discolored, loss, longitudinal striations, white spots *Cattle, Horses:* deformity, coronary band inflammation cracks *Pigs:* deformed
Zinc deficiency	Goats, sheep	Hoof soft, deformed, lack surface striations

Nasal discharge

Element	Species	Comments
Cesium toxicity	Rats	Blood stained
Iodine toxicity	Cats	

Nasal lesions

Element	Species	Comments
Sulfur toxicity	Humans	Hyperplasia, rhinitis, secretions
Vanadium toxicity	Mice, rats	Hyperplasia, metaplasia

Nasal papilloma

Element	Species
Chromium toxicity	Mice

Nausea

Element	Species
Antimony toxicity	Humans
Arsenic toxicity	Humans
Barium toxicity	Humans
Boron toxicity	Humans
Cadmium toxicity	Humans
Cesium toxicity	Humans
Calcium toxicity	Reptiles
Chlorine deficiency	Humans
Chromium toxicity	Humans
Copper toxicity	Various
Europium toxicity	Humans
Gadolinium toxicity	Humans (*After injection for MRI scan*)
Gallium toxicity	Humans
Germanium toxicity	Humans
Iron toxicity	Humans
Lanthanum toxicity	Humans
Lithium toxicity	Humans
Lutetium toxicity	Humans
Magnesium deficiency	Humans
Magnesium toxicity	Humans
Mercury toxicity	Humans
Molybdenum deficiency	Humans
Nickel toxicity	Humans
Selenium toxicity	Humans
Silver toxicity	Humans
Sodium deficiency	Humans
Sulfur toxicity	Humans
Tellurium toxicity	Humans
Thallium toxicity	Humans

Tin toxicity	Humans
Tungsten toxicity	Humans
Vanadium toxicity	Humans
Zinc deficiency	Humans
Ionizing radiation	Humans

Neck extension

Element	Species
Phosphorus deficiency	Buffaloes

Neck pain

Element	Species
Selenium deficiency	Horses

Nephrocalcinosis

Element	Species
Calcium toxicity	Birds, cats, dogs (usually due to nutritional or renal hyperparathyroidism)
Phosphorus toxicity	Rats, guinea pigs
Selenium deficiency	Dogs
Selenium toxicity	Fish
Tin toxicity	Rats

Nephrogenic diabetes insipidus

Element	Species
Lithium toxicity	Humans

Nephrogenic systemic fibrosis

Element	Species	Comments
Gadolinium toxicity	Humans	After Gd contrast for MRI scan

Nephrotic syndrome

Element	Species
Lithium toxicity	Humans

Nervousness

Element	Species	Comments
Iodine deficiency	Dogs	Timidity
Mercury deficiency	Dogs	
Sulfur toxicity	Cattle	

Neurasthenic signs

Nonspecific emotional signs such as fatigue, headache, irritability, lassitude.

Element	Species
Nickel toxicity	Humans

Neuropathy

Nerve dysfunction often causing numbness, pain, tingling sensation, muscle weakness.

Element	Species	Comments
Aluminum toxicity	Dogs	Diffuse cerebral neuropathy and diffuse peripheral neuropathy
Antimony toxicity	Humans	Peripheral neuropathy
Arsenic toxicity	Humans	Peripheral neuropathy in Chronic toxicity, sensory neuropathy (hands and feet initially)
Bromine toxicity	Humans	Neurological signs including coma, confusion, hallucinations, headaches, slurred speech, stupor
Cesium toxicity	Humans	tingling in lips, cheeks, hands, feet
Calcium deficiency	Nonhuman primates	Cynomolgus monkeys: developed motor neuron damage after 3.5 years on 0.32% Ca with high Al (Garruto et al., 1989)
Chromium deficiency	Humans	Peripheral neuropathy
Cobalt (cobalamin) deficiency	Humans, cattle (calves)	Humans: neurological lesions Cattle: demyelination of peripheral nerves
Copper deficiency	Guinea pigs	Delayed myelination, cerebellar agenesis, cerebral edema
Lead toxicity	Humans	Peripheral; demyelination, reduced nerve conduction speed
Mercury toxicity	Humans	Reduced nerve conductance speed
Potassium deficiency	Humans, cats	Loss of reflexes

Neuropsychiatric signs

Element	Species
Bismuth toxicity	Humans

Neuropsychological changes

Element	Species
Mercury toxicity	Humans

Neurotoxic effects

Element	Species	Comments
Aluminum toxicity	Cats (*experimental*), Rabbits (*experimental*)	Neurofilament formation.
Arsenic toxicity	Humans	
Chromium toxicity	Humans	Nerve damage
Europium toxicity	Mice	Ataxia, disorientation, loss motor activity, tremor (infrequent)
Iodine toxicity	Birds	Central nervous system signs
Lithium toxicity	Humans	Pyramidal, extrapyramidal and cerebellar, nystagmus
Mercury toxicity	Humans, cats, dogs, horses, mice, pigs, rats	Loss of neurones, necrosis
Molybdenum toxicity	Humans	Psychosis, seizures,
Phosphorus toxicity	Humans	Neurological damage—atrophy
Selenium toxicity	Humans	Anesthesia, confusion, coma, hemiplegia, paresthesia, tremors
Thallium toxicity	Humans	Peripheral neuropathy: feels like walking on coals
Tin toxicity	Humans	CNS edema
Zinc toxicity	Humans, dogs	
Ionizing radiation	Various	Exposure—confusion, coma, elevated intracranial pressure, incoordination

Neutropenia

Element	Species	Comments
Arsenic toxicity	Humans	Chronic
Lead toxicity	Frogs	
Zinc toxicity	Humans	Due Cu deficiency

Neutrophil activity—Inhibition

Element	Species
Copper deficiency	Cattle

Neutrophilia

Element	Species
Barium toxicity	Rats
Cerium toxicity	Rats
Gallium toxicity	Mice
Zinc toxicity	Humans

Night blindness

Element	Species
Molybdenum deficiency	Humans
Zinc deficiency	Humans

Nitric oxide—High

Element	Species
Copper deficiency	Buffaloes

Nitrogen balance reduced

Element	Species
Chromium deficiency	Humans

Nodule formation

Element	Species	Comments
Dysprosium toxicity	Guinea pigs	
Lutetium toxicity	Guinea pigs	At injection site

Nucleated erythrocytes

Element	Species
Vanadium toxicity	Rats

Numbness

Element	Species	Comments
Barium toxicity	Humans	Face
Phosphorus deficiency	Humans	

Nystagmus

Repetitive, involuntary eye movements, which can be in a horizontal or vertical plane and may move faster in one direction than another.

Element	Species
Lithium toxicity	Humans
Mercury toxicity	Cats (vertical), dogs (horizontal)

Ocular discharge

Element	Species	Comments
Lead toxicosis	Mink	Mucopurulent discharge

Oliguria

Element	Species	Comments
Bismuth toxicity	Humans	
Iron toxicity		Acute toxicity

Ophthalmoplegia

Impaired function of the ocular muscles.

Element	Species
Thallium toxicity	Humans

Opisthotonos

Spasm of the muscles causing backward arching of the head, neck, and spine.

Element	Species
Calcium deficiency	Birds
Magnesium deficiency	Buffaloes
Mercury toxicity	Dogs
Molybdenum deficiency	Humans

Optical atrophy

Death (atrophy) of retinal ganglion cell axons that constitute the optic nerve.

Element	Species
Antimony toxicity	Humans
Lead toxicity	Humans
Thallium toxicity	Humans

Oral lichen planus

Chronic inflammation of the mucous membranes of the mouth which visually looks like white, lacy patches, red, swollen tissues; or open sores.

Element	Species
Palladium toxicity	Humans

Oral pain

Element	Species	Comments
Zirconium toxicity	Humans	Burning sensation

Oral ulcers

Element	Species
Zinc deficiency	Humans

Organ of corti damage

Part of the cochlea (in inner ear) that produces nerve impulses in response to sound vibrations.

Element	Species	Comments
Mercury toxicity	Marine mammals	Harp seals

Organic brain syndrome

Element	Species
Thallium toxicity	Humans

Osteoarthritis

Element	Species	Comments
Calcium toxicity	Cattle	
Silicon deficiency	Humans	Proposed, not confirmed. Si is nonessential in humans

Osteoblast activity—Reduced

Element	Species	Comments
Copper deficiency	Pigs	Resulting in impaired endochondral ossification + Other species with developmental skeletal problems

Osteochondrosis

Element	Species	Comments
Calcium toxicity	Dogs	
Copper deficiency	Horses, red deer possibly also in dogs	? due low copper containing metalloenzymes in ossification
Zinc toxicity	Horses	Due induced Cu deficiency

Osteodystrophy

Element	Species	Comments
Aluminum toxicity	Humans	Especially in patients with renal failure

Otolith malformation

Element	Species
Manganese deficiency	Guinea pig, mice, mink, rabbit, rat

Ovarian cysts

Element	Species	Comments
Selenium deficiency	Cattle	Se supplementation reduced occurrence rate

Ovarian necrosis

Element	Species
Cadmium toxicity	Hamsters, mice, rats

Ovulation—Abnormal

Element	Species	Comments
Manganese deficiency	Cattle, goats, rats	Cattle, goats—depressed or delayed estrus
Zinc deficiency	Cattle, hamsters, monkeys, sheep	Abnormal estrus cycles or cessation

Packed cell volume (PCV)—Low

Element	Species	Comments
Cobalt deficiency	Cattle, sheep	Vit B12 cobalamin deficiency
Iron deficiency	Poultry (rare)	
Potassium toxicity	Cattle	Experimental

Packed cell volume (PCV)—High

Element	Species
Cobalt toxicity	Cattle

Pain—Peripheral

Element	Species
Selenium toxicity	Humans

Pain sensation loss

Element	Species
Mercury toxicity	Cats

Pain—Substernal

Element	Species
Cadmium toxicity	Humans

Pallor

Pale mucous membranes.

Element	Species	Comments
Cobalt deficiency	Cattle, sheep	Vit B12 cobalamin deficiency
Copper deficiency	Humans	
Iron deficiency	Humans, cats, cattle (calves), dogs, pigs, horses, reptiles	
Iron toxicity	Pigs	
Lead toxicity	Humans (children), reptiles	
Zinc deficiency	Ferrets	
Zinc toxicity	Dogs, reptiles	

Palpitations

Element	Species
Iron deficiency	Humans
Magnesium toxicity	Humans
Nickel toxicity	Humans

Pancreatic aplasia/hypoplasia

Element	Species
Manganese deficiency	Guinea pigs

Pancreatic atrophy

Element	Species
Tin toxicity	Rats

Pancreatic islet cell necrosis

Element	Species
Aluminum toxicity	Rats

Pancreatic lesions

Element	Species	Comments
Aluminum toxicity	Rats	Islet cell necrosis
Copper deficiency	Humans	High pancreatic leucine enkephalin-containing peptides
Manganese deficiency	Guinea pigs	Pancreatic aplasia/hypoplasia
Selenium deficiency	Humans	Pancreatic degeneration
Tin toxicity	Rats	Pancreatic atrophy

Pancreatitis

Element	Species	Comments
Aluminum toxicity	Humans	
Antimony toxicity	Humans, guinea pigs	
Calcium toxicity	Reptiles	Acute pancreatitis
Zinc toxicity	Humans, dogs	

Pancytopenia

Element	Species	Comments
Arsenic toxicity	Humans	Acute—severe

Panting/gasping

Element	Species	Comments
Cobalt toxicity	Cattle	Shortness of breath
Iodine toxicity	Dogs	
Magnesium deficiency	Poultry (rare)	
Zinc deficiency	Poultry	Tachypnoea

Papilledema

Swelling of the optic disc.

Element	Species
Lead toxicity	Humans (children)

Paresthesia

Abnormal sensation: tingling or pricking (pins and needles).

Element	Species	Comments
Bismuth toxicity	Humans	
Magnesium toxicity	Humans	Hands and feet
Selenium toxicity	Humans	Peripheral paresthesia

Parakeratosis

Excess keratinization of epidermis, crusts, horny dry scale, cracks, fissures, and brown exudate with alopecia.

Element	Species	Comments
Lithium deficiency	Goats, minipigs	Maybe due interference with Zn
Calcium toxicity	Pigs	Due interference with Zn
Zinc deficiency	Cats, cattle, goats, horses, monkeys (*squirrel monkeys*) pigs	*Squirrel monkeys:* tongue

Paralysis

Element	Species	Comments
Barium toxicity	Humans, dogs	*Dogs:* flaccid paralysis
Calcium deficiency	Birds, buffaloes, pigs, poultry, squirrels	May be due vertebral fractures
Chlorine deficiency	Birds	limbs
Copper deficiency	Sheep	Spastic paralysis
Gallium toxicity	Humans	
Iron toxicity	Pigs	Posterior paralysis
Lead toxicity	Birds	
Magnesium deficiency	Dogs	Hindleg
Magnesium toxicity	Humans	Flaccid paralysis hands/feet
Mercury toxicity	Mink, pigs	Partial
Phosphorus deficiency	Buffaloes, pigs, poultry	*Buffaloes:* hindleg paralysis
Potassium deficiency	Humans, dogs, goats, poultry (chicks)	
Selenium deficiency	Birds	Oral paralysis in cockatiels

Selenium toxicity	Cattle	
Sodium deficiency	Birds	
Strontium toxicity	Pigs	Posterior paralysis
Tellurium toxicity	Rats	*Experimental:* hindleg (temporary)

Paralytic ileus

Functional intestinal obstruction due to paralysis of the intestinal muscles.

Element	Species
Bismuth toxicity	Humans
Uranium toxicity	Humans

Parasitism—Increased risk

Element	Species	Comments
Cobalt deficiency	Cattle, sheep	Vit B12 cobalamin deficiency

Paresthesia

Abnormal skin sensation, e.g., prickling, burning, or tingling.

Mineral	Species	Comments
Gadolinium toxicity	Cattle, sheep	Vit B12 cobalamin deficiency
Mercury toxicity	Humans	

Paresis

Muscular weakness. Paraparesis: partial paralysis of lower limbs. Tetraparesis: partial paralysis all four limbs.

Element	Species	Comments
Aluminum toxicity	Dogs	Paraparesis, tetraparesis
Calcium deficiency	Cattle, reptiles	*Reptiles:* hindleg paresis *Cattle:* milk fever
Mercury toxicity	Cattle, dog, sheep	*Dog:* tetraparesis

Pericardial effusion

Abnormal accumulation of fluid in the pericardial sac around the heart.

Element	Species
Cobalt toxicity	Humans

Parkinsonism

A combination of movement abnormalities seen in Parkinson's disease—such as tremor, slow movement, impaired speech, or muscle stiffness.

Element	Species	Comments
Bismuth toxicity	Humans	

Parosmia

Abnormal sense of smell.

Element	Species
Gadolinium toxicity	Humans

Performance reduction

Element	Species
Iron deficiency	Cattle (calves)
Magnesium toxicity	Goats
Potassium deficiency	Pigs
Sodium deficiency	Pigs
Sulfur deficiency	Goats

Peritonitis

Element	Species	Comments
Iron toxicity	Humans	Gastrointestinal perforation
Mercury toxicity	Humans	
Palladium toxicity	Rats	After intraperitoneal administration
Sulfur toxicity	Cattle	
Zinc toxicity	Birds	

Perosis

Element	Species	Comments
Manganese deficiency	Birds, poultry	*Birds, Chicks:* slipped tendon *Birds (gray cranes and others) Chicken, Duck:* slipped gastrocnemius muscle

Petechial hemorrhages

Element	Species	Comments
Sulfur toxicity	Cattle	Organs, e.g., kidney

Pharyngitis

Element	Species	Comments
Antimony toxicity	Humans	
Beryllium toxicity	Humans	
Mercury toxicity	Humans	Severe
Zirconium toxicity	Humans	Painful

Phospholipid concentrations in blood increased

Element	Species
Copper deficiency	Rats

Phosphorus absorption reduced from intestinal tract

Element	Species	Comments
Cadmium toxicity	Rats	Ca absorption also reduced

Phosphorus deficiency

Element	Species	Comments
Aluminum toxicity	Mice, poultry (chicks), sheep	Al chelates P.

Phosphorus—Plasma low

Element	Species
Boron toxicity	Cattle

Pica

Depraved appetite-eating unusual items, foreign objects, rocks, bones (ruminants).

Element	Species
Chlorine deficiency	Cattle
Iron deficiency	Humans
Phosphorus deficiency	Buffaloes, cattle, dogs (young), goats, horses, sheep
Potassium deficiency	Cattle
Sodium deficiency	Cattle, goats, horses, sheep
Zinc toxicity	Sheep

DD *Humans:* malnutrition, pregnancy, mental health disorders.

Placental abruption

Premature separation of the placenta—prior to birth.

Element	Species
Zinc deficiency	Humans

Placental retention

Element	Species
Selenium deficiency	Cattle

Plasma cell hyperplasia

Element	Species	Comments
Gallium toxicity	Mice	In lymph nodes

Platelet count—High

Element	Species
Gallium toxicity	Mice

Platelet count—Reduced

Element	Species
Cerium toxicity	Mice
Gadolinium toxicity	Rats

Pleural effusion

Excess free fluid accumulation in the pleural cavity outside the lungs.

Element	Species	Comments
Titanium toxicity	Humans	Part of "yellow nail syndrome"

Pneumoconiosis

Fibrotic pneumoconiosis nodules in the lungs due chronic inhalation.

Element	Species	Comments
Aluminum toxicity	Humans	Inhalation bauxite
Antimony toxicity	Humans	
Cerium toxicity	Humans	
Lanthanum toxicity	Humans	Exposure to carbon arc light
Lutetium toxicity	Humans	
Neodymium toxicity	Humans	

Praseodymium toxicity	Humans	
Samarium toxicity	Humans	
Terbium toxicity	Humans	
Ytterbium toxicity	Humans	

Pneumomediastinum

Element	Species	Comments
Selenium toxicity	Humans	Inhaled

Pneumonia

Element	Species	Comments
Aluminum toxicity	Humans	Granulomatous pneumonia, desquamative interstitial pneumonia
Cerium toxicity	Humans	
Manganese toxicity	Humans	
Mercury toxicity	Humans	

Pneumonitis

Inflammation of lung tissue.

Element	Species	Comments
Antimony toxicity	Humans	
Beryllium toxicity	Humans	
Bromine toxicity	Humans	
Chlorine toxicity	Humans	Inhalation: toxic pneumonitis
Mercury toxicity	Humans	
Nickel toxicity	Humans	Interstitial pneumonitis
Palladium toxicity	Rats	
Selenium toxicity	Humans	
Silicon toxicity	Rats	Inhalation—inflammation
Sulfur toxicity	Humans	
Zinc toxicity	Humans, guinea pigs	Inhalation
Zirconium toxicity	Hamsters, guinea pigs, rats	Interstitial pneumonitis after inhalation

Pneumothorax

Element	Species	Comments
Zinc toxicity	Humans	Inhalation

Polioencephalomalacia

Element	Species
Mercury toxicity	Sheep
Selenium toxicity	Cattle

Polioencephalomalacia-like syndrome

Element	Species
Sulfur toxicity	Cattle

Poliomyelomalacia

Element	Species
Selenium toxicity	Pigs

Polyarthritis

Element	Species
Mercury toxicity	Humans

Polycythemia

Element	Species
Cobalt toxicity	Humans, cattle, rats (*experimental*)

Polydipsia

Excessive thirst.

Element	Species
Calcium deficiency	Birds
Calcium toxicity	Humans
Chlorine deficiency	Humans, cattle
Copper deficiency	Birds
Lead toxicity	Birds
Lithium toxicity	Humans, birds
Phosphorus deficiency	Birds
Potassium toxicity	Cattle
Sodium deficiency	Cats, dogs, poultry

Sodium toxicity	Birds, cattle
Zinc toxicity	Birds, dogs
Zirconium toxicity	Humans

Polyneuritis

Element	Species
Arsenic toxicity	Humans

Polyneuropathy

Element	Species	Comments
Mercury toxicity	Humans	
Phosphate toxicity	Humans	Sensorimotor polyneuropathy
Thallium toxicity	Humans	Both motor and sensory
Zinc toxicity	Humans	Sensory. Due to Cu deficiency

Polyuria

Element	Species
Calcium deficiency	Birds
Calcium toxicity	Humans
Chlorine deficiency	Cattle
Lead toxicity	Birds
Lithium toxicity	Humans
Molybdenum toxicity	Rats
Phosphorus deficiency	Birds
Potassium deficiency	Poultry (chicks)
Potassium toxicity	Cattle
Sodium deficiency	Cats, dogs
Sodium toxicity	Birds
Zinc toxicity	Birds, dogs

Poor survivability of young

Element	Species	Comments
Arsenic deficiency	Goats, minipigs, rats	
Chlorine deficiency	Poultry	
Cobalamin deficiency	Cattle, pigs, sheep	
Iodine deficiency	Cattle	
Iron deficiency	Pigs	

Lanthanum toxicity	Fish	Embryotoxic: low hatchability and survival, tail deformities
Magnesium deficiency	Poultry	
Manganese deficiency	Rats	
Molybdenum deficiency	Humans, poultry	*Humans:* genetic defect
Nickel deficiency	Goats, minipigs, rats	
Phosphorus deficiency	Cattle	
Potassium deficiency	Poultry (chicks)	
Selenium toxicity	Pigs, poultry	

Postural reflexes—Abnormal

Element	Species
Manganese deficiency	Mink (pastel)

Premature birth

Element	Species
Manganese deficiency	Guinea pigs
Zinc deficiency	Guinea pigs

Progesterone levels—Low (in serum)

Element	Species
Copper deficiency	Buffaloes
Chromium toxicity	Amphibians

Prolonged labor

Element	Species
Zinc deficiency	Humans

Prolonged recovery after exercise

Element	Species	Comments
Iron deficiency	Humans (women)	Due anemia

Proprioceptive deficits

Element	Species
Mercury toxicity	Cats

Prostate problems

Element	Species	Comments
Aluminum toxicity	Gerbils	Impaired development
Cadmium toxicity	Humans	Prostatic hyperplasia
Zinc deficiency	Humans	

Prostration

Element	A
Phosphorus deficiency	Buffaloes

Protein—Total in blood—Increased

Element	Species
Cerium toxicity	Rats

Protein losing enteropathy (gastroenteropathy)

Protein loss from the body through the gastrointestinal tract.

Element	Species
Zirconium toxicity	Humans

Proteinuria

Abnormally large amounts of protein in urine.

Element	Species
Antimony toxicity	Humans
Chromium toxicity	Humans, rats
Mercury toxicity	Humans cattle, dogs
Palladium toxicity	Rats

Prothrombin time—Prolonged

The time it takes for the plasma to clot is too long.

Element	Species	Comments
Cerium toxicity	Dogs	
Gadolinium toxicity	Rats	
Lutetium toxicity	Dogs	And clotting time
Praseodymium toxicity	Dogs	

Protoporphyria IX

Excess protoporphyrin accumulated in tissues.

Element	Species	Comments
Lead toxicity	Birds	In blood, brain, liver

Pruritus

Severe itching of the skin.

Element	Species	Comments
Gadolinium toxicity	Humans	
Mercury toxicity	Humans, sheep	
Selenium toxicity	Humans	Contact
Sulfur toxicity	Humans	Contact

Psychiatric abnormalities

Element	Species
Manganese toxicity	Humans

Psychomotor disturbance

Movements that serve no purpose, in humans: aimless pacing, tapping toes, rapid talking.

Element	Species
Tin toxicity	Humans

Psychosis

When someone has lost contact with reality.

Element	Species
Bismuth toxicity	Humans
Molybdenum toxicity	Humans
Thallium toxicity	Humans
Tin toxicity	Humans

Pterygium

Fibrovascular growth of the bulbar conjunctiva and subconjunctival tissue that may cause blindness.

Element	Species	Comments
Arsenic toxicity	Humans	Chronic toxicity

Ptosis

Drooping of the upper eyelid.

Element	Species
Phosphorus toxicity	Humans

Puberty—Onset delayed

Element	Species
Molybdenum toxicity	Sheep

Pulmonary alveolar proteinosis

Element	Species
Aluminum toxicity	Humans

Pulmonary congestion

Accumulation of fluid in the lungs, causing impaired gas exchange, and arterial hypoxemia.

Element	Species
Boron toxicity	Mice

Pulmonary edema

Excess fluid in the lungs.

Element	Species	Comments
Aluminum toxicity	Humans	After ingestion
Antimony toxicity	Humans	Rare
Arsenic toxicity	Humans	
Boron toxicity	Mice	
Bromine toxicity	Humans	
Cadmium toxicity	Humans	
Chlorine toxicity	Humans	Inhalation
Gallium toxicity	Humans	
Iron toxicity	Sheep	
Mercury toxicity	Humans	
Nickel toxicity	Humans	
Selenium toxicity	Humans	Inhalation
Tellurium toxicity	Humans	
Zirconium toxicity	Humans	

Pulmonary embolism

A blockage in one of the pulmonary arteries, e.g., by a blood clot.

Element	Species	Comments
Neodymium toxicity	Humans	Inhalation of dust
Praseodymium toxicity	Humans	

Pulmonary fibrosis

Replacement of normal lung tissue with fibrous tissue rendering it nonfunctional—called scarring.

Element	Species	Comments
Aluminum toxicity	Humans	Following inhalation Restrictive, obstructive airway disease
Beryllium toxicity	Humans, dogs	Interstitial fibrosis
Cerium toxicity	Humans	
Mercury toxicity	Humans	
Praseodymium toxicity	Humans	
Samarium toxicity	Humans	Inhalation
Silicon toxicity	Humans, rats	
Sulfur toxicity	Humans	
Tungsten toxicity	Humans	Suspected—but may be due to cobalt exposure
Zinc toxicity	Humans, guinea pigs	
Zirconium toxicity	Humans, rabbits	

Pulmonary granulomatosis

Element	Species
Aluminum toxicity	Humans

Pulmonary hemorrhage

Element	Species	Comments
Boron toxicity	Mice	
Cerium toxicity	Mice	And tracheal hemorrhage
Iron toxicity	Horses (foals)	
Palladium toxicity	Rats	

Pulse weak

Element	Species
Arsenic toxicity	Dogs

Pupillary constriction (miosis)

Element	Species
Phosphate toxicity	Humans

Pupillary dilation

Element	Species
Fluorine toxicity	Cattle

Pupillary reflex loss

Element	Species
Mercury toxicity	Dogs

Pyloric ceca epithelium degeneration

Located at the junction of the stomach and the intestines.

Element	Species
Magnesium deficiency	Fish

Pyloric stenosis

Narrowing of the pyloric canal in the stomach which lies proximal to the first section of the small intestine (duodenum).

Element	Species
Zirconium toxicity	Humans

Pyramidal signs

Pyramidal tract dysfunction signs including a positive Babinski sign (big toe goes up when sole of foot is stimulated), hyperreflexia, slowing of rapid alternating movements, spasticity, weakness.

Element	Species
Bismuth toxicity	Humans

Quick temperedness

Element	Species
Mercury toxicity	Humans

Reactive airway dysfunction syndrome (RADS)

Airway compromise due to inhalational injury from fumes, gas, mists, vapors.

Element	Species
Sulfur toxicity	Humans

Recumbency

Element	Species
Aluminum toxicity	Dogs
Arsenic toxicity	Humans *Chronic toxicity*
Selenium deficiency	Cattle (*calves*), goats (*kids*), horses (*foals, donkeys*) sheep (*lambs*),
Selenium toxicity	Buffaloes, cattle
Sulfur toxicity	Cattle

Red cell fragility

Element	Species
Selenium deficiency	Humans

Reflexes depressed or lost

Element	Species	Comments
Aluminum toxicity	Dogs	Menace response, patellar reflexes, pupillary light reflex, withdrawal reflexes
Barium toxicity	Dogs	Limb reflexes impaired
Lead toxicity	Cattle	
Manganese deficiency	Nonhuman primates	Righting responses inadequate
Mercury toxicity	Cats, dogs	Face

Regurgitation

Element	Species
Calcium deficiency	Birds
Iodine deficiency	Birds
Zinc toxicity	Birds

Renal failure (acute)

Element	Species
Mercury toxicity	Humans, cattle, horses

Selenium toxicity	Humans, sheep
Thallium toxicity	Humans

Renal failure (chronic)

Element	Species
Aluminum toxicity	Koalas (*suspected not confirmed*)
Mercury toxicity	Horses

Renal hypertension

Element	Species
Calcium toxicity	Reptiles

Renal tubular acidosis

Element	Species
Arsenic toxicity	Humans

Renal tubular necrosis

Element	Species	Comments
Bismuth toxicity	Humans	
Chromium toxicity	Humans, hamsters, rats	*Rats:* proximal tubules
Mercury toxicity	Humans, dogs, horses	

Reproductive performance—Reduced

Element	Species	Comments
Antimony toxicity	Rats	Reproductive failure
Aluminum toxicity	Mice, rats	*Mice:* delayed puberty and adulthood *Rats:* fetal resorption, reduced progeny rate, reduced viability of offspring
Arsenic toxicity	Humans	Abortions, low birth weight, stillbirths, neonatal and infant mortality
Boron toxicity	Rats	impaired ovarian development,
Calcium deficiency	Birds, buffaloes, cattle, dogs, nonhuman primates poultry	Reduced fertility *Nonhuman primates:* Tarsiers
Chromium deficiency	Rats	Infertility, reduced sperm count
Chromium toxicity	Amphibians, hamster, mice, mongooses	*Amphibians:* ovarian dysfunction *Hamsters:* fetal abnormalities (cleft palate, hydrocephalus,

		skeletal abnormalities), resorption *Mice:* fetal abnormalities (anencephaly, exencephaly), reduced litter size, reduced sperm count and abnormal sperm
Cobalt deficiency	Pigs, poultry, sheep	*Pigs:* reduced litter size *Poultry:* reduced hatchability *Sheep:* reduced fertility
Copper deficiency	Camels, cats, cattle, pigs, poultry, sheep	*Camels:* infertility *Cattle:* delayed/depressed estrus, reduced conception rate, retained placenta *Sheep:* infertility Poultry: reduced hatchability
Copper toxicity	Various	
Fluorine toxicity	Cattle, sheep, fish	Reduced fertility
Iodine deficiency	Cattle, elephants, goats, horses, marsupial (wombat), pigs, poultry (*experimental*) sheep	*Cattle, Sheep, Horses:* abnormal estrus cycle *Cattle:* prolonged gestation *Goats:* sterility
Iodine toxicity	Foxes	Reduced number successfully breeding, reduced number of offspring, low birth weights, reduced viability
Iron deficiency	Mice	Low birth weight, low litter size
Lead toxicity	Humans, birds, sheep	*Sheep:* sterility
Lithium deficiency	Rats	
Manganese deficiency	Cattle, dogs, goats, guinea pigs, sheep, rabbits, rats	Cattle, Goats: abortion, low conception and birth rates, delayed/depressed estrus. Rats, Rabbits: male sterility, lack sperm and libido Rats: defective ovulation Rabbits: low litter size Guinea pigs: premature births
Molybdenum deficiency	Goats	Reduced conception rate
Molybdenum toxicity	Cattle, rats, sheep	*Cattle:* reduced fertility, delayed puberty *Rats:* reduced fertility, seminiferous tubular degeneration, abnormal oocysts, prolonged estrus *Sheep:* delayed puberty
Nickel deficiency	Goats minipigs	Reduced conception rate Delayed estrus
Phosphorus deficiency	Birds, buffaloes, cattle, dogs, goats, horses, poultry	*Buffaloes:* reduced fertility *Goats:* low conception rate, silent heats due low P and Ca:P ratio
Selenium deficiency	Cattle, sheep	*Cattle:* abortions, metritis, retained placenta, stillbirths *Sheep:* failure to implant, embryonic deaths
Selenium toxicity	Cattle, pigs, poultry, rats	Reproductive failure
Sodium deficiency	Poultry, rats	Gonadal inactivity (poultry) Infertility (rats—males) Delayed sexual maturity (rats—females)
Uranium toxicity	Mice, rats	Impaired fertility

Vanadium deficiency	Goats	
Zinc deficiency	Humans, cats, cattle, goats, nonhuman primates, pigs, poultry, rabbits, rats, sheep	*Humans, Cats, Cattle, Goats, Pigs, Nonhuman primates:* abortions, stillbirths, fetal malformations *Rats, Sheep:* Male sex organs not developed *Cattle, Rabbit, Poultry, Sheep:* all phases of reproduction failure *Rats, Sheep:* aspermiogenesis *Pigs:* stillbirths

Respiratory distress/failure/labored breathing

Element	Species	Comments
Arsenic toxicity	Humans	*Acute:* respiratory failure
Barium toxicity	Humans, dogs	*Respiratory failure:* Barium carbonate
Cadmium toxicity	Humans	
Calcium deficiency	Reptiles	Respiratory muscle paralysis
Cerium toxicity	Rats	
Chlorine deficiency	Cattle	Reduced respiratory rate
Dysprosium toxicity	Mice	
Erbium toxicity	Mice	Labored breathing
Europium toxicity	Mice	
Gadolinium toxicity	Rats	Labored breathing
Iron deficiency	Cattle (calves), horses, pigs	Pigs: pulmonary edema, free thoracic fluid and lung collapse Horses: pneumonia
Iron toxicity	Sheep	Respiratory failure
Lutetium toxicity	Cats	Acute toxicity: respiratory paralysis
Magnesium toxicity	Humans	Respiratory depression
Mercury toxicity	Humans	Restrictive lung disease Chronic respiratory insufficiency
Nickel toxicity	Humans	Shortness of breath
Phosphorus deficiency	Humans	Irregular breathing
Phosphorus toxicity	Humans	Respiratory paralysis
Potassium deficiency	Humans, poultry	
Selenium deficiency	Horses	White muscle disease
Selenium toxicity	Humans, pigs	*Humans Inhalation:* acute respiratory distress syndrome
Sodium deficiency	Humans	Respiratory failure
Sulfur toxicity	Cattle, horses	Horses: accidental poisoning
Thallium toxicity	Humans	
Vanadium toxicity	Cattle, mice, rats	*Cattle:* pulmonary and tracheal hemorrhage (after ingestion) *Mice, Rats:* alveolar/bronchiolar hyperplasia, inflammation and fibrosis *Rats:* dyspnoea, tachypnoea
Ytterbium toxicity	Guinea pigs	Respiratory paralysis
Zinc deficiency	Rats, poultry	*Rats:* Tachypnoea *Poultry:* labored breathing
Zinc toxicity	Humans	Acute respiratory distress syndrome, reduced vital capacity
Zirconium toxicity	Humans	

Respiratory tract inflammation (mucous membranes)

Element	Species	Comments
Antimony toxicity	Humans	
Arsenic toxicity (chronic)	Cattle	
Bismuth toxicity	Humans	Inhalation
Boron toxicity	Humans	Inhalation
Chromium toxicity	Mice, rats	Ulceration and perforation of nasal septum
Fluorine toxicity (gas)	Humans and others	
Gadolinium toxicity	Humans	Respiratory signs
Indium toxicity	Hamsters, mice, rats	*Hamsters:* severe inflammation, alveolar cell hyperplasia, adenomas *Mice, rats:* adenoma, carcinomas
Molybdenum toxicity	Mice, rats	*Mice, rats:* hyaline degeneration nasal septum squamous metaplasia epiglottis *Rats:* chronic inflammation
Nickel toxicity	Humans	Pneumonitis, upper airway irritation
Palladium toxicity	Humans	
Selenium toxicity	Humans	*Inhalation:* pneumonitis, restrictive and obstructive respiratory disease
Vanadium toxicity	Various	

Respiratory noise

Element	Species	Comments
Iodine deficiency	Birds	Expiratory noise: chirp, wheeze
Lead toxicity	Horses	Due paralysis of larynx/pharynx called "roaring"

Respiratory quotient reduced

The volume of carbon dioxide released over the volume of oxygen absorbed during respiration, used as an indirect measure of basal metabolic rate.

Element	Species
Chromium deficiency	Humans

Restlessness

Inability to rest or relax.

Element	Species
Iodine toxicity	Dogs
Lithium toxicity	Humans
Potassium deficiency	Dogs
Selenium toxicity	Humans
Tin toxicity	Humans (*Organotin compounds*)

Retained placenta

A placenta that has not been fully expelled from the uterus after birth.

Element	Species	Comments
Copper deficiency	Cattle	
Iodine deficiency	Cattle, sheep	
Selenium deficiency	Goats	And Vit E

Retching

Element	Species
Zirconium toxicity	Humans

Reticulocytopenia

An abnormal decrease in reticulocyte numbers.

Element	Species	Comments
Cobalt deficiency	Cattle, sheep	Vit B12 cobalamin deficiency
Cerium toxicity	Mice	

Reticulocytosis

An increase in number of reticulocytes.

Element	Species
Cobalt toxicity	Rats
Vanadium toxicity	Rats

Retinol-binding protein in urine

Element	Species
Cadmium toxicity	Humans

Retrobulbar neuritis

Element	Species
Thallium toxicity	Humans

Rhabdomyolysis

Element	Species
Barium toxicity	Humans
Magnesium deficiency	Dogs

Phosphorus deficiency	Dogs
Potassium deficiency	Goats
Uranium toxicity	Humans

Rhinitis

Element	Species
Antimony toxicity	Humans
Beryllium toxicity	Humans
Sulfur toxicity	Humans

Rhinorrhoea

Element	Species	Comments
Selenium toxicity	Humans	Inhaled

Riboflavinuria

Element	Species	Comments
Boron toxicity	Humans	Pinto et al. (1978)

Rickets

Imperfect endochondral ossification with softening and deformity of the bones typically resulting in angular deformity of the legs. Typically due to Ca, P, or vitamin D deficiencies.

Element	Species
Calcium deficiency	Humans, bears (polar), birds, camels, dogs, goats, guinea pigs, sheep
Beryllium toxicity	Rats
Molybdenum toxicity	Horses
Phosphorus deficiency	Birds

Rigidity—Trunk

Element	Species
Lithium toxicity	Humans
Manganese toxicity	Humans

Rumen papille degeneration

Element	Species
Magnesium toxicity	Cattle

Ruminal stasis

Element	Species
Fluorine toxicity	Cattle

Salivation (ptyalism)/hypersalivation

Element	Species
Copper toxicity	Various
Lithium toxicity	Dogs
Magnesium deficiency	Cattle, sheep
Potassium toxicity	Cattle
Selenium toxicity	Cattle
Zinc deficiency	Cattle

Sarcopenia/muscle wastage

Element	Species	Comments
Cobalt deficiency	Cattle, sheep	Vit B12 cobalamin deficiency

Screwneck

Element	Species
Manganese deficiency	Mink (Pastel)

Seborrhoea

Element	Species
Zinc deficiency	Humans

Sedation

Element	Species
Lutetium toxicity	Mice
Praseodymium toxicity	Mice

Seizures (fits)

A sudden onset, uncontrolled electrical disturbance in the brain causing changes in behavior, movements, feelings or level of consciousness. (**See also convulsions.**)

Element	Species	Comments
Antimony toxicity	Humans	
Aluminum toxicity	Humans, cats (*experimental*), rabbits (*experimental*)	Cats and rabbits: associated with neurofilament formation

Arsenic toxicity	Humans	Acute and chronic toxicities
Barium toxicity	Humans	
Bismuth toxicity	Humans	
Boron toxicity	Humans	
Cesium toxicity	Humans	
Calcium deficiency	Humans (child), birds, dogs, reptiles, squirrels	
Chlorine deficiency	Humans	
Copper deficiency	Humans	Menkes disease
Copper toxicity	Various	
Iron toxicity	Humans, pigs	
Lead toxicity	Human (adults), birds, cats, dogs	Seizures
Lithium toxicity	Humans	
Magnesium deficiency	Humans, cats, cattle, dogs, gerbils, horses, rabbits, rats, sheep	Seizures
Manganese deficiency	Rats	Epileptiform
Mercury toxicity	Humans, cats, cattle, dogs, ferrets, mink	Seizures *Ferrets, Mink:* Convulsions
Molybdenum deficiency	Humans	Seizures
Molybdenum toxicity	Humans	Seizures
Phosphorus deficiency	Dogs	Seizures
Potassium deficiency	Poultry (chicks)	
Sodium deficiency	Birds	
Sulfur toxicity	Horses	Accidental poisoning
Thallium toxicity	Humans	
Tungsten toxicity	Humans	Rare
Zinc toxicity	Dogs	
Zirconium toxicity	Humans	

Selenosis

Element	Species	Comments
Selenium toxicity	Humans, reptiles	High levels in tissues including hair and nails

Sensory disturbances

Element	Species	Comments
Mercury toxicity	Humans	
Tungsten toxicity	Humans	Sensory deficits

Sensory polyneuritis

Element	Species
Thallium toxicity	Humans

Sexual maturity delayed

Element	Species	Comments
Zinc deficiency	Humans	Hypogonadism

Shivering

Element	Species
Iron toxicity	Pigs
Sodium deficiency	Cattle

Shock

Element	Species	Comments
Antimony toxicity	Humans	
Barium toxicity	Humans	
Chromium toxicity	Humans	
Iron toxicity	Humans, dogs	*Dogs:* hypovolemic Acute
Mercury toxicity	Humans, horses	
Ionizing radiation	Various	Exposure

Shortness of breath

Element	Species
Barium toxicity	Humans
Mercury toxicity	Humans
Sulfur toxicity	Humans
Tellurium toxicity	Humans
Yttrium toxicity	Humans *Industrial exposure*

Shyness—Excessive

Element	Species
Mercury toxicity	Humans

Silicosis

Accumulation of silica dust in lungs usually as a result of occupational exposure.

Element	Species
Silicon toxicity	Humans

Sinusitis

Inflammation of the nasal sinuses.

Element	Species
Titanium toxicity	Humans

Skeletal lesions/disorders

In vertebrates bone forms from a cartilaginous skeleton by the process of endochondral ossification. The conformation of the skeleton is partially dictated by genetic inheritance but also by biomechanical forces exerted on the bone—under Wolff's Law, this results in the more dense bone where forces are greater. Once formed bone is in a continual flux of removal and formation through the actions of osteoclasts and osteoblasts respectively and these are controlled primarily by the parathyroid hormone (increases osteoclast activity) and calcitonin (increases osteoblast activity). Other factors are gastrointestinal absorption of minerals under the influence of vitamin D.

Skeletal disease linked to minerals is due to lack of minerals required for the structure, interference with normal endochondral ossification resulting in abnormal bone or cartilage formation and development, or removal of minerals from the skeleton resulting in weak bones.

Element	Species	Comments
Aluminum toxicity	Humans	Bone pain, fractures, deformity
Boron toxicity	Mice, rabbits poultry (chicks)	Congenital deformities Poultry: due riboflavin deficiency
Cadmium toxicity	Mice, rats	Dysplastic tail
Calcium deficiency	Humans, amphibians, bears, birds, buffaloes, camels, cats, cattle, dogs, ferrets, foxes, horses, mink, nonhuman primates, pigs, reptiles, sheep, squirrels	Rickets, poor bone density, spontaneous fractures, angular deformities, swollen joints, vertebral compression Undermineralization of bone, thin cortices, metabolic bone disease, nutritional secondary hyperparathyroidism, osteomalacia, osteopenia, spontaneous fractures *Ferrets:* angular deformities limbs, metabolic bone disease, weak bones, fractures, osteoporosis, skeletal deformities, can't weight bear *Foxes:* angular deformities, enlarged cranial bones, gum swelling, muzzle swollen, metabolic bone disease, walk on pasterns, *Horses:* delayed growth plate closure, enlargement of facial bones (big head disease) *Sheep:* osteomalacia (adults), rickets (lambs) *Squirrels:* facial deformity, metabolic bone disease *Reptiles:* mandibular swelling, swollen limbs, spinal deformities *Rhesus macaques:* osteoporosis
Calcium toxicity	Dogs, cattle	

Chromium deficiency	Humans, mice, poultry (turkeys), rats	
Chromium toxicity	Hamsters	Skeletal abnormalities and failure of normal endochondral ossification
Copper deficiency	Humans, camels, cats, cattle, dogs, horses, pigs, poultry, rabbits	*Humans:* undercalcified bone, Skeletal abnormalities *Cats:* kinked tails, abnormal carpi *Cattle:* rickets, undermineralised bone, spontaneous fractures *Dogs:* hyperextension carpi *Horses:* osteochondrosis *Pigs:* hyperflexion hocks, impaired endochondral ossification, reduced osteoblast activity *Red Deer:* osteochondrosis *Rabbits:* abnormal cartilage formation *Poultry:* undermineralised bone, spontaneous fractures
Fluorine deficiency	Goats	Skeletal and joint deformity
Fluorine toxicity	Humans, cattle, pigs	Skeletal deformities, exostoses, hypermineralization bone, joint swellings, spinal and joint ankylosis, tendon and ligament mineralization
Iodine deficiency	Humans, dogs, pigs	
Lead toxicity	Birds	Skeletal abnormalities
Magnesium deficiency	Fish	Spinal deformities
Manganese deficiency	Humans, dogs, fish, goats, guinea pigs, pigs, poultry, rabbits, rats	*Humans:* skeletal abnormalities, *Dogs, Guinea pigs, Rabbits, Rats:* reduced growth *Dogs, Goats, Guinea pigs, Pigs, Rabbits, Rats:* angular limb deformities *Dogs, Guinea pig, Pigs, Rabbits, Rats:* shortened limbs *Guinea pigs, Poultry Rabbits, Rats:* skull deformities
Molybdenum deficiency	Poultry	Impaired ossification and cartilage formation
Molybdenum toxicity	Cattle, horses, sheep, rabbits, rats	*Cattle, Rat:* exostosis *Cattle:* deformities, osteoporosis *Horses:* rickets *Rabbit, Rats:* deformities *Rats:* loss bone strength, impaired endochondral ossification, wide growth plates, short limbs, hemorrhage and tears to ligaments and tendons *Sheep:* fractures, subperiosteal hemorrhages
Nickel deficiency	Goats, minipigs	Poor calcification and zinc
Phosphorus deficiency	Humans, birds, buffaloes, cattle, Fish, goats, mink, poultry, sheep	*Buffaloes:* fractures, retarded growth, joint stiffness, neck extension, paralysis Rickets, poor bone density, fractures, angular deformities, swollen joints *Goats:* Skeletal deformities *Fish:* abnormal calcification of ribs, bone weakness, frontal bones, spinal deformity (curvature, scoliosis, twisted), soft fins *Sheep:* skeletal deformities
Phosphorus toxicity	Fish, horses, squirrels	*Fish:* skeletal deformities *Horses:* enlargement of facial bones due replacement of bone with fibrous tissue (big head disease) *Squirrels:* metabolic bone disease

Selenium deficiency	Horses	Locomotor problems due myopathy
Selenium toxicity	Fish	Congenital malformations
Silicon deficiency	poultry (chicks) rats	Reduced bone strength, cartilage damage, low collagen, low glycosaminoglycans Retarded growth, skull deformity
Vanadium deficiency	Goats	Bone abnormalities
Zinc deficiency	Humans, cattle, nonhuman primates, pigs, poultry	*Humans and others:* Dwarfism, shortened long bones, delayed skeletal development *Cattle:* angular deformity (bowing) hind legs *Pigs:* loss of bone density and strength *Poultry:* deformities (vertebral fusion, curvature of spine, swollen joint, short thick legs), abnormal posture

Skin lesions

Element	Species	Comments
Arsenic toxicity	Humans, dogs	Chronic toxicity *Dogs -contact:* bleeding, blisters, cracking, infections, swelling
Beryllium toxicity	Humans	Rash
Bismuth toxicity	Humans	Lichen planus-like rash
Boron toxicity	Rats	Scaly tails
Bromine toxicity	Humans	Burns
Calcium toxicity	Elephants	Due Zn deficiency
Cerium toxicity	Humans	Itchiness, skin lesions
Copper deficiency	Goats, rabbits	*Goats:* depigmentation skin and hair *Rabbits:* dermatitis
Europium toxicity	Humans, guinea pigs, rabbits	*Humans:* Scarring of abraded skin *Guinea pigs:* Nodules at injection site *Rabbits:* Inflamed and fibrotic skin, abraded skin irritation
Fluorine toxicity	Horses	Loses elasticity
Gadolinium toxicity	Humans Rabbits	*Humans:* Skin rash—after injection for MRI scan *Rabbits:* Severe skin irritation abraded skin
Gold toxicity	Humans	Skin rashes
Holmium toxicity	Rabbits, rats	*Rabbits:* severe skin irritation abraded skin *Rats:* inflammation, fibrous changes
Iodine deficiency	Goats	
Iodine toxicity	Humans, amphibians, cats, cattle	*Humans:* skin ulcers *Amphibians:* external gill shedding in axolotls

		Cats: scales *Cattle:* scales, sloughing
Iron toxicity	Dogs	Discoloration and edema at injection site
Lead toxicity	Frogs	Skin sloughs
Lutetium toxicity	Humans	Scarring
Manganese deficiency	Humans	Skin rash
Molybdenum toxicity	Sheep	Depigmentation skin (and hair/wool)
Mercury toxicity	Humans, sheep	*Sheep:* scabs around anus and vulva *Humans:* skin irritation
Nickel toxicity	Rats	*Dermal contact:* epidermal acanthosis, epidermal atrophy, hyperkeratosis
Palladium toxicity	Humans, rabbits, rats	*Humans:* dermatitis, skin sensitivity (contact allergy) *Rabbits:* eschar *Rats:* exudates,
Praseodymium toxicity	Humans, rabbits	Severe skin irritation on contact
Selenium toxicity	Humans, buffaloes, pigs, reptiles	*Humans Contact:* erythema, burns, heat, pain, pruritus, rash, swelling *Buffaloes:* cracks, sloughing *Pigs:* rough skin
Silver toxicity	Humans	Gray—and internal organs
Sodium deficiency	Birds, horses	*Birds:* dry, pruritic skin *Horses:* reduced skin turgor
Tellurium toxicity	Humans	Dry, cracks, erythema
Thallium toxicity	Humans	Dry, crusty scales
Thulium toxicity	Rats	Skin ulcers
Titanium allergy	Humans	Hypersensitivity reaction to implant
Ytterbium toxicity	Humans	Skin irritation
Zinc deficiency	Humans, cats, cattle, dogs, elephants, goats, horses (*foals*) nonhuman primates (*squirrel monkeys*), pigs, poultry	*Humans:* rash *Cats, Cattle, Goats, Monkeys, Pigs:* Parakeratosis *Cattle:* inflamed scrotal skin, nose, mouth; crusting and cracks hoof; ears dry scaly *Dogs, Poultry:* hyperkeratosis *Dogs:* crusting around eyes, abdomen, extremities, hyperkeratosis foot pads *Elephants:* hyperkeratosis *Foals:* rough coat, crusting, flaking, desquamation, hoof, nose, face, muzzle *Nonhuman primates:* unkempt hair, alopecia *Poultry:* changed comb color
Ionizing radiation	Various	Burns

Sleep disturbance

Element	Species	Comments
Antimony toxicity	Humans	Sleeplessness
Bromine toxicity	Cat	*Experimental*—paradoxical sleep
Calcium deficiency	Squirrels	Excessive sleeping

Sluggishness

Slow to respond to stimulation and very slow in movement.

Element	Species
Lead toxicity	Humans, frogs
Mercury toxicity	Humans
Palladium toxicity	Rabbits, rats

Small intestine degeneration

Element	Species	Comments
Aluminum toxicity	Rats	And goblet cell proliferation

Smell—Loss of sensation

Element	Species
Bismuth toxicity	Humans
Sulfur toxicity	Humans
Zinc deficiency	Humans

Sneezing

Element	Species
Sulfur toxicity	Humans

Sodium: Potassium ratio (Na:K)—Decreased

Element	Species
Cerium toxicity	Rats

Soft tissue calcification

Element	Species	Comments
Calcium deficiency	Nonhuman primates	Lemurs

Calcium toxicity	Humans, marsupials guinea pigs	Wombats
Magnesium deficiency	Cats	
Phosphorus toxicity	Humans	

Somesthetic sensory loss

Includes the somatic senses of touch, proprioception (sense of position and movement), and haptic perception.

Element	Species
Mercury toxicity	Monkeys

Somnipathy

Abnormal sleep patterns that affect normal emotional, mental, and physical functioning.

Element	Species	Comments
Tin toxicity	Humans	Organotin compounds

Somnolence

Excess drowsiness or sleepiness.

Element	Species
Bismuth toxicity	Humans

Sorbitol dehydrogenase (SDH)—Increased

A cytosolic enzyme.

Element	Species
Cerium toxicity	Rats
Copper toxicity	Goats, sheep
Gallium toxicity	Mice

Speech problems

Element	Species	Comments
Aluminum toxicity	Humans	
Bromine toxicity	Humans	Slurred speech

Sperm count reduction

Element	Species	Comments
Aluminum toxicity	Humans, rats	*Humans:* reduced numbers, reduced motility, increased abnormalities *Rats:* reduced numbers
Barium toxicity	Rats	Reduced numbers, motility, and osmotic resistance

Boron toxicity	Dogs, rats	Reduced spermiogenesis
Chromium deficiency	Humans, rats	
Chromium toxicity	Mice	Reduced number, abnormalities
Gallium toxicity	Mice	Reduced number and motility
Manganese deficiency	Mice, rabbits	
Mercury toxicity	Humans	
Molybdenum toxicity	Cattle, rats	
Zinc deficiency	Humans, cats, cattle, goats, pigs, rats, sheep	*All:* lack of development of male reproductive organs *Rats, Sheep:* Aspermiogenesis

Spinal ankylosis

Element	Species
Fluorine toxicity	Cattle

Spinal curvature

Element	Species
Calcium deficiency	Amphibians, camels, cats, reptiles
Magnesium deficiency	Fish
Zinc deficiency	Poultry

Splenic degeneration

Element	Species	Comments
Gadolinium toxicity	Rats	Splenic necrosis (and liver)
Iron toxicity	Sheep	
Lanthanum toxicity	Mice	Splenic degeneration
Thulium toxicity	Mice	

Splenomegaly

Element	Species	Comments
Antimony toxicity	Guinea pigs	Hyperplasia
Cerium toxicity	Mice	Hypertrophy
Iron deficiency	Rats	
Selenium toxicity	Humans	

Staggering

Element	Species
Arsenic toxicity	Dogs

Standing difficulty

Element	Species
Manganese deficiency	Cattle (calves)

Steatitis

Inflammation of fat, causing yellow discoloration of adipose tissue with saponification and fat necrosis. Called "yellow fat disease."

Element	Species	Comments
Selenium deficiency	Horses	And donkeys. Linked to vitamin E deficiency

Stepping syndrome

Element	Species
Magnesium deficiency	Pigs

Stiffness

Element	Species	Comments
Calcium deficiency	Camels, foxes	Difficulty moving
Copper deficiency	Sheep	
Fluorine toxicity	Cattle, sheep	
Lead toxicity	Mink	
Magnesium deficiency	Cattle, guinea pigs	
Manganese toxicity	Humans	
Mercury toxicity	Sheep	
Phosphorus deficiency	Humans, buffaloes	*Humans:* joint stiffness
Potassium deficiency	Cattle, goats, sheep	
Selenium deficiency	Cattle (*calves*), goats (*kids*), horses (*foals*), sheep (*lambs*)	And donkeys (foals)
Zinc deficiency	Cattle, horses, poultry	

Stillbirths

Element	Species	Comments
Arsenic toxicity	Humans	Water contamination
Iodine deficiency	Humans, cattle, goats, horses, pigs sheep	
Manganese deficiency	Cattle, goats	
Nickel toxicity	Marine mammals	Ringed Seals—airborne route
Selenium deficiency	Cattle	
Zinc deficiency	Nonhuman primates, pigs	

Stomatitis

Element	Species
Iron deficiency	Humans
Mercury toxicity	Humans (severe), cattle (oral ulcers), horses
Palladium toxicity	Humans
Thallium toxicity	Humans
Zinc deficiency	Humans

Stridor

A harsh or grating, vibrating noise when breathing, caused by upper airway obstruction.

Element	Species
Zirconium toxicity	Humans

Stroke

Element	Species	Comments
Aluminum toxicity	Humans	Ischemia stroke due thrombus

Stupor

Element	Species
Lithium toxicity	Humans

Subperiosteal hemorrhages

Element	Species	Comments
Molybdenum toxicity	Sheep	Especially humerus

Substernal pain

Element	Species
Nickel toxicity	Humans

Succinyl dehydrogenase activity (reduced)

Element	Species	Comments
Iron deficiency	Rat	Heart, kidneys

Suckling difficulties—Young

Element	Species	Comments
Selenium deficiency	Horses (foals)	Myopathy

Suckling—Failure to suckle young

Element	Species
Magnesium deficiency	Rats

Superoxide dismutase—Decreased activity

Element	Species	Comments
Copper deficiency	Buffaloes	
Manganese deficiency	Fish	Cu-Zn and Mn in liver and heart
Samarium toxicity	Rats	In liver and brain

Survival time reduced/life expectancy reduced

Element	Species	Comments
Barium toxicity	Mice, rats	
Gallium toxicity	Rats	*Experimental* Females
Germanium toxicity	Mice, rats	*Experimental*
Manganese deficiency	Humans	*Humans:* genetic deficit
Molybdenum toxicity	Rats	
Tungsten toxicity	Mice, rats	
Vanadium deficiency	Goats	Kids
Vanadium toxicity	Rats	
Ytterbium toxicity	Fish	
Zirconium toxicity	Rats	Mild effect

Swayback syndrome (human)

Abnormal gait, sensory ataxia, spasticity.

Element	Species	Comments
Zinc toxicity	Humans	Due Cu deficiency

Swaying of hindquarters

Element	Species
Copper deficiency	Sheep

Sweating

Element	Species	Comments
Beryllium toxicity	Humans	Night sweats
Magnesium deficiency	Horses	
Phosphorus toxicity	Humans	
Zinc toxicity	Humans	

Sweating inhibition

Element	Species
Tellurium toxicity	Humans
Magnesium deficiency	Cattle, horses pigs, sheep
Sodium deficiency	Humans

Tachycardia

Element	Species	Comments
Arsenic toxicity	Dogs	
Cadmium toxicity	Humans	
Cesium toxicity	Humans	Ventricular tachycardia
Europium toxicity	Mice	
Gallium toxicity	Humans	Acute
Iodine toxicity	Cats, dogs	
Iron toxicity	Dogs	Due hypovolemia
Lithium toxicity	Humans	
Lutetium toxicity	Dogs	
Magnesium deficiency	Cats	
Mercury toxicity	Humans	
Molybdenum deficiency	Humans	
Selenium toxicity	Humans, cattle	
Sodium deficiency	Dogs	
Thallium toxicity	Humans	
Zinc toxicity	Dogs	

Tachypnoea

Rapid breathing rate.

Element	Species	Comments
Barium toxicity	Dogs	
Chlorine toxicity	Humans	Inhalation
Selenium toxicity	Humans	
Vanadium toxicity	Rats	
Zinc toxicity	Dogs	

Tail drooping

Element	Species
Iron deficiency	Pigs

Taste—Loss of sensation

Element	Species
Bismuth toxicity	Humans
Zinc deficiency	Humans

Teeth discoloration

Element	Species	Comments
Fluorine toxicity	Horses	
Selenium toxicity	Humans	Mottled

Teeth grinding

Element	Species
Lead toxicity	Cattle

Tendon mineralization

Element	Species	Comments
Fluorine toxicity	Cattle	And ligaments

Tenesmus

Element	Species	Comments
Zirconium toxicity	Humans	Acute toxicity

Teratological changes

Element	Species	Comments
Selenium toxicity	Poultry, sheep	Chick embryo deformaties Lambs eye deformities
Tellurium toxicity	Rats	Exophthalmia, hydrocephalus, ocular hemorrhage, small kidneys, edema, umbilical hernia, retained testes
Ytterbium toxicity	Hamsters	

Testicular disease

Element	Species	Comments
Aluminum toxicity	Dogs, rabbits	Reduced teste weight, sperm quality
Boron toxicity	Dogs, gerbils, mice, rats	Testicular degeneration
Cadmium toxicity	Various: rats	Retarded development and degeneration. Rats: testicular necrosis
Gallium toxicity	Mice	Testicular atrophy and weight loss. Reduced sperm count and motility
Iodine deficiency	Cattle, horses, pigs, poultry, sheep	Poor semen production Poultry—small testes
Manganese deficiency	Rabbits, rats	Testicular tubular degeneration
Molybdenum toxicity	Humans, cattle, rats	Testicle damage, reduced spermatogenesis, *Humans:* reduced testosterone *Rats:* seminiferous tubular degeneration
Nickel deficiency	Goats	Low testes weight
Palladium toxicity	Mice, rats	Reduced weight
Tellurium toxicity	Rats	*Teratological effect:* retained testicles
Tin toxicity	Rats	Testicular degeneration
Zinc deficiency	Humans, cats, cattle, goats, pigs, rats, sheep	Failure of male sex organs to develop properly. *Rats, Sheep:* Aspermiogenesis *Cats:* seminiferous tubular degeneration

Testosterone formation reduced

Element	Species
Aluminum toxicity	Rats

Tetany

Element	Species	Comments
Calcium deficiency	Humans (child), amphibians, birds, cattle, dogs, marsupials (sugar-gliders), reptiles	Due neuromuscular hyperexcitability
Chlorine deficiency	Poultry	
Fluorine toxicity	Cattle	
Magnesium deficiency	Humans, cats, goats, guinea pigs, horses, pigs	
Potassium deficiency	Fish	

Thirst reduced

Element	Species	Comments
Fluorine toxicity	Pigs	
Palladium toxicity	Mice, rats	
Sodium deficiency	Horses	
Sodium toxicity	Cattle	If very saline water

Throat pain

Element	Species	Comments
Arsenic toxicity	Humans	Chronic
Beryllium toxicity	Humans	Throat soreness after inhalation
Selenium toxicity	Humans	
Sulfur toxicity	Humans	Throat soreness
Vanadium toxicity	Humans	Soreness
Zirconium toxicity	Humans	Burning sensation

Thrombocytopenia

Element	Species
Bismuth toxicity	Humans
Zinc toxicity	Humans

Thrombus formation

Element	Species	Comments
Aluminum toxicity	Humans	Left ventricular thrombus, stroke due embolism from thrombus

Thymus atrophy

Element	Species
Nickel toxicity	Mice, rats

Thymus hypoplasia

Element	Species	Comments
Fluorine deficiency	Goats	Kids
Fluorine toxicity	Goats	

Thyroid hyperplasia

Element	Species
Cobalt toxicity	Human (children)
Iodine deficiency	Dogs, cats

Thyroxine T4—Reduced

Element	Species
Molybdenum toxicity	Rabbits
Iodine deficiency	Dogs, cats

Tibial chondrodysplasia

Element	Species
Chlorine deficiency	Birds (ratites)

Tingling sensation

Element	Species
Mercury toxicity	Humans

Tinnitus

Element	Species	Comments
Chromium toxicity	Human	
Tin toxicity	Humans	Organotin compounds

Tip Toe gait

Element	Species
Palladium toxicity	Rats
Praseodymium toxicity	Mice

Tongue coating (white)

Element	Species
Zinc deficiency	Humans

Tongue papille lesions: Atrophy/parakeratosis

Element	Species	Comments
Iron deficiency	Cattle (*calves*)	Atrophy
Zinc deficiency	Monkeys (*Squirrel monkeys*)	Parakeratosis

Tooth loss

Element	Species	Comments
Calcium deficiency	Cats, dogs, foxes	Due poor bone mineralization in hyperparathyroidism
Mercury toxicity	Sheep	
Phosphorus toxicity	Dogs	

Torsade de pointes

Torsade de pointes is a polymorphic ventricular tachycardia, with prolonged QT interval changed amplitude and twisting of QRS complexes around the isoelectric line.

Element	Species
Chromium deficiency	Humans, mice, rats

Tracheobronchitis

Element	Species	Comments
Antimony toxicity	Humans	Bronchitis, tracheitis
Beryllium toxicity	Humans	
Sulfur toxicity	Humans	Inhalation

Transaminase (liver) activity increased

Element	Species
Bismuth toxicity	Humans

Transferrin—Low blood concentration

Element	Species
Iron deficiency	Dogs

Tremors

Element	Species	Comments
Aluminum toxicity	Dogs	
Bismuth toxicity	Humans	
Calcium deficiency	Squirrels	
Lead toxicity	Birds, dogs, mink	
Lithium toxicity	Humans	
Magnesium deficiency	Humans	
Manganese toxicity	Humans	Hands
Mercury toxicity	Humans, cattle, ferrets, horses, mink, nonhuman primates	*Humans:* legs, arms, tongue, lips
Phosphorus deficiency	Dogs	
Selenium toxicity	Humans	
Sodium deficiency	Marine mammals	
Tin toxicity	Humans	

Triglyceride concentrations (blood)—Low

Element	Species	Comments
Cerium toxicity	Rats	In liver

Trousseau sign

A sign of latent tetany induced by applying an inflated blood pressure cuff to the arm for 3 min to occlude blood supply. This causes spasms of the hand and forearm muscles. The wrist and metacarpophalangeal joints flex, the DIP and PIP joints extend, and the fingers adduct.

Element	Species
Calcium deficiency	Humans
Magnesium deficiency	Humans

Tumors

Element	Species	Comments
Cadmium toxicity	Mice, rats	Mice: lung tumors Rats: hematopoietic system, lung, prostate, testes,
Ionizing radiation	Various	Exposure

Tunnel vision

Element	Species
Mercury toxicity	Humans

Twitching

Element	Species	Comments
Calcium deficiency	Cattle, squirrels	Muscles under skin, toes, legs, tail
Phosphorus toxicity	Humans	
Tin toxicity	Humans	Organotin compounds

Umbilical hernia

Element	Species	Comments
Tellurium toxicity	Rats	*Teratogenic effect*

Unthriftiness

Element	Species
Calcium deficiency	Buffaloes
Cobalt deficiency	Buffaloes, cattle. Goats, sheep
Cobalt toxicity	Goats, sheep
Copper deficiency	Buffaloes (*calves*), deer
Fluorine toxicity	Cattle, horses, sheep
Iron deficiency	Buffaloes (*calves*)
Lead toxicity	Sheep
Molybdenum toxicity	Cattle
Phosphorus deficiency	Cattle
Potassium deficiency	Horses
Selenium deficiency	Buffaloes, deer
Selenium toxicity	Cattle
Sodium deficiency	Cattle
Zinc deficiency	Buffaloes, cattle

Upper respiratory tract obstruction

Element	Species	Comments
Zinc toxicity	Cats	Inhalation

Uremia

High blood urea.

Element	Species
Aluminum toxicity	Rats
Antimony toxicity	Rats
Chlorine deficiency	Cattle
Mercury toxicity	Humans, dogs, marine mammals (*seals*), pigs
Selenium deficiency	Sheep

Uric acidemia—Lowered

Element	Species
Lanthanum toxicity	Mice

Urine β-glucuronidase—Increased

Element	Species
Chromium toxicity	Humans, rats

Urine concentrating ability—Lost

Element	Species
Lithium toxicity	Humans

Urine production decreased

Element	Species
Palladium toxicity	Rabbits, rats

Urine retention

Element	Species
Magnesium toxicity	Humans

Urine specific gravity—Low

Element	Species	Comments
Selenium toxicity	Humans	Renal insufficiency
Sodium deficiency	Cats	

Urolithiasis

Element	Species	Comments
Calcium toxicity	Poultry, reptiles	*Poultry:* calcium urate in ureters *Reptiles:* renal nephroliths
Magnesium toxicity	Cattle (calves), cats,? dogs	*Cattle:* calcium apatite nephroliths *Cats/? Dogs* struvite uroliths—if pH over 6.6
Molybdenum deficiency	Sheep	Xanthine uroliths
Molybdenum toxicity	Humans	Urate and cystine uroliths
Phosphorus toxicity	Giraffes, goats	
Potassium deficiency	Sheep	
Silicon toxicity	Humans dogs ruminants	Silicosis, mesothelioma in lungs Si Uroliths—eating soil Si uroliths and nephroliths

Uterine hemorrhage

At parturition—due to ruptured uterine artery.

Element	Species
Copper deficiency	Horses

Uveitis

Element	Species
Antimony toxicity	Humans

Vaginal development delayed

Element	Species
Manganese deficiency	Rats

Vascular disease (peripheral)

Element	Species	Comments
Arsenic toxicity	Humans	Gangrene of extremities

Vasodilation

Element	Species
Antimony toxicity	Humans
Cobalt toxicity	Humans
Magnesium deficiency	Rats
Magnesium toxicity	Humans

Ventricular dysrhythmia

Element	Species
Selenium toxicity	Humans

Ventricular tachycardia

Element	Species	Comments
Arsenic toxicity	Humans	Ventricular arrhythmia
Barium toxicity	Humans	
Selenium toxicity	Humans	

Ventroflexion of the neck

Element	Species
Cadmium deficiency	Goats
Iron deficiency	Pigs
Potassium deficiency	Dogs, cats

DD: causes of muscle weakness, thiamine deficiency, taurine deficiency (cats).

Vestibular syndrome/loss of balance/head tilt

Element	Species	Comments
Manganese deficiency	Rats, sheep, guinea pigs	Otolith abnormality
Mercury toxicity	Dogs	Head tilt

Vertebral ankylosis

Element	Species	Comments
Calcium toxicity	Cattle	
Zinc deficiency	Poultry	Vertebral fusion

Vertigo

Element	Species
Gallium toxicity	Humans
Magnesium deficiency	Humans
Mercury toxicity	Humans
Nickel toxicity	Humans

Violent acts

Element	Species
Manganese toxicity	Humans

Violent movements—Uncontrolled

Element	Species
Lead toxicity	Buffaloes

Visual problems

Element	Species	Comments
Bismuth toxicity	Humans	Blurred vision
Mercury toxicity	Humans, dogs, pigs, monkeys	*Humans, dogs, pigs:* loss of sight, blindness *Humans:* loss of color vision, tunnel vision *Monkeys:* loss spatial vision; loss sensitivity; loss of visual field
Selenium toxicity	Humans, cattle	

Vital capacity (lungs)—Reduced

Element	Species
Zinc toxicity	Humans

Vitiligo

Pigment is lost from areas of skin causing whitish patches.

Element	Species
Copper deficiency	Buffaloes

Vocalization—Uncontrolled

Element	Species	Comments
Mercury toxicity	Cats, mink	*Cats:* howling *Mink:* high pitched
Zinc toxicity	Dogs	

Vomiting

Element	Species	Comments
Antimony toxicity	Humans	
Arsenic toxicity	Dogs	
Barium toxicity	humans, rats	
Bismuth toxicity	Humans	
Boron toxicity	Humans	
Cadmium toxicity	Humans	
Calcium toxicity	Humans	
Chlorine deficiency	Humans	
Chlorine toxicity	Humans	Inhalation
Chromium toxicity	Humans	May contain blood
Gadolinium toxicity	Humans	After injection for MRI scan
Gallium toxicity	Humans	
Gold toxicity	Humans	
Iron toxicity	Humans, cats	
Lanthanum toxicity	Humans	
Lead toxicity	Humans, dogs	
Lithium toxicity	Humans	
Magnesium deficiency	Humans, dogs	
Magnesium toxicity	Humans	
Mercury toxicity	Humans, cats, dogs, mink	
Molybdenum deficiency	Humans	
Nickel toxicity	Humans, dogs	
Phosphorus toxicity	Humans	
Selenium toxicity	Humans, nonhuman primates	*Macaques*
Silver toxicity	Humans	
Sodium deficiency	Humans	
Tellurium toxicity	Humans	
Thallium toxicity	Humans	
Tin toxicity	Humans	
Uranium toxicity	Dogs	After inhalation exposure
Vanadium toxicity	Humans	

Zinc deficiency	Dogs, pigs	
Zinc toxicity	Humans, dogs	
Zirconium toxicity	Humans	
Ionizing radiation	Various	Exposure

Wandering

Element	Species
Cobalt deficiency	Sheep
Mercury toxicity	Pigs

Water intake—Reduced

Element	Species
Chlorine deficiency	Cattle
Potassium deficiency	Cattle

Weakness

Element	Species	Comments
Aluminum deficiency	Goats	Hindlegs (Anke et al., 2005a)
Aluminum toxicity	Humans, dogs	
Arsenic toxicity	Humans, dogs	
Barium toxicity	Humans	
Beryllium toxicity	Humans	
Cadmium deficiency	Goats	
Calcium deficiency	Amphibians, birds, camels, cattle	
Chlorine deficiency	Humans, dogs	
Cobalt deficiency	Cattle (calves)	Cobalamin deficiency
Copper deficiency	Birds, camels	
Fluorine toxicity	Cattle	
Iodine deficiency	Horses (foals), pigs, sheep (lambs)	
Iron deficiency	Cats, dogs, horses	
Lead toxicity	Birds	Neck, limbs
Lithium toxicity	Humans	
Manganese deficiency	Birds	Neck and limbs
Manganese toxicity	Humans	
Mercury toxicity	Humans, mice (hindlegs), horses	
Magnesium deficiency	Cats, pigs, poultry (rare)	
Magnesium toxicity	Cattle	
Nickel toxicity	Humans	

Phosphorus deficiency	Humans, buffaloes, horses, poultry	
Phosphorus toxicity	Humans	
Potassium deficiency	Humans, cats, cattle, dogs, poultry	*Humans:* skeletal and intestinal
Selenium deficiency	Humans, goats, horses	
Selenium toxicity	Humans	
Sodium deficiency	Cattle, marine mammals	*Marine mammals:* seals
Sulfur deficiency	Sheep	
Tellurium toxicity	Humans	
Thallium toxicity	Humans	
Zinc deficiency	Ferrets (*hindleg*), poultry	
Zinc toxicity	Birds, dogs, nonhuman primates	
Ionizing radiation	Various	Exposure

Weak offspring

Element	Species	Comments
Cobalt deficiency	Cattle, sheep	Vit B12 cobalamin deficiency

Weight gain

Element	Species
Lithium toxicity	Humans

Weight gain—Reduced

Element	Species
Bromine deficiency	Goats
Calcium deficiency	Mice
Calcium toxicity	Poultry
Copper deficiency	Cats
Fluorine toxicity	Pigs, poultry
Iron deficiency	Cattle (calves), pigs, poultry (rare)
Lithium deficiency	Goats
Magnesium deficiency	Cats, dogs, fish, guinea pigs, rabbits
Magnesium toxicity	Pigs, Poultry
Molybdenum toxicity	Sheep
Phosphorus deficiency	Goats, sheep
Sodium deficiency	Goats, rabbits
Strontium toxicity	Poultry (chicks)
Sulfur deficiency	Pigs
Uranium toxicity	Mice
Zinc deficiency	Humans, dogs, mice, rabbits
Zirconium toxicity	Hamsters, guinea pigs, rats

Weight loss

Element	Species	Comments
Antimony toxicity	Humans, rats	
Arsenic toxicity (Chronic)	Cattle	
Barium toxicity	Mice, rats	
Beryllium toxicity	Humans	
Bismuth toxicity	Humans	
Boron toxicity	*Fetal:* rabbits, rats	
Cadmium toxicity	Mice	
Calcium deficiency	Squirrels	
Chlorine deficiency	Cattle	
Cobalt deficiency	Goats, sheep	
Cobalt toxicity	Humans, sheep	
Copper deficiency	Deer	
Gallium toxicity	Mice	
Iodine toxicity	Dogs	
Iron deficiency	Cats	
Lanthanum toxicity	Rats	Or weight gain
Lithium toxicity	Humans	
Lead toxicity	Birds	Slow weight loss
Magnesium deficiency	Dogs	
Mercury toxicity	Horses, ferrets, marine mammals (*seals*), mink, pigs	
Molybdenum toxicity	Cattle, guinea pigs, rabbits, rats	
Nickel toxicity	Cattle (calves)	
Palladium toxicity	Rats	
Phosphorus deficiency	Cattle, horses	
Potassium deficiency	Cattle, mice, sheep	
Selenium deficiency	Cattle, nonhuman primates (*Squirrel monkeys*)	
Selenium toxicity	Deer, nonhuman primates (*macaques*), pigs, poultry	
Sodium deficiency	Cattle, poultry (laying hens)	
Sulfur deficiency	Sheep	
Tungsten toxicity	Rats	
Vanadium toxicity	Rats	Pregnant mothers
Zinc deficiency	Humans, hamsters, mice, nonhuman primates, rabbits	
Zinc toxicity	Cattle	
Zirconium toxicity	Birds, mice	

Wheezing

Element	Species
Aluminum toxicity	Humans
Antimony toxicity	Humans
Phosphorus toxicity	Humans
Selenium toxicity	Humans (inhalation)
Sulfur toxicity	Humans

White blood cell count—Low

Element	Species
Antimony toxicity	Guinea pigs
Cerium toxicity	Mice

Withdrawal reflex loss

Indicates loss of deep pain sensation.

Element	Species
Mercury toxicity	Dogs

Wool abnormalities

Element	Species	Comments
Molybdenum toxicity	Sheep	Loss of crimp, loss color
Sulfur toxicity	Sheep	Wool loss
Zinc deficiency	Sheep	Brittle, crimp lost, loose,

Wound healing—Impaired

Element	Species
Zinc deficiency	Humans, cattle, pigs

Wrinkled skin

Element	Species	Comments
Iron deficiency	Pigs	
Zinc deficiency	Sheep	Thickened as well

Wrist drop

Due to opposing action of flexor muscles the wrist and fingers cannot extend at the metacarpophalangeal joints so the wrist stays partially flexed.

Element	Species
Lead toxicity	Humans

Writhing

Element	Species
Dysprosium toxicity	Mice
Erbium toxicity	Mice
Europium toxicity	Mice
Holmium toxicity	Mice
Lutetium toxicity	Mice
Praseodymium toxicity	Mice

Xanthine oxidase—Increased activity

Element	Species	Comments
Molybdenum toxicity	Humans	In tissues

Xanthinuria—Increased

Element	Species
Molybdenum deficiency	Humans

Xerosis

Abnormally dry skin.

Element	Species	Comments
Selenium toxicity	Nonhuman primates	*Macaques*
Zinc deficiency	Humans	Dry, scaly skin

Yellow nail syndrome

A syndrome including yellow discolored, thickened nails, cough, bronchial obstruction, lymphedema, pleural effusion, sinusitis.

Element	Species	Comments
Titanium toxicity	Humans	Linked to dental implants

Zinc—Liver content decreased

Element	Species
Iron toxicity	Horses

Zinc—Serum concentrations reduced

Element	Species
Copper deficiency	Buffaloes
Iron toxicity	Horses
Zinc deficiency	Cats, cattle, dogs, sheep

Zinc superoxide dismutase—Reduced activity

Element	Species
Zinc toxicity	Humans

Section 2

Minerals, trace elements, and rare earth elements

In this section, the individual minerals, trace elements, rare earth, and other elements are listed in alphabetical order. Clinical signs are listed in alphabetical order and include abnormal signs identified from the history, physical examination, diagnostic tests including imaging and blood tests, or postmortem examination.
In mineral studies, there are often confounding factors that obscure the significance of clinical observations, for example, with Industrial exposure workers are often exposed to more than one potentially toxic substance—so determining the significance of the effect of one mineral over another can be difficult or impossible. Clinical signs have only been included when scientific studies/clinical reports demonstrated a likely cause and effect relationship or a strong association. This does not mean that other observations, for example, the occurrence of cataracts in rats exposed to high doses of antimony are not valid.
Following exposure whether or not an individual will develop clinical signs due to toxicity is dependent on numerous factors including species, breed, the exposure dose, duration, route (ingestion, drinking, inhalation, or skin contact), other chemical exposures, age, sex, nutritional status, genetics, lifestyle, and state of health.
Official tolerance levels are not listed because these are subject to change and for any individual person or animal signs may occur at much lower or much higher doses than those currently held as valid guidelines.

Aluminum (Al)

Al is a metal which resists corrosion and it has many uses including the aerospace industry, cans, car industry, cosmetics, medicines, transportation, and building industries.

Essentiality

There is evidence for essentiality in goats and Al may be essential for other species as well.

Human	No
Other species Goats	 Yes

Body distribution

Parameter	Location
Site(s) of absorption	Gastrointestinal tract (poor), lungs, skin
Route(s) of excretion	Feces, urine
Transportation in blood	Bound to transferrin
Tissue accumulation	Bone, liver, spleen. In uremia more Al is in bone.

Clinical Signs in Humans and Animals Associated with Minerals, Trace Elements, and Rare Earth Elements. https://doi.org/10.1016/B978-0-323-89976-5.00006-2

Other minerals whose absorption is blocked by Al	Ca, F, Fe, P, Na, Sr, cholesterol
Al availability blocked by	Fe, Si
Nutritional sources	In foods, from utensils

Medical uses

Al is used to reduce F in fluorosis, and reduce P absorption in renal patients. Excess supplementation can result in hypophosphatemia and osteomalacia.

Biological activity

Not reported.

Biological effects

Al stimulates enzyme systems in succinate metabolism; accumulates in regenerating bone; decreases intestinal motility; is higher (x9) in cerebral cortex DNA and associated with neurofibrillary tangles seen in patients with Alzheimer's (McLachlan et al., 1990); is deposited in bone mitochondria and inhibits bone phosphate formation (Lieberherr et al., 1982). Al reduces the efficacy of erythropoietin by reducing Fe bioavailability (Donnelly and Smith, 1990); Al causes anemia by reducing heme synthesis, it decreases globulin synthesis and increases hemolysis.
Negative biological effects include synthesis of amyloid, apoptosis, dysplasia, impaired enzyme function, impaired immune response, altered iron homeostasis, genotoxicity, altered membrane stability, metabolic changes, necrosis, oxidative stress, a proinflammatory effect, denaturation, or transformation of peptides.

Deficiency

Species	Clinical signs	Comments
Goats	Abortion, impaired growth, incoordination, life expectancy reduced, weakness (hindlegs)	
Poultry	*Chicks:* impaired growth	

Toxicity

Aluminum toxicity is linked to inhalation of aerosols or particles, ingestion of food, water, or medicines, skin contact, vaccination (Aluminum hydroxide used as an adjuvant), dialysis and infusions, the use of Al-containing antacids and phosphate-binders and patients receiving parenteral nutrition. Al utensils have been linked to Al toxicity in people (Bichu et al., 2019).
In dogs, Al toxicity has been reported in renal failure patients given aluminum hydroxide as a phosphate binder, and following ingestion of Al-containing foreign objects.
Osteomalacia seems to be due to impaired osteoblast activity (Sedman et al., 1987).

Species	Clinical signs
Humans	Amyotrophic lateral sclerosis, anemia, asthma, autism, bone disease (osteomalacia serum Al $>30\,\mu g/L$; due interference with P, fractures and pain), brain disease (confusion, dementia, encephalopathy (linked to dialysis and serum $>80\,\mu g/L$), directional disorientation, hallucinations, myoclonus, seizures), conjunctivitis, constipation, contact dermatitis, granulomatous enteritis, delayed gastric emptying, dyspnoea, fractures; growth delayed (children), hepatorenal disease, reduced Fe absorption, lung problems (desquamative interstitial pneumonia, pneumoconiosis, pulmonary alveolar proteinosis, pulmonary fibrosis, granulomatous pneumonia, pulmonary granulomatosis after inhalation), impaired lung function typically assessed by measuring forced expiratory volume (FEV1) and forced volume capacity (FVC), macrophagic myofasciitis, multiple sclerosis, muscle weakness (myopathy); toxic myocarditis, myocardial dysfunction myocardial hypokinesia, osteomalacia, pancreatitis, pulmonary edema (after ingestion) skeletal deformities, speech problems (dyspraxia), decreased daily sperm production, reduced sperm count and motility and increase in abnormal spermatozoa ischemic stroke due to thrombosis, left ventricular thrombosis, wheezing Al accumulation linked to Alzheimer's, breast cancer, and Parkinson's disease, linked to Crohn's disease and inflammatory bowel disease (IBD)

Cats	Neurotoxic
Dogs	Altered mentation, ataxia, decreased activity, convulsions, low hemoglobin, acute kidney injury, lethargy, decrease in mean corpuscular volume (MCV), microcytosis, muscle twitches, neurological signs (depressed menace response, normal cranial reflexes, obtundation, decreased pupillary light response, paraparesis, reduced patellar reflexes, poor pupillary light reflex, reduced withdrawal reflexes), diffuse cerebral and peripheral neuropathy or junctionopathy, recumbency, testes weight, reduced sperm quality, tetraparesis, tremor, weakness
Gerbils	Disrupted prostate development
Horses	Equine granulomatous enteritis
Koalas	? Renal failure
Mice	Less active, slowed reflexes, behavioral changes (altered escape behavior, delayed puberty, and adulthood), impaired fetal bone development, fetal micronucleated erythrocytes, increased fetal abnormalities, intestinal inflammation, learning and memory disorders, reduced estradiol production, motor disturbances (grip strength), reduced neural stem cells, impaired cell proliferation, and neuroblast differentiation, increased startle responses Signs of P deficiency as Al binds to P
Pigs	Growth retardation, muscle atrophy—due hypophosphatemia, osteomalacia
Poultry	*Chicks:* Cardiac teratogenesis in embryonic chick heart: defects in ventricular septation and ventricular myocardium Signs of P deficiency as Al binds to P. See P
Rabbits	Reduced testes weights, reduced sperm quality Neurotoxic: ataxia, seizures, motor inhibition, neurofibrils form
Rats	High ALP, ALT, AST, reduced androgen receptor protein expression, low bilirubin, bone disease, suppressed erythropoiesis, congenital heart defects in offspring: septal defects, conotruncal defects, and right ventricular outflow, high blood cortisol, high creatinine, high blood epinephrine, hepatotoxicity, impaired glucose tolerance, hyperglycemia, hyperinsulinemia, hypoparathyroidism, hypoproteinemia, hypothyroidism (T3 and T4), hyperlipidemia, hypercholesterolemia and hypertriglyceridemia, low luteinizing hormone, muscle necrosis at the injection site, neuronal death in the hippocampus, pancreatic islet cell necrosis, small intestine epithelial degeneration, goblet cell proliferation, and lymphocyte infiltration in the mucosa of the small intestine, low pregnancy rate, impairment of spermatogenesis and increase in sperm malformation rate. Low testosterone
Sheep	Signs of P deficiency as Al binds to P. See P.

There is insufficient evidence to support toxic effects of Al in vaccine adjuvants or in antiperspirants.

Carcinogenesis, mutagenicity, embryotoxicity

Al containing underarm cosmetics have been associated with breast cancer incidence as Al levels in breast tissues were significantly higher in breast cancer cases than controls (5.8 versus 3.8 nmol/g) (Linhart et al., 2017).
Teratogenic effects resulting in cardiac defects reported in chicks and rats, and genotoxicity resulting in mitodepressive effect in the bone marrow of mice. However, the reports of mutagenesis, cell proliferation and impaired mitosis in Al toxicosis require further studies to confirm the relationship.

Diagnosis

Al can be measured in blood, bone, feces, urine, and hair (if environmental contamination can be ruled out). Plasma Al concentrations (humans) usually $>100\,\mu g/L$.
Bone has wide osteoid seams on X-ray and fails to take up tetracycline markers.
Several high-tech methods can be employed including spectroscopy (accelerator mass spectroscopy) or various forms of spectrometry (graphite furnace atomic absorption, flame atomic absorption, electrothermal atomic absorption, neutron activation analysis, inductively coupled plasma atomic emission, inductively coupled plasma mass, and laser microprobe mass spectrometry).
Mean corpuscular volume (MCV) and microcytosis can be used to identify Al toxicity early in dogs given Al-containing phosphate binders.

Prevention

Avoid exposure through contaminated foods, intravenous solutions, e.g., dialysate or industrial exposure—skin contact or inhalation. Also, avoid Al-containing antacids or antiperspirants, reduce Al absorption, increase Al excretion, maintain renal function, reducing Al load by chelation, and ameliorate toxic effects with antioxidants and other agents.

Chelators: desferrioxamine mesylate. Deferoxamine Side effects include: allergic reactions (pruritus, wheals, and anaphylaxis), abdominal discomfort, cataract, cramps (leg), diarrhea, dysuria, fever, tachycardia.
Malic acid may also increase fecal and urinary excretion of Al reducing tissue concentrations.
(*See Appendix 1 and 2*).

References

NIOSH Aluminium (2020), Tietz et al. (2019), Klein (2019), Igbokwe et al. (2019), Bichu et al. (2019), Linhart et al. (2017), Segev et al. (2016), CDC (2015), Exley (2013), Segev et al. (2008), Krewski et al. (2007), Anke et al. (2005a), Haynes et al. (2004), Golub and Germann (2001), Garruto et al. (1989), Léonard and Gerber (1988), McDowell (1992o), Donnelly and Smith (1990), Alfrey (1986), Ganrot (1986).

Antimony(Sb)

Sb is a metalloid used for hardening lead, the manufacture of batteries and cables, it is used in smelters, mining, glass working, soldering, and brazing.

Clinical applications

Historically Sb compounds were widely used topically or by parenteral injection to treat epilepsy, gout, leprosy, mania, syphilis, whooping cough; as an emetic (tartar emetic—antimony potassium tartrate), as a decongestant, a sedative and to treat leishmaniasis and schistosomiasis—latter now largely replaced by more modern drugs: sodium stibogluconate and praziquantel, respectively.

Essentiality

Human	No
Other species	No

Body distribution

Parameter	Location
Site(s) of absorption	Gastrointestinal tract (poor), inhalation, skin, eye
Route(s) of excretion	Urine mainly, Bile, Feces
Transportation	Red blood cells (trivalent compounds); plasma (pentavalent compounds)
Site(s) of metabolism	Liver—pentavalent reduced to trivalent compounds
Tissue accumulation	Mainly in lung, liver, bone, skin, thyroid, and adrenal glands

Biological activity

None reported.

Biological effects

Sb containing compounds can be emetics and have antiparasitic properties.

Deficiency

Not reported.

Toxicity

The pentavalent antimonials are most toxic. Human cases are often associated with the use of medicinal products or follow industrial exposure, especially contact or inhalation of fumes.

Species	Clinical signs	Comments
Humans	Anorexia, arthralgia, bone marrow hypoplasia, bronchitis, cardiotoxicity (~9% of patients—ECG changes: decrease T wave height, T-wave inversion, prolonged QT, ventricular ectopics, ventricular tachycardia, torsade de pointes, ventricular fibrillation), cerebral edema, coma, conjunctivitis, convulsions, corneal burns, chronic bronchitis, cough, death, dermatitis (Antimony spots: papules, pustules around sweat and sebaceous glands with eczema and lichenification), diarrhea, dizziness, electrolyte disturbances, chronic emphysema, epistaxis, eye irritation, fluid loss, garlic odor, gastrointestinal signs (gastric pain, ulcers), hypotension, impaired immunity (low IgG, IgA, and IgE levels), insomnia, laryngitis, liver damage (transient elevation of hepatocellular enzyme levels), metallic taste, myalgia, myocardial depression, nausea, neuropathy (peripheral), nose bleeds, optic atrophy, palmar keratosis, pancreatitis, pharyngitis, pleural adhesions, pneumoconiosis, pneumonitis, proteinuria, pulmonary edema (rare), acute renal failure, respiratory irritation (chronic coughing), and upper airway inflammation, retinal hemorrhages, rhinitis, shock, skin spots, obstructive sleep apnea, sleeplessness, smell reduced sensitivity, stomach cramps, throat irritation, tracheitis, uveitis, vasodilation, vomiting, weight loss, wheezing	After treatment Pancreatitis seen when HIV and visceral leishmaniasis coinfections
Guinea pigs	Death, hepatic lipidosis, interstitial pneumonitis, splenic hyperplasia, decreased white blood cell count	
Pigs	*Pulmonary lesions:* Focal fibrosis, adenomatous hyperplasia, multinucleated giant cells, cholesterol clefts, pneumonocyte hyperplasia, and pigmented macrophages.	Inhalation studies
Rabbits	Interstitial pneumonia	Inhalation
Rats	High creatinine, high blood urea, low body weight, kidney damage, high potassium, reproductive failure, high sodium *Inhalation*: Lung tumors, interstitial inflammation, fibrosis, granulomatous changes, alveolar-wall cell hypertrophy and hyperplasia, and cuboidal and columnar cell metaplasia. degenerating macrophages, alveolitis, multinucleated giant cells, high aspartate transaminase concentrations	Antimony trioxide and antimony trisulfide Inhalation

Carcinogenesis, mutagenicity, embryotoxicity

Sb causes cancer in rats. It may cause lung tumors and skin cancer in people, especially in smokers. Also implicated in bladder cancer in patients treated with antimony products (Bradberry et al., 2020; Winship, 1987).

Antimony trioxide is classified as possibly carcinogenic to humans (Group 2B) by the International Agency for Research on Cancer.

Based on in vitro testing using the embryonic stem cell test (EST) Sb demonstrated weak embryotoxicity (Imai and Nakamura, 2006).

Diagnosis

Urine testing is considered the most accurate, reliable, and valid test method for measuring antimony levels in the body. Hair analysis is not recommended.

Prevention

Avoid contact with environmental or medicinal sources of Sb. Protective clothing in workplace.

Treatment

Irrigate eyes, wash skin, gastric lavage, and activated charcoal, otherwise symptomatic and supportive treatment. The use of Dimercaprol or the thiol chelating agents DMSA and DMP are possible treatment modalities but only dimercaprol has been successfully used in humans (Bradberry et al., 2020).
(See Appendix 1 and 2).

References

NIOSH Antimony (2020), Bradberry et al. (2020), Wu and Chen (2017), Scinicariello et al. (2017), Sundar and Jaya Chakravarty (2010), De Boeck et al. (2003), Léonard and Gerber (1996), IRIS (1995), Bailly et al. (1991), Inchem (1989), Winship (1987), Nielson (1986), Brieger et al. (1954).

Arsenic (As)

As is considered an essential nutrient in some species.
As is used in the manufacture of ammunitions, cotton desiccants, feed additives, fungicides, glass, herbicides, insecticides, metal adhesives, paints, paper, pesticides, pharmaceuticals, pigments, tanning, textiles, and wood preservatives. As is also found in lasers, light-emitting diodes, semiconductors, and transistors.

Clinical applications

Historically As (Fowler's mixture) was recommended for treatment of gingivitis and stomatitis in infants leukemia, skin conditions (dermatitis herpetiformis, and eczema and psoriasis), and Vincent's angina. Long-term intake caused cancers. Arsenic is still present in many Eastern and Traditional Chinese Medicines and it is a common contaminant in Herbal remedies in the US. In addition:

1. Arsenic may be beneficial in the management of some cancers (Frost, 1983), e.g., multiple myeloma, myelodysplastic syndromes. It reduced mammary tumor growth in mice but As can also increase growth rate
2. As may protect against Se toxicity
3. Arsenic trioxide is used to treat acute promyelocytic leukemia
4. As-containing supplements are sometimes added to feed for perceived health benefits or as growth promoters in production animals. They must be discontinued before slaughter

Essentiality

Human	No
Other species	
Goat	Yes (min <50 ppb)
Hamster	Yes
Pigs (Minipig)	Yes
Poultry (chicks)	Yes
Rat	Yes

Body distribution

Parameter	Location
Site(s) of absorption	Small intestine
Site(s) of excretion	Mainly in urine, some in feces
Site(s) of metabolism	Liver
Site(s) of accumulation	Mainly: Liver, kidneys, heart, lung, and other tissues. Long-term stored in hair and nails
Nutritional sources	Mainly in fish, seafood, and some rice

Biological activity

As is involved in the metabolism of cysteine, polyamines, and taurine from methionine and it inactivates over 200 enzymes, notably those involved in energy pathways and DNA synthesis and repair.
Intake of fish-containing pet foods is associated with renal and hair accumulation of arsenic but the clinical significance of this is unclear (Alborough et al., 2020).

Deficiency

For species in which arsenic is essential, the impact of deficiency is affected by other nutritional components including arginine, choline, guanidinoacetic acid, methionine, taurine, and zinc.
Deficiency affects myocardial mitochondria and impairs enzyme activity involved in amino acid and protein synthesis.
As deficiency increases Cu content of tissues within normal limits, increases organ Mn in kids and piglets, and Mn in milk.
As deficiency increases Mn in the liver of rats.

Mineral	Clinical signs
Goats:	Abortion, death (sudden death in lactating goats), dermatitis, reduced food intake, poor growth, poor milk production, mitochondrial damage (in cardiac muscle, skeletal muscle, and liver), poor survivability of kids, skeletal lesions
Poultry (Chicks):	Increased arginase activity in kidney, depressed hematocrit (PCV), depressed hemoglobin, reduced plasma uric acid
Rats	High mortality rate

Toxicity

Inorganic As is highly toxic and sources include naturally contaminated water or foods, pesticides, and industrial environmental contamination. Organic salts which are common in seafood and other foodstuffs are regarded as being relatively safe.

Species	Clinical signs
	Acute toxicity
Humans	*Ingestion:* Abdominal pain (colic), basophil stippling, blood volume depletion, bone marrow depression, cardiomyopathy, coma, convulsions, death, dehydration (severe), depression, dermatitis (increased capillary permeability and necrosis), diarrhea (can be bloody), fever, fluid loss (severe), gastritis, intestinal hemorrhage, hemoglobinuria, hemolysis, hepatic steatosis, hypotension, intravascular coagulation, jaundice, muscle spasms, nausea, esophagitis, oral pain (burning of the mouth), pancytopenia (severe), pulmonary edema, renal failure, renal tubular acidosis, respiratory failure, ventricular arrhythmia, vomiting
Dogs	Abdominal pain, collapse, death, diarrhea, lethargy, hypersalivation, hypothermia, staggering, vomiting, weakness, rapid weak pulse *Dermal contact:* blisters, swelling, cracking, bleeding, and infections
Fish	Death (increased mortality rate), reduced food utilization, impaired growth, increased liver enzymes activity—SGOT and SGPT
Rabbits	Death
Toads	Death
	Chronic toxicity
Humans	Abortion, anemia (low Fe), behavior changes, low birth rate, cataract, cerebrovascular infarction, cirrhosis, cognitive impairment, confusion, convulsions, deep cracks palms and soles of feet, demyelination, dermatitis (exfoliative), diabetes mellitus risk, drowsiness, encephalopathy, erectile dysfunction, eye congestion, fever, gangrene, garlic smell on the breath, headaches, hearing loss (children), hepatic fibrosis (noncirrhotic portal fibrosis), hepatomegaly, hyperkeratosis, hyperpigmentation skin, hypersalivation, hypertension, infant mortality, the mortality rate increased, ischemic heart disease (*possibly*), limb edema (nonpitting), memory loss, mucosal lesions (inflammation throat), muscle pains, myocardial injury, nail pigmentation (linear—Mee's lines), neonatal death, neutropenia, pericapillary hemorrhages within the white matter, nephritis, peripheral neuropathy, peripheral vascular disease, polyneuritis, recumbency, skin lesions, stillbirths, weakness.

	Cancers including bone, colon, larynx, lung, prostate, skin, stomach, basal cell carcinoma, hemangiosarcoma, angiosarcoma of the liver, lymphoma, and nasopharyngeal carcinoma. Transition cell carcinomas of kidney, bladder, ureter. All urethral tumors. Adenocarcinoma of the bladder *Inhalation:* Respiratory disease, bronchiectasis, chronic bronchitis, lung cancer, chronic obstructive pulmonary disease, restrictive pulmonary disease, throat soreness
Cattle (rare)	Ataxia, hair coat changes, hypocalcemia, hypoglycemia, inflammation of upper respiratory tract membranes, inflammation of eyes, weight loss
Dogs	Renal tubular lesions (mild degeneration and vacuolization)
Poultry (chicks)	Increased erythrocyte formation, reduced leukocyte production (Turkeys): reduced weight gain, death
Rats	Impaired growth

Carcinogenesis, mutagenicity, embryotoxicity

As is a recognized carcinogen in humans associated with cancer in the following: Bladder adenocarcinoma and transition cell carcinoma (TCC), bone, colon, kidney (TCC), larynx, liver, lung, lymphoma, nasal cavity, prostate, skin (Bowen's disease; basal cell carcinoma), ureter (TCC), urethra (all types).

Diagnosis

1. High As in blood, hair, nails, and urine. Human hair 0.1–0.5 mg As/kg = chronic; >1 mg/kg = acute.
2. Radiography may show radiopaque As in the gastrointestinal tract lumen.
3. Urinary As concentration is considered best for confirmation of acute toxicity, hair analysis for long-term chronic exposure.

Prevention

Prevent access and exposure to arsenic in products and emissions in the workplace, the environment (especially water supplies in many countries).

Treatment

No antidote or effective specific treatment. Treat symptomatically and possibly use antioxidants.
(*See Appendix 1 and 2*).

References

NIOSH Arsenic (2020), Alborough et al. (2020), Kuivenhoven (2019), Ratnaike (2003), Quansah et al. (2015), Long (2019), Barai et al. (2017), McDowell (1992o), Anke (1986).
NIOSH Arsenic, 2020. Arsenic in NIOSH Pocket Guide to Chemical Hazards. The National Institute for Occupational Safety and Health (NIOSH), Centres for Disease Control and Prevention (CDC). Available from: https://www.cdc.gov/niosh/topics/arsenic/ (Accessed 11 September 2020).

Barium (Ba)

Ba is a metal used in heated cathodes, electroceramics, fireworks, iron and steel industry, oil well drilling, rodenticides, vacuum tubes, YBCO superconductors.

Clinical applications

Barium sulphate is used in radiography as a positive contrast agent.

Essentiality

Human	No
Other species	No

Body distribution

Parameter	Location
Site(s) of absorption	Gastrointestinal tract, lungs,skin
Site(s) of excretion	Feces
Site(s) of metabolism	Not reported
Site(s) of accumulation	Lungs after inhalation

Biological activity

None reported.

Biological effects

None reported.

Deficiency

Not reported.

Toxicity

Water-soluble barium salts including barium acetate, chloride, hydroxide, nitrate, sulfide may cause signs whereas insoluble barium sulfate does not. Orally administered barium carbonate also can cause harm. Gastrointestinal signs may occur rapidly whereas cardiovascular and neuromuscular signs may be delayed.
A systematic review (Nigra et al., 2016) failed to find sufficient evidence to confirm Ba toxicity as a risk factor for cardiovascular disease.

Species	Clinical signs
Humans	Abdominal pain, anxiety, baritosis (accumulation of Ba in the lungs after inhalation of dust) blindness, cardiac arrhythmias, cardiac arrest, death, watery diarrhea, dizziness, dyspnoea, gastric pain, gastrointestinal hemorrhage, hematuria, hypertension or hypotension, hypokalemia, hypophosphatemia, mydriasis, nausea, numbness, paralysis, renal damage (chronic toxicity), renal degeneration, renal insufficiency, acute renal failure, respiratory failure, rhabdomyolysis, seizures, sensitization, shock, shortness of breath, tremors, ventricular tachycardia, vomiting, muscle weakness
Dogs	Acidemia, arrhythmias (ventricular), collapse, severe hypokalemia, loss of muscle tone, muscle twitches, tachypnea, impaired limb reflexes, flaccid paralysis, respiratory failure
Mice	Death, kidney damage, decreases in body weight, decreased survival, renal tubule dilatation, hyaline cast formation, interstitial fibrosis, glomerulosclerosis
Rats	Death, gastroenteritis, kidney damage, decreases in body weight, decreased survival, decreased number of sperm, decreased percentage of motile sperm and decreased osmotic resistance of sperm, *Blood changes:* erythropenia, leukocytosis, eosinophilia, neutrophilia

Carcinogenesis, mutagenicity, embryotoxicity

Not reported.

Diagnosis

Measure Ba in body tissues and fluids (bones, blood, urine, and feces).

Prevention

Avoid unnecessary exposure to Ba compounds.

Treatment

No specific treatment. Treat symptomatically.
(*See Appendix 1 and 2*).

References

Nigra et al. (2016), Adam et al. (2010), ATSDR Barium (2007), Tao et al. (2016), Johnson and Van Tassell (1991), Nielsen (1986), Doig (1976).

Beryllium (Be)

Be is a metal used in electronics, high-technology ceramics, metals extraction, and dental alloys.

Clinical applications

Dental alloys.

Essentiality

Human	No
Other species	No

Body distribution

Parameter	Location
Site of absorption	Poorly absorbed from gastrointestinal tract. Inhalation, Dermal contact
Site(s) of excretion	Feces, urine
Site(s) of accumulation	Bone, liver, kidney, spleen, blood—lung if inhaled

Biological activity

Not reported.

Deficiency

Not reported.

Toxicity

Inhalation or skin exposure to dust particles, fumes or vapors are the usual methods of human exposure to Be. Buried shrapnel has been linked to Be-associated lung disease.

Species	Clinical signs	Comments
Humans	Appetite loss, bronchioalveolar lavage positive, chest pain, conjunctivitis, contact dermatitis, dry cough, death, difficulty breathing (dyspnoea)/shortness of breath, fatigue, fever, granuloma formation (inflammatory nodules), hypersensitivity reaction, joint pain, liver damage, lung disease (interstitial edema, cellular infiltration, high plasma cells, alveolar cell proliferation or desquamation, interalveolar edema, hyaline membrane formation), lymphadenopathy, pharyngitis, interstitial pulmonary fibrosis, night sweats, pneumonia, pneumonitis, rhinitis, skin rash, sore throat, tracheobronchitis, ulcers (skin), weakness, weight loss	Industrial exposure—inhalation Deficient and dysfunctional T-cells in the lungs contribute to persistent inflammation
Dog	Granulomatous lung disease	*Inhalation* Beryllium oxide (Haley et al., 1989)
Guinea pigs	Granulomatous lung disease, fibrosis	*Intratracheal*
Hamsters	Death, Bronchiolar-alveolar tumors, adenomas Granulomatous lung lesions	*Inhalation*-Be ores
Mice	Bone tumors	*Injection intravenous: Be salts*

Nonhuman primates	Death, inflammatory cells in lungs (macrocytes, lymphocytes, plasma cells) Lung tumors	*Inhalation*: Be ores Squirrel monkeys *Inhalation*: Be sulphate Rhesus monkeys
Rabbits	Death, hypoglycemia, liver necrosis, convulsions Osteosarcomas Bone tumors	*Intravenous* Be metal in water *Intravenous*: Be salts *Intramedullary bone injection or subperiosteal implantation*
Rats	*Inhalation:* Lung tumors (mainly bronchioalveolar, adenocarcinomas and adenomas) *Intratracheal:* Lung neoplasia (mainly adenocarcinomas and adenomas) *Intravenous:* Death, hypoglycemia *Intranasal:* Necrotizing, hemorrhagic pulmonitis, intraalveolar fibrosis, chronic inflammation *Oral in food:* Rickets, inhibits alkaline phosphatase	*Inhalation*—Be ores *Intratracheal:* Be metal and various salts *Oral* Be carbonate: Severe rickets not ameliorated by vitamin D

Carcinogenesis, mutagenicity, embryotoxicity

According to the WHO IARC Monologue Be and Be compounds are carcinogenic to humans (Group 1) and there is sufficient evidence for the carcinogenicity of Be and Be compounds in guinea pigs, hamsters, rabbits, rats, and possibly monkeys.

In vitro: soluble Be chloride and sulfate compounds cause gene mutations and cell transformations in microorganisms and mammalian cells. Be compounds bind to nucleoproteins and inhibit enzymes in DNA synthesis, for nucleic acids binding to cell membranes and inhibit microtubule polymerization. So Be is likely mutagenic.

It is not known if Be is teratogenic.

Diagnosis

History of exposure. Radiography, measure beryllium lymphocyte proliferation test [BeLPT] in blood sample.

Prevention

Avoid exposure to dust and inhalation or direct contact with skin. Use PPE in workplace.

Treatment

No specific treatment. Treat symptomatically, ventilator use if necessary.
(*See Appendix 1 and 2*).

References

Boffetta et al. (2016), Fontenot et al. (2016), Fireman et al. (2012), Strupp (2011a,b), Van Dyke et al. (2011), Middleton and Kowalski (2010), NORD Beryllium (2009), Cummings et al. (2009), Scott et al. (2008), Culver and Dweik (2003), Newman et al. (1996a), Newman et al. (1996b), Rossman (1996), IARC (1993), Haley et al. (1989), Léonard and Lauwery (1987), Nielsen (1986).

Bismuth (Bi)

Bi is a heavy metal used in cosmetics, pigments, and a few pharmaceuticals.

Clinical applications

Bi salts are used as an ointment for burns, warts, as a fungicide, and for gastrointestinal symptoms: bismuth subsalicylate for diarrhea, dyspepsia, nausea, and stomach ulcers. Prescribed long term as surgical packing pastes for colostomy patients.

Essentiality

Human	No
Other species	No

Body distribution

Parameter	Location
Site of absorption	Poorly absorbed from gastrointestinal tract, inhalation, transdermal
Site(s) of excretion	Bile and urine
Site(s) of metabolism	Not reported
Site(s) of accumulation	Kidney mainly, also in lung, spleen, liver, brain and muscle

Biological activity

None reported.

Deficiency

Not reported in any species.

Toxicity

Due to inhalation, ingestion of over the counter Bi-containing preparations, Bi-impregnated surgical packing pastes.

Species	Clinical signs	Comments
Humans	Abdominal pain, albuminuria, aminoaciduria, anemia, angiedema, anxiety, apathy, appetite loss, ataxia, balance loss, blurred vision, bone marrow hypoplasia, buccal hemorrhage, cognitive deficits, coma, concentration difficulty, confusion, constipation, death, delirium, delusions, dementia, depression, dermatitis, diarrhea, diplopia, dysarthria, encephalopathy [confusion, disorientation, seizures (rare)], epistaxis, erythematous rash (maculopapular), eye irritation, falls, fever, gait abnormalities, gastrointestinal symptoms, gingivitis, glomerular filtration rate (GFR) reduced, glycosuria, hematuria, halitosis, hallucinations, headache, hyperreflexia, incoordination, insomnia, irritability, joint pain, lethargy, malaise, melena (black stools), memory loss, metallic taste, muscle cramps, nausea, neuropsychiatric symptoms, memory loss, myoclonus, kidney damage (acute renal failure, glomerular and tubular necrosis), liver damage, oliguria, paraesthesia, Parkinsonism, petechial hemorrhages, poor oral hygiene with a blue-black gum line, proteinuria, psychosis, pyramidal signs, rectal bleeding, respiratory irritation, seizures, sleeplessness, smell sensation reduced, somnolence skin lesions (lichen planus-like rashes), taste sensation reduced, thrombocytopenia, tremor, ulcerative stomatitis, vomiting, weight loss	
Rats	Foreign body inflammatory response in lungs	*Inhalation*

Carcinogenesis, mutagenicity, embryotoxicity

Not reported.

Diagnosis

No specific diagnostic test.

Prevention

Avoid excessive use of Bi-containing preparations, e.g., ointments, oral preparations.

Treatment

No specific treatment. Avoid further exposure and treat symptomatically.
The chelator Dimercaprol may reduce Bi in the brain, kidneys and liver and increase renal elimination—given by deep intramuscular injection 2.5–5 mg/kg four hourly for two days followed by 2.5 mg/kg bd for up to 14 days. May get gastrointestinal signs and hypertension.
Based on animal studies DMPS and DMSA are the chelating agents of choice in bismuth poisoning. Either may be administered orally in a daily dose of 30 mg/kg body weight. May get gastrointestinal signs and increased liver transaminase activity.
(*See Appendix 1 and 2*).

References

NORD (2020a), Lenntech (2020a), Borbinha et al. (2019), Hogan (2018), DiPalma (2001), Bradberry et al. (1996), Gordon et al. (1995), Nielsen (1986).

Boron (B)

B is a trace element that is essential for some species of mammal.
B compounds are used in disinfectants, fertilizers, fiberglass, textiles, herbicides, insecticides, insulation, Pyrex glass, pyrotechnic flares, rocket fuel and, as boric (or boracic) acid, borax (sodium borate), and boric oxide, in eye drops, mild antiseptics, washing powders, and tile glazes.

Clinical applications

B compounds are used in some mouthwashes, denture cleaners, eye drops, and antiseptics.
B supplementation may aid in the management of rheumatoid arthritis (Newnham, 1990)—humans and animals. B may also be of benefit in managing some cancers including prostate, cervical, and lung cancers, and multiple and non-Hodgkin's lymphoma and may help ameliorate the adverse effects of chemotherapeutic agents (Pizzorno, 2015).
B compounds are used for optical and nuclear imaging and as therapeutic agents with anticancer, antiviral, antibacterial, antifungal, and other disease-specific activities (Das et al., 2013).

Essentiality

Human	No
Other species	
Rat	Yes—0.3–0.4 ppm
Poultry (*Chicks*)	Yes—Min 1 ppm

Body Distribution

Parameter	Location
Site of absorption	Small intestine
Site(s) of excretion	Urine
Interference with minerals	Ca and Mg metabolism
Site(s) of accumulation	Bone, hair, nails

Biological activity

In the body, B is important for (after Pizzorno, 2015):

(1) Bone growth and maintenance by inducing mineralization of osteoblasts by regulating genes expression related to tissue mineralization and the actions of key hormones (17β-estradiol [E2], testosterone, and vitamin D) (Hakki et al., 2010)
(2) Optimal wound healing by stimulating the activity of elastase, trypsin-like enzymes, collagenase, and alkaline phosphatase in fibroblasts
(3) Metabolism of estrogen, testosterone, and vitamin D
(4) Optimizing magnesium absorption
(5) Reduces inflammatory biomarkers [C-reactive protein (hs-CRP), tumor necrosis factor α (TNF-α)]
(6) Increases activity of antioxidant enzymes (e.g., catalase, glutathione peroxidase, and superoxide dismutase (SOD)]
(7) Protects against pesticide-induced oxidative stress and heavy metal toxicity
(8) Improves brain electrical activity, cognitive performance, and short-term memory for older people
(9) Affects the formation/activity of key biomolecules, such as S-adenosyl methionine (SAM-e) and nicotinamide adenine dinucleotide (NAD+)

Deficiency

Species	Clinical signs	Comments
Humans	Impaired behavior: attention, dexterity, memory (short term), motor speed	Observations on low B intake, but B is not considered an essential nutrient
Rat	Growth impairment, low hemoglobin, impaired immune function, reduced red cell production (red cell count)	
Poultry	(Vitamin D deficient chicks): growth impairment, elevated Alkaline phosphatase activity	

Toxicity

Routes of exposure to B resulting in natural cases of toxicity are ingestion of food or water, use of pesticides, inhalation of powders or dusts, or use of cosmetics or medical preparations containing B.

Species	Clinical signs	Comments
Humans	*Ingestion:* anorexia, death, dermatitis (exfoliative dermatitis), diarrhea, mucosal exfoliation, inflammation, congestion, exfoliation gastrointestinal effects, kidney lesions (cloudy swelling and granular degeneration of renal tubular cells), lethargy, liver lesions, nausea, nervous system signs (irritability, seizures), respiratory failure, riboflavinuria, vomiting *Inhalation:* breathlessness, conjunctivitis, cough, eye irritation, nose irritation, acute respiratory irritation, throat irritation *Skin contact:* erythema	Diborane gas is very irritant
Cattle	Food intake reduced, growth impairment, edema (limbs), low hematocrit (PCV), low hemoglobin, low plasma phosphate	
Dogs	Testicular degeneration, arrested spermatogenesis, vomiting	1170 μg/L B for 2 years
Gerbils	Testicular degeneration	
Mice	Pulmonary congestion, bleeding, and edema, hepatic necrosis, gastric hyperplasia and dysplasia, renal lesions, testicular degeneration *Embryo:* skeletal malformations (e.g., rib agenesis, fused rib, cervical rib, reduced rib length) Reductions in fetal body weights	Inhalation diborane gas Oral
Poultry	*Chicks*: curled-toe paralysis, skeletal abnormalities	Due riboflavin deficiency
Rabbits	*Fetuses*: Decreases in the number of live fetuses and litters, increase in resorptions, decreases in body weight, and increases in the occurrence of external visceral, and cardiovascular malformation Adults: conjunctivitis	44 mg boron/kg/day on gestation days 6–19
Rats	Hair changes (coarse), delayed maturation of prepubescent hair, incisors lack pigmentation, scaly tails, desquamation pads on feet, abnormal nails (long), aspermia, impaired ovarian development, testicular atrophy, hemorrhage from eyes, low hemoglobin, low hematocrit. *Fetuses:* decreases in body weight and increases in the occurrence of skeletal malformations	1170 μg/L B for 2 years

Carcinogenesis, mutagenicity, embryotoxicity

Boron is not regarded as being carcinogenic, mutagenic or embryotoxic.

Diagnosis

Measurement of B in tissues and body fluids (blood, urine).

Prevention

Ensure adequate nutritional intake (as cobalamin); avoid excess exposure.

Treatment

Ensure adequate intake, avoid excess exposure, treat symptomatically.
(*See Appendix 1 and 2*).

References

ATSDR Boron (2020), NIOSH Boron (2020), Pizzorno (2015), Das et al. (2013), Hakki et al. (2010), Bakirdere et al. (2010), Bai and Hunt (1996), Naghii and Samman (1993), McDowell (1992p), Newnham, 1990, Nielsen (1986).

Bromine (Br)

Br-containing products are used in agriculture, sanitation, and as fire retardants

Clinical applications

Historically Br (as potassium bromide) has been used as a sedative, antilibido or anticonvulsant agent.

Essentiality

Human	No
Other species Goats	Possibly

Bromine is not currently regarded as being an essential nutrient however there is growing evidence that it is essential for all life (McCall et al., 2014) and in studies feeding a low Br ration can cause signs such as impaired growth. Many scientists now consider Br to be essential in goats.

Body distribution

Parameter	Location
Site of absorption	Small intestine
Site(s) of excretion	Urine
Sites of metabolism	Is taken up by thyroid in absence of iodine

Biological activity

Bromide and chloride are interchangeable in body metabolic processes.
Recently, Br has been shown to be essential for assembly of collagen IV scaffolds in tissue development and architecture (McCall et al., 2014).

Deficiency

Species	Clinical signs	Comments
Humans	Insomnia	Low Br diet
Goats	Abortions, conception rate reduced, fatty liver, fat in skeletal muscle, reduced food intake, reduced fertility, high gamma glutamyl transferase, impaired growth, poor weight gain, *Blood changes:* low hematocrit, low hemoglobin, reduced life expectancy, high triglycerides	Not fully corroborated
Mice	impaired growth	Low Br diet
Poultry	impaired growth (chicks)	Low Br diet

Toxicity

Exposure routes to Br in solid, liquid, or gaseous forms are: eye contact, ingestion, inhalation, skin contact.

Species	Clinical signs	Comments
Humans	Abdominal pain, confusion, coma, cough, diarrhea, dizziness, epistaxis, hallucinations, headache, lacrimation, neurological symptoms, pneumonitis, pulmonary edema, slurred speech, skin burns, skin lesions, stupor	Bromism (1999)
Cat	Paradoxical sleep	*Experimental:* Br as 1-methylheptyl-gamma-bromoacetoacetate given IV

Carcinogenesis, mutagenicity, embryotoxicity

Bromate is considered to be a genotoxic carcinogen and is categorized as "possibly carcinogenic to humans" by the International Agency for Research on Cancer (IARC).

Diagnosis

History of exposure. Analysis of serum bromide (human reference level is 50–100 mg/L).

Prevention

Avoid exposure to Br compounds especially in the workplace.

Treatment

No specific treatment. Treat symptomatically.
(*See Appendix 1 and 2*).

References

NIOSH Bromine (2020), Anke et al. (2005b), Bromism (1999), Nielsen (1986, 1998).

Cadmium (Cd)

Cd used for electroplating, storage batteries, vapor lamps, and some solders.

Clinical applications

Cd is a component in some medical devices such as cadmium zinc telluride devices used in imaging.

Essentiality

Human	No
Other species Rats	 Possible
Goats	Yes—(Anke et al., 1983, 2005c)

Body distribution

Parameter	Location
Site of absorption	Poorly absorbed after ingestion or transdermally It is absorbed after inhalation.
Site(s) of excretion	Urine (mainly), also Bile
Transported	Mainly in erythrocytes in the blood and metallothionein
Sites of metabolism	Cd is tightly bound to metallothionein and accumulates in tissues especially kidney and liver
Site(s) of accumulation	Kidneys, liver

Biological activity

The radioisotope has a biological half-life of 20–30 years, Cd affects biosynthesis of cytochromes, hem, and porphyrins.

Interferences

Cd interferes with Cu, Fe, and Zn metabolism.

Deficiency

Species	Clinical signs	Comments
Goats	Abortion, conception impaired, death, lethargy, mitochondrial damage (especially in liver and kidneys), mobility reduced, muscle weakness, myasthenia, ventroflexion of the neck	Low Cd diet fed during pregnancy

Toxicity

Routes of exposure: ingestion (water, food), inhalation, contact with soil.

Species	Clinical signs	Comments
Humans	Abdominal cramps, anemia (low Fe), anosmia (loss of sense of smell) breathlessness, cancer (liver, lung, prostatic), death, chest tightness, chills, cough, cyanosis, diarrhea, discoloration of teeth (yellow), dyspnoea, emphysema, fatigue, fever, headaches, liver function impairment, muscle aches, nausea, pain (substernal), prostatic hyperplasia, pulmonary edema,	

	renal damage: tubular epithelial cell degeneration cellular atrophy, interstitial fibrosis, and glomerular sclerosis, hypercalciuria, itai-itai disease (osteomalacia—linked to renal disease), proteinuria, proteins of low relative molecular mass in the urine (beta2-microglobulin and retinol-binding protein), glucosuria, aminoaciduria, enzymuria (*N*-acetyl-beta-glucosaminidase), nausea, pulmonary edema, increased salivation, tachycardia, vomiting, and urinary concentration of cadmium and metallothionein	
Dogs	Osteomalacia	
Fish	Death (increased mortality rate), reduced food utilization impaired growth, increased liver enzymes activity—SGOT and SGPT	
Hamsters	Alveolar hyperplasia, hemorrhagic ovarian necrosis, interstitial fibrosis Fetal deformities (craniofacial, eyes, limbs)	Inhalation Injection during pregnancy
Horses	Chronic interstitial nephritis	
Mice	Alveolar hyperplasia, death, fetal developmental abnormalities if high Cd during pregnancy (included exencephaly, Fetal deformities (craniofacial, eyes, limbs), hemorrhagic ovarian necrosis, hydrocephaly, cleft lip and palate, microphthalmia, micrognathia, club foot, and dysplastic tail), lung tumors, pulmonary interstitial fibrosis, testicular necrosis, osteomalacia (femur calcium content), weight loss	
Rabbits	Glomerular sclerosis, tubular-cell degeneration, and interstitial fibrosis	
Rats	Increased albuminuria, anemia, behavioral changes (reduced activity, decreased acquisition of avoidance behavior), carcinogenicity (*Cd orally*: leukemia, prostate, testis, and hematopoeitic system; *Cd inhalation*: lung; *Cd injection:* at the injection site (rhabdomyosarcomas; fibrosarcomas) and interstitial-cell tumors of the testes, pancreatic Islet cell tumors), fetal developmental abnormalities if high Cd during pregnancy(included exencephaly, hydrocephaly, cleft lip and palate, microphthalmia, micrognathia, club foot, and dysplastic tail) death, reduced calcium and phosphorus absorption from the intestine, lung tumors, inhibition of the renal conversion of 25-hydroxycholecalceferol to 1,25-dihydroxycholecalceferol (only if on a low ca diet), reduced femoral bone density, desquamation, and necrosis of gastric and intestinal mucosa, hemorrhagic ovarian necrosis, ovarian necrosis, periportal liver-cell necrosis, proteinuria, pulmonary edema, renal tubular necrosis, testicular necrosis.	Sarcomas and carcinomas formed after injection or inhalation of Cd salts Tumor formation could be inhibited if there was Zn deficiency
Rhesus monkeys	Renal tubular dysfunction	1.2 mg/kg bw per day for 9 years

Carcinogenesis, mutagenicity, embryotoxicity

The IARC of the WHO consider that cadmium and cadmium compounds are carcinogenic to humans and cadmium compounds are carcinogenic to animals.

Cd causes chromosome aberrations in animals and in animal and human cells in vitro.

Cd is teratogenic.

Diagnosis

Cadmium levels can be measured in the blood, hair, nail, saliva, and urine.

Prevention

Avoid exposure especially in the workplace.

Treatment

No specific antidote so treat cases symptomatically and avoid further exposure. The use of chelators is controversial as there is an absence of proven efficacy and safety and they can increase the risk of nephrotoxicity (Rahimzadeh et al., 2017). New products may be more efficacious and safer but whenever possible use licensed medications.
(See Appendix 1 and 2).

References

NIOSH Cadmium (2020), Rahimzadeh et al. (2017), Nigra et al. (2016), IARC (1993), WHO (1992), Inchem (2020), McDowell (1992o), Kostial (1986).

Cesium (Cs)

Cs is used in drilling fluids, production of electricity, electronics, and chemistry. It is radioactive isotopes present a great health risk.

Clinical application

Over-the-counter supplements containing Cs are widely taken by people, usually in an attempt to treat cancers but efficacy and safety studies are lacking and toxic effects are commonly reported.

Essentiality

Human	No
Other species	No

Body distribution

Parameter	Location
Site of absorption	Small intestine, transdermal, inhalation
Site(s) of excretion	Feces, urine

Biological activity

Cs can be interchangeable with K in lower animals. Cs is present in higher concentrations in erythrocytes than in plasma.

Deficiency

Not reported.

Toxicity

Nonradioactive Cs toxicity in humans is often due to consumption of cesium chloride to treat cancer. This practice can also induce hypokalemia with sometimes fatal outcomes.

Species	Clinical signs
Humans	Anorexia, cardiac arrest, cardiotoxicity, death, diarrhea, ECG changes: QT interval prolonged to 650 ms headache, hypokalemia, hypomagnesemia, hypotension, hypothyroidism (Harari et al., 2015), nausea, neurological signs (tingling sensations in lips, cheeks, hands, and feet), seizures, syncope, polymorphic ventricular tachycardia (Lyon and Mayhew, 2003)

Radioactive Cs is much more toxic: *Dogs:* a single dose of 3.8 mCi (140 MBq, 4.1 μg of cesium-137) per kilogram is lethal within three weeks; smaller amounts cause infertility and cancer (Redman et al., 1972). Radioactive Cs was one of the main pollutants resulting from the Chernobyl and Fukushima nuclear accidents.

Carcinogenesis, mutagenicity, embryotoxicity

No evidence for nonradioactive Cs.
Radioactive Cs is carcinogenic, mutagenic, and embryotoxic.

Diagnosis

History of exposure. Measurement of Cs in body tissues, environmental samples.

Prevention

Avoid exposure, especially to radioactive isotopes in the workplace or contaminated environment.

Treatment

There is a report of successful treatment using amiloride (Horn et al., 2015) 10mg amiloride daily, increased to 20mg. After 1 week of amiloride treatment, the patient no longer required extra-dietary potassium supplementation, and his urine K+ excretion decreased to 36mmol/day.
(*See Appendix 1, 2, and 3*).

References

Aaseth et al. (2019), Harari et al. (2015), Horn et al. (2015), Sessions et al. (2013), Melnikov and Zanoni (2009), ATSDR Cesium (2004), Lyon and Mayhew (2003), Nielsen (1986), Redman et al. (1972).

Calcium (Ca)

Ca is an important micromineral that is present in many foodstuffs, but absent in muscle meat. It has many uses including in bleaching agents as calcium hypochlorite.

Clinical applications

Calcium supplements are widely used but care is needed to avoid excess intake, interference with the bioavailability of other minerals, and to avoid an inappropriate balance, e.g., Ca:P ratio, all of which can cause clinical signs.

Essentiality

Humans	Yes
Other species	
All	Yes
Fish	Yes—but may not be essential for salmonids

Function	Site(s)	Comments
Site(s) of absorption	Gastrointestinal tract (small intestine) by transcellular and paracellular pathways, gills (fish)	
Transportation	In blood as free ions or bound to proteins, e.g., albumin	Blood concentrations strictly regulated by hormones—especially calcitonin and parathyroid hormone *Rabbits:* blood concentrations reflect dietary intake
Excretion	Urine, feces, and sweat	*Rabbits:* Unusually mainly excrete Ca in bile
Body storage	Mainly in bones and teeth, but also other tissues, e.g., muscle	

Excretion: Rabbits urine (unusual) others excrete Ca in the bile, also in rabbits blood calcium reflects dietary intake (unusual) in other species it is maintained within very narrow References limits.

Biological activity

Closely linked to P and vitamin D activity.

1. A structural component of bone
2. Important for thrombin formation from prothrombin in blood clotting
3. A cofactor in enzyme reactions
4. Hormone secretion
5. Neuromuscular activity
6. Cardiac activity
7. Membrane permeability

Deficiency

Ca bioavailability reduced by the presence of oxalate, phytate, dietary fiber and high amounts of other minerals. Also by low vitamin D.
Low Ca causes neuromuscular hyperexcitability and if very low tetany and death.
In vertebrates, skeletal abnormalities due to impaired endochondral ossification include poor stunted growth, rickets with poor mineralization of bone, developmental abnormalities, swollen joints, angular deformities, lameness, and spontaneous fractures. (Common in dogs and cats fed all meat rations).

Species	Clinical signs	Comments
Humans	Bone demineralized, fractures (spontaneous), joint enlargement, lameness, muscle weakness, rickets, seizures (child), skeletal deformity (bowed legs), spinal curvature, tetany (child), tooth enamel defects	
Amphibians	Poor bone mineralization, limb fractures, metabolic bone disease, spinal deformities (e.g., scoliosis) tetany, weakness	
Birds	Angular deformity of limbs, anorexia, beak malformations, bone demineralization, carpal rotation (waterfowl) collapse, egg abnormalities (shape, soft, thin shell), egg binding, falling from perches, dermatitis, feather abnormalities (brittle, frayed), fractures, poor growth, hypocalcemia, incoordination, leukocytosis, metabolic bone disease (hyperparathyroidism), opisthotonos, osteomalacia, osteoporosis, paralysis, polydipsia, polyuria, regurgitation, reproductive problems, rickets, seizures, skeletal deformities, syncope, tetany, weakness, (especially African gray parrots; Timneh parrots)	
Buffaloes	Bone fragility, fractures, reduced fertility, delayed maturity, stunted growth, reduced milk production, osteoporosis (adults), paralysis, rickets (calves), unthriftiness,	
Camels	*Calves* (1–12 *months*): angular deformity of limbs, anorexia, arched back, death (due to starvation), epiphyseal plate widened, facial deformity, fibrous osteodystrophy, fractures (spontaneous), joint pain, difficulty to move, joint swelling (forelimbs), lameness, osteoporosis, stiff gait, weakness (*and P deficiency*) *Adults*: reduced appetite, bone fragility, emaciation, lameness, osteomalacia, pica (*and P deficiency*)	
Cats	*Common in cats fed an all meat diet*: bone cortices thin, bone pain, bones undermineralized, fractures (spontaneous), lameness, nutritional secondary hyperparathyroidism, rickets, spinal curvature	Includes domesticated and captive wild cats
Cattle	*Dairy:* fractures (spontaneous), retarded growth, osteoporosis, osteomalacia	
Deer	*Antler abnormalities*: deformity, delayed velvet shedding	
Dogs	*Common in dogs fed an all meat diet*: bone cortices thin, bone undermineralized, fractures, feed efficacy reduced, growth rate slowed, nutritional secondary hyperparathyroidism, osteomalacia, rickets, tooth loss, reproductive failure, seizures, tetany, vertebral compression	
Elephants	Hypocalcemia, metabolic bone disease, poor bone calcification, osteodystrophy	
Foxes	Angular deformity of long bones, bones poorly calcified, bones soft, bones vascularized, facial bones enlarged, gums swollen, hypocalcemia, joint swelling, lameness, muzzle enlarged, rickets, spasms, stiffness (rear legs), tooth loss, walk-on pasterns	
Goats	Osteomalacia (adults), rickets (kids)	
Guinea pigs	Dental enamel hypoplasia, rickets	(P deficiency
Horses	*Foals:* rickets (poor mineralization of osteoid, enlarged joints, crooked long bones (angular deformity)) *Adults*: osteomalacia, lameness	
Marsupials	*Opossums:* Nutritional secondary hyperparathyroidism, undermineralized bone, fractures	
Meercats	Fractures, metabolic bone disease, nutritional secondary hyperparathyroidism, undermineralized bone	
Mice	Poor weight gain, poor bone mineralization	
Nonhuman Primates	Lemurs: hyperphosphatemia, hypocalcemia, increased alkaline phosphatase activity, impaired mobility, angular deformity of the long bones, undermineralized skeleton, soft tissue mineralization Tarsiers: poor reproductive performance	Tomson and Lotshaw (1978) Roberts and Kohn (1993)
Rats	Bone undermineralized, food intake reduced, growth impaired, hemorrhage (internal), joint swelling, lactation failure, paralysis (hindlegs), rickets, reproduction impaired	

Reptiles	Anorexia, poor bone mineralization, prolapsed cloaca, death (due asphyxiation respiratory failure), dystocia, eggs poorly mineralized, hypocalcemia, limb fractures (spontaneous), muscle twitches (toes and legs), neuromuscular signs, osteopenia, hind limb paresis, limb deformity (bowing due weakness, healing fractures), paralysis respiratory muscles, poor shell development (chelonia), shell deformity, swollen limbs, mandibular swelling, muscles flabby, seizures, spinal deformity, tetany	
Sheep	Milk fever, osteomalacia (adults), rickets (lambs)	
Squirrels	Aggressiveness, angular deformity of bones (limbs, spine, tail), anorexia, bones poorly calcified (weak), growth impairment, reluctance to move (lethargy), labored breathing, metabolic bone disease (abnormal bone growth, facial deformities (malocclusion) fractures, swollen joints) muscle spasms, paralysis, seizures, excessive sleeping, tremors, twitching (muscles under skin, toes, legs, tail), weight loss	
Fish	Food intake reduced; growth depressed	

Toxicity

Ca in the blood is maintained within a narrow References range and Ca absorption is regulated by vitamin D, parathyroid hormone (PTH) and calcitonin. High Ca intake inhibits PTH and increases osteoblast activity so mineral is deposited in bone.

Ca toxicity is associated with bone disorders (especially in young growing animals), reduced food intake and reduced weight gain.

Excess Ca intake reduces the efficacy of utilization of P and other minerals and cause deficiency of Fe, I, Mn mg, and Zn. The maximum tolerance levels vary between species:

Species	Maximum tolerance	Comments
Human	USL: 2500 mg/day	
Dog	6.5 g/100 g food	For adult
Cat	—	Not specified
Cattle, sheep, horses, rabbits	2% of ration	
Pigs	1% of ration	
Poultry (most) Laying hen	1.2% of ration 4% of ration	 Due high Ca requirement to form eggs

Species	Clinical signs	Comments
Humans	Abdominal pain, anxiety, arrhythmias, bone pain, confusion, depression, fatigue, lethargy, gastric pain, hypercalcemia, hypertension, lethargy, muscle weakness, polydipsia, polyuria, soft tissue calcification, vomiting	
Birds (Captive)	Visceral gout (uric acid crystal deposition in organs), nephrocalcinosis, nephropathy, renal fibrosis	
Camels	Urolithiasis (and P excess)	
Cats	Increased bone mineral content, serum calcium concentrations increased (total and ionized), reduced food intake, growth retardation,	
Dogs	Angular deformities, bone increased-mineralization, growth impaired, osteochondrosis, skeletal abnormalities, stunted growth, tooth loss	
Elephants	Immunosuppression, skin lesions (hyperkeratosis)	
Guinea pigs	Soft tissue calcification	And P toxicity
Marsupials	*Wombats*: osteodystrophy, soft tissue calcification	Associated with hypervitaminosis D:

Pigs	Hemorrhage (young pigs—due to interference with vitamin K), parakeratosis	
Poultry (growing)	Food intake reduced, death, gout (visceral), hypercalcemia, hypophosphatemia, nephrosis, sexual maturity delayed, urate urolithiasis (calcium urate), weight gain poor	Scott et al. (1982)
Poultry (laying)	Conflicting studies about effects on egg laying	
Reptiles	Abdominal pain, cardiac disorders, fatigue, muscle weakness, nausea, acute pancreatitis, renal hypertension, renal urolithiasis	

Carcinogenesis, mutagenicity, embryotoxicity

Evidence is conflicting but there is some evidence that Ca supplementation can be carcinogenic (Kasprzak and Waalkes, 1986).
Calcium hypochlorite IARC: Not classifiable as to carcinogenicity to humans.

Diagnosis

Total blood calcium is not helpful as an indicator dietary intake (except rabbits) as blood Ca is maintained within narrow limits, and also depends on other factors such as protein content. Ionized Ca is a more useful measurement.

Prevention

Avoid feeding rations low in Ca content especially all meat rations.
Avoid feeding rations high in Ca or with an abnormal Ca:P ratio.
Avoid feeding rations low in Ca and high in components that might inhibit Ca bioavailability such as P, dietary fiber, oxalate, and phytate.

Treatment

Ensure adequate dietary intake of Ca correctly balanced to other minerals especially P and Zn.
Treat toxicity symptomatically.
(*see Appendix 1 and 2*).

References

ATSDR Calcium (2011) Calcium hypochlorite, NRC (2006) Dogs and Cats, Weaver and Heaney (1999), McDowell (1992b), Kasprzak and Waalkes (1986).

Cerium (Ce)

Ce is a rare earth metal and, as cerium (ceric) oxide (CeO2), uses include: a polishing agent for mirrors, plate glass, television tubes, ophthalmic lenses, and precision optics, to prevent solarization and discoloration in glass in TV screens, cinematography, photoengraving, and to control emissions from gasoline and diesel engines.

Clinical applications

Cerium nitrate has been used as a topical treatment for burn wounds.

Essentiality

Human	No
Other species	No

Parameter	Location
Site of absorption	Poorly absorbed from gastrointestinal tract or lungs
Site(s) of excretion	Mainly fecal via bile. From lungs by ciliary clearance.
Sites of accumulation	Bone, liver, kidney, lung, spleen

Biological activity

Not reported.

Deficiency

Not reported.

Toxicity

Ce has low toxicity. At risk are workers as optical lens grinders, in cinematography and photoengraving.

Species	Clinical signs	Comments
Humans	Dendriform pulmonary ossification, emphysema, endomyocardial fibrosis, itching, heat sensitivity, myocardial infarction, interstitial lung disease or pneumoconiosis, pneumonia, pulmonary fibrosis skin lesions	Industrial exposure—often carbon arc lamps
Dogs	Prothrombin levels and blood coagulation time were significantly increased (Graca et al., 1964)	
Mice	Altered hepatic levels of some cytochrome (CYP) P450, congestion of abdominal organs, death, gastroenteritis including mucosal necrosis, hepatocellular necrosis, pulmonary and tracheal hemorrhage, hepatic congestion, renal damage, splenic hypertrophy, lower white blood cell, and platelet counts, reticulocytes reduced. CD3 and CD8 lymphocytes decreased, IgM decreased. Liver fatty degeneration, alanine aminotransferase high, increased liver metallothionein and glutathione	
Rats	Ataxia, death, dyspnoea, lethargy, prostration, Neutrophilia, lymphoid hyperplasia *Post-IV injection:* increased ALT, AST, and SDH fatty liver, fatty degeneration, and necrosis, damage to mitochondria, and invaginations of the nuclear membrane in hepatocytes, decrease cell viability, increased lactate dehydrogenase, total protein, and alkaline phosphatase. leukocytosis, percentage neutrophils, and concentrations of proinflammatory cytokines elevated, liver damage, decreased liver weight, dose-dependent hydropic degeneration, hepatocyte enlargement, sinusoidal dilation, and nuclear enlargement. Low albumin, the sodium-potassium ratio, and triglyceride levels were decreased Nalabotu et al. (2011)	Inhalation challenge

Carcinogenesis, mutagenicity, embryotoxicity

There is "inadequate information to assess the carcinogenic potential" of cerium in humans (IRIS) and the data is insufficient to ascertain the mutagenicity of cerium compounds.

Diagnosis

Known exposure.

Prevention

Avoid exposure, especially in the workplace.

Treatment

No specific treatment. Treat symptomatically and avoid further exposure.
(*See Appendix 1 and 2*).

References

Damian (2014), Nalabotu et al. (2011), Srinivas et al. (2011), EPA (2009a), IRIS (2009), Gómez-Aracena et al. (2006), Eapen (1998), Kutty et al. (1996), Valiathan et al. (1989).
Eapen, J., 1998. Elevated levels of cerium in tubers from regions endemic for endomyocardial fibrosis (EMF). Bull. Environ. Contam. Toxicol. 60, 168–170. https://doi.org/10.1007/s001289900606.

Chlorine (Cl)

As chloride ion Cl is an essential electrolyte and it is a very important industrial chemical used in the manufacture of thousands of products including bleach and cleansers. It is used in water purification.

Clinical applications

Cl-based products are used in the manufacture of medicines and in hospital and other medical premises as an antibacterial agent (hypochlorite).

Essentiality

Human	Yes
Other species	Yes—all

Parameter	Location
Site of absorption	Proximal small intestinal (very efficient)
Site(s) of excretion	Kidneys (95%), gastrointestinal tract, sweat
Sites of metabolism	88% is extracellular
Interaction with other minerals	Not reported

Biological activity

Biologically available as chloride salts—notably sodium chloride. Cl is an essential electrolyte responsible for maintaining acid/base balance, nerve impulse transmission, and regulating fluid balance.

Deficiency

Cl deficiency is usually associated with NaCl (salt) deficiency—see Sodium.

Species	Clinical signs
Humans	Behavioral problems dehydration, dyspnoea, diarrhea, fatigue, growth impairment, hypertension, language deficits, metabolic alkalosis, vomiting, weakness
Cats	Hypochloridemia, hypokalemia
Cattle	Anorexia, cardiovascular depression, constipation, emaciation, reduced food intake, lethargy, milk yield reduced, pica, polydipsia (mild), polyuria (mild), eye defects (severe) and reduced respiration rates, blood and mucus in feces, reduced water intake, weight loss, alkalosis (severe), hypochloremia, secondary hypokalemia, hyponatremia, and uremia.
Dogs	Ataxia, growth impairment, hypochloremia, hypokalemia, metabolic acidosis, weight gain ceased, weakness
Poultry	Dehydration, poor growth rate, high mortality rate, hypochloridemia, tetany
Rats	Reduced appetite, reduced growth, renal lesions

Toxicity

Chlorine gas is highly irritant to airways and respiratory tract, chlorinated liquid (hypochlorite) is highly toxic (caustic) to the gastrointestinal tract if ingested.
Manifest as hyperchloridemia which is caused by:

- Loss of body fluids from prolonged vomiting, diarrhea, sweating or high fever causing sweating

- High blood sodium
- Renal failure, or renal disease
- Diabetes insipidus or diabetic coma
- Medications including: androgens, corticosteroids, estrogens, some diuretics

Species	Clinical signs
Humans	Dehydration, diarrhea, vomiting, hyperchloridemia *NaCl toxicity:* appetite loss, cerebral hemorrhage, confusion, jitteriness; muscle twitches, nausea, seizures, coma. Death, thirst, weakness *Inhalation chlorine gas*: airway irritation (nose, throat), chest pain, cough, death, eye irritation, immediate chest pain, pneumonitis (toxic), pulmonary edema, tachypnoea, vomiting *Ingested hypochlorite* (*caustic*): death, esophagitis, severe gastrointestinal damage *Dermal contact hypochlorite:* irritation and inflammation

Carcinogenesis, mutagenicity, embryotoxicity

The International Agency for the Research of Cancer (IARC) has concluded there is insufficient evidence for the carcinogenicity of chlorine in drinking water in animals and humans. It has classified chlorine as neither a carcinogen nor a possible carcinogen.

Diagnosis

History of exposure and clinical signs.

Prevention

Avoid exposure especially in the workplace and home – prevent children and animas from accessing toxic (caustic) chlorine-based household/agricultural products such as hypochlorite cleaning agents.

Treatment

Treat symptomatically and avoid further exposure.
(*see Appendix 1 and 2*).

References

ATSDR Chlorine (2010), McDowell (1992c), Malloy et al. (1991), Felder et al. (1987).

Chromium (Cr)

Cr used in manufacture of cars, glass, pottery, and linoleum.

Clinical applications

Cr supplementation (as chromium tripicolinate) has been successfully used in dogs with diabetes to reduce blood glucose (Spears et al., 1998; Lindemann et al., 2000).
Cr supplementation reduced glucose concentrations and improved glucose tolerance in normal-weight cats in one study (Appleton et al., 2002) but failed to affect blood glucose in obese and normal cats in another (Cohn et al., 1999).

Essentiality

Humans	Yes in US; No in the EU
Other species Cattle	Yes
Horses	Yes
Pigs	Yes
Poultry	Yes
Rats	Yes
Sheep	Yes

Parameter	Location
Site of absorption	Poorly absorbed from the gastrointestinal tract, jejunum main site
Transportation	In blood bound to β-globulin and to tissues bound to transferrin
Site(s) of excretion	Mainly in urine; also in milk, bile, and sweat

Biological activity

Cr is part of an oligopeptide—chromodulinpotentiating insulin binding to receptors at the cell surface and so potentiates insulin activity and is essential for normal glucose tolerance. It is involved in protein synthesis, nucleic acid metabolism, and lipid metabolism. Cr is also a component of GTF an organic complex which has many biological actions.

Deficiency

Species	Clinical signs	Comments
Humans	Appetite increase, body fat increased, corneal lesions, encephalopathy, fertility reduced, glucose tolerance impaired, glycosuria, growth impairment, hypercholesterolemia, hyperglycemia, hypoglycemia, hyperinsulinemia, hypertriglyceridemia, insulin-binding reduced, insulin receptor numbers reduced increased intraocular pressure, lean muscle mass reduced, neuropathy (peripheral), reduced nitrogen balance, reduced respiratory quotient, sperm count reduced	Anderson (1987a,b)
Cattle	Humoral immune response decreased, hypercholesterolemia, morbidity increased	
Guinea pigs	Glucose intolerance	
Mice	Aortic plaques, appetite increase, Glucose intolerance, growth disorders, hypercholesterolemia, longevity reduced	
Nonhuman primates	Corneal lesions, Glucose intolerance	

Pigs	Body fat increased, hypercholesterolemia, hyperinsulinemia, reduced lean muscle mass	
Poultry: Turkeys	Growth disorders	
Rabbits	Aortic plaques	
Rats	Aortic plaques, appetite increase, corneal lesions, fertility reduced, glucose intolerance, glycosuria, growth disorders, hypercholesterolemia, hyperinsulinemia, lean muscle mass reduced, longevity reduced, sperm count reduced	

Toxicity

Exposure to Cr can occur from food, air and water and in humans occupational exposure to Cr VII is high during chromate production, welding, chrome pigment manufacture, chrome plating and spray painting. Exposure to other forms of Cr during mining, ferrochromium and steel production, welding, cutting, and grinding of chromium alloys.

Species	Clinical signs	Comments
Humans	Acidosis, allergic dermatitis (irritation, ulcers), asthma, death, diarrhea (bloody), electrolyte imbalance, liver lesions (necrosis), lung cancer (bronchiogenic carcinoma), myocardial lesions, proteinuria, renal disease (tubular necrosis), reduced respiratory function, shock, skin ulcers, increased urine ß-glucuronidase, vomiting (bloody)	Industrial exposure
Amphibians	Ovarian dysfunction, estradiol and progesterone levels depressed; follicle stimulating hormone (FSH) and luteinizing hormone levels significantly elevated.	
Cats	*Inhalation:* Bronchitis, pneumonia	
Cattle	Inflammation and ulceration of abomasum, stomach and rumen	
Hamsters	*Fetus:* cleft palate, increased resorption rate, growth retardation, hydrocephalus, skeletal ossification impaired, *Maternal:* death, renal tubular necrosis, reduced weight gain	
Mice	Death (post inhalation), fetal abnormalities (anencephaly, exencephaly) (*after intraperitoneal injection,*) granuloma formation (*inhalation*), growth impaired, kidney damage, liver damage, nasal papillomas, nasal septum perforation, pulmonary adenomas, sarcomas at intramuscular implant site, *Intraperitoneal injection*: reduced litter size, reduced sperm count, abnormal sperm	
Mongoose	Impaired reproductive performance	Andleeb et al. (2019)
Poultry	*Egg embryos:* short and twisted limbs, microphthalmia, exencephaly, short and twisted neck, everted viscera, edema and reduced body size	
Rats	Bronchial squamous metaplasia, growth impaired, kidney damage (proximal convoluted tubules, increases in urinary β-glucuronidase, lysozyme, glucose, and protein), liver damage, lung tumors (implantation) Sarcomas (rhabdomyosarcomas, spindle-cell sarcomas, and fibrosarcomas *at subcutaneous or intramuscular injection site*), nasal septum necrosis and perforation	

Carcinogenesis, mutagenicity, embryotoxicity

According to the IARC there is sufficient evidence in humans for the carcinogenicity of chromium(VI) (Group 1) and there is sufficient evidence for the carcinogenicity of some chromium compounds in animals.
In humans, chromosomal aberrations and sister chromatid exchanges have been found in workers exposed to chromium(VI) compounds and chromium VI compounds show embryotoxicity or fetotoxicity and teratogenicity in mice and hamsters. Based on in vitro testing using the embryonic stem cell test (EST) Cr demonstrated strong embryotoxicity (Imai and Nakamura, 2006).

Diagnosis

Measurement of Cr in body or environmental samples.

Prevention

Ensure adequate dietary intake, avoid exposure to large concentration.

Treatment

Deficiency: Cr supplements or (better) change to a balanced diet.
Toxicity: treat symptomatically.

References

Andleeb et al. (2019), Pechova and Pavlata (2007), Imai and Nakamura (2006), Appleton et al. (2002), Lindemann et al. (2000), Cohn et al. (1999), Spears et al. (1998), Anderson (1997, 1987), McDowell (1992p), Léonard and Lauwerys (1980).

Cobalt (Co)

Co used in the manufacture of jet engines.

Clinical applications

Cobalt supplementation has been used in humans to stimulate erythropoiesis in patients with nephritis or infection-associated anemias. Cobalt supplementation has been found to help prevent Phalaris staggers (in Australia) which is due to alkaloids in the pasture.

Essentiality

Co is essential in the chemical form of cobalamin (vitamin B12) which is produced by microbes in the intestinal tract—especially in ruminants as cobalamin does not occur in plants. Monogastric animals cannot synthesize enough of it from Co in the diet as their gastrointestinal microbiome has limited capability to synthesize cobalamin, so they require dietary intake. All-plant rations (vegetarian/vegan) contain almost no cobalamin.
Rabbits need to perform coprophagy to obtain their vitamin B12 requirement.

Human	Yes
Other species	Yes—all animal species

Function	Site(s)	Comments
Site(s) of absorption	Small intestine, lungs (inhaled)	Cobalamin requires the presence of hydrochloric acid and intrinsic factor to be absorbed
Site(s) of excretion	Urine mainly, also feces	
Site(s) of accumulation	Liver, kidneys, heart, pancreas, skeleton, and muscle	

Biological activity

Cobalamin stimulates erythropoiesis by increasing the renal synthesis of erythropoietin. Low cobalt intake (as well as low I) has been linked to goiter.
High cobalt intake may inhibit Fe absorption.
Cobalamin is important in several enzyme systems and for metabolic processes including carbohydrate, fat, nucleic acids, purines, and proteins. It promotes red blood cell synthesis and neurological function.
In vitamin B12 (cobalamin) deficiency methylmalonic acid increases in urine.

Deficiency

Lack of intrinsic factor will reduce absorption of cobalamin in the gastrointestinal lumen. Vegetarians and vegans are at high risk.

Species	Clinical signs
Humans	anemia (megaloblastic/pernicious anemia), goiter, neurological lesions
Buffaloes	Unthriftiness
Cattle	Anemia, emaciation, listlessness, pallor, muscle wastage (marasmus), reduced appetite, poor growth, rough haircoat, loss of hair color, unthriftiness, thickening of the skin, body weight loss, increased susceptibility to infections/parasites, death, reduced milk production, weak calves, poor survivability, low blood glucose, low red cell count, low hemoglobin, impaired neutrophil activity, fatty liver, hemosiderin in spleen, bone marrow hypoplasia *Calves*: poor appetite, poor growth, poor condition, muscle weakness, demyelination of peripheral nerves

Deer	Death, emaciation
Horses	Not reported—even when on pastures so low in Co that ruminants die
Pigs	Reduced body weight gain, reduced litter size, poor survivability, mild anemia (uncommon)
Poultry	Reduced hatchability, embryonic death
Rabbits	Not reported
Rats	Reduce thyroid hormone synthesis
Sheep	Anemia (normocytic, normochromic), low body condition score, death, depression, emaciation, lachrymation, lethargy, listlessness, muscle wastage (marasmus), reduced appetite, poor growth, head pressing, increased susceptibility to parasites, neonatal deaths, pallor, reduced milk production, poor survivability, low red cell count, low hemoglobin, fatty liver, poor fertility, hemosiderin in spleen, bone marrow hypoplasia, hypocobalminemia (lambs), impaired immune response, low liver B12 (lambs), tear staining, unthriftiness, poor response to vaccination, aimless wandering, weak lambs, body weight loss

Toxicity

Species	Clinical signs
Humans	Anorexia, cardiac disease, high cardiac enzymes, cyanosis, death, ECG changes (low voltage), flushing, goiter, heart failure, hypotension, kidney disease, high lactic acid, lethargy, nausea, nerve damage, pericardial effusion, respiratory disease, tinnitus, vasodilation, weight loss Following industrial exposure: cardiomyopathy Beer drinkers (possibly with other nutritional problems, e.g., thiamine deficiency): cardiomyopathy colloid depletion, polycythemia, thyroid hyperplasia *Children*: congestive heart failure, myxedema, thyroid hyperplasia
Cattle	Anaemia,reduced appetite, reduced body weight gain, death, polycythemia, polyuria, salivation, shortness of breath, increased hemoglobin, increased packed cell volume
Dogs	Cardiomyopathy
Goats	Anemia, impaired growth, hepatic fatty degeneration, unthriftiness, weight loss
Mice	Hypothermia
Rats	(*Experimental*) polycythemia, bone marrow hyperplasia, reticulocytosis, increased blood volume, reduced myocardial hypoxic contracture, death LD50 200–500 mg/kg body weight.
Sheep	anemia, reduced appetite, death, growth impairment, fatty degeneration of liver, unthriftiness, reduced body weight gain, weight loss

Carcinogenesis, mutagenicity, embryotoxicity

Co metal is genotoxic in vitro. Co metal dust is linked to an increased lung cancer risk, and is proven to be genotoxic in vitro and in vivo.

Diagnosis

Deficiency: based on clinical signs and inappropriate diet.
Toxicity: Based on known exposure and body tissue or environmental sample measurement of Co levels.

Prevention

Avoid exposure.

Treatment

Deficiency: Ensure adequate dietary intake
Toxicity: Treat symptomatically and avoid further exposure.
(*See Appendix 1 and 2*).

References

Packer (2016), De Boeck et al. (2003), McDowell (1992i), Smith (1987).

Copper (Cu)

Cu used in the manufacture of electrical wires, in motors, in construction (plumbing and roofing) and in industrial machinery.

Clinical applications

Copper injections are routinely given to newborn piglets because they have little stored Cu at birth and are very susceptible to develop Cu deficiency.

Essentiality

Human	Yes
Other species	Yes—all animal species

Biological activity

Cu is important for:

- Bone formation
- Connective tissue formation and maintenance
- Mobilization of iron stores
- Mitochondria preservation
- Cardiac function
- Spinal cord myelination
- Keratinization
- Pigmentation (melanin formation)
- Hemoglobin synthesis (with Fe)
- Red cell maturation and life span
- Lipid metabolism
- Superoxide detoxification

Copper is a component of several metalloenzymes of importance in biochemical pathways:

Enzyme system	Biological Activity
Ceruloplasmin (ferrous oxidase)	Hematopoiesis
Monoamine oxidase	Achromotrichia due lack of tyrosinase (polyphenol oxidase)—tyrosine needed for melanin synthesis. Neurotransmitters and neuropeptides
Lysyl oxidase	Important for connective tissue integrity and collagen cross-linking—cardiorespiratory and bone. Cu deficiency causes aortic aneurism and rupture.
Cytochrome oxidase	Protection from free radicals following cellular respiration Important for myelin (phospholipid) formation. Deficiency causes ataxia (swayback) in lambs
Superoxide dismutase (a Cu, Mn, Zn) dependant enzyme	Reduced immune function Impaired cardiac function
Dopamine-beta-hydroxylase	Deficiency associated with neurological signs
Tyrosinase	Melanin synthesis

Parameter	Location
Site of absorption	Most sections of gastrointestinal tract including stomach (humans) and large intestine (sheep)—but mainly from the small intestine. Involves metallothionein Active and passive processes involved
Site(s) of excretion	Bile and Feces
Transportation	Bound to ceruloplasmin—mainly Bound to albumin Bound to amino acids
Sites of metabolism	Liver
Factors blocking bioavailability	Fiber, phytate, Ca, Fe, S, Mo—high protein (ruminants), ascorbic acid
Nutritional sources	Meats, especially liver, plants
Interferences with other nutrients	Ag, Cd, Fe, Mo, Pb, S, Se, Zn

Cu affects Fe absorption and use.
Cu deficiency also reduces neurotransmitter concentrations including norepinephrine and dopamine which can cause neurological signs.

Deficiency

Pastures may be deficient in copper for example in a survey of 1106 New Zealand pastures (Knowles and Grace, 2014) 75% would not meet the needs of cattle.
In humans with Wilson's disease and in Bedlington Terriers with inherited copper toxicosis they lack ceruloplasmin transfer protein and Cu accumulates in the liver.
Copper oxide (CuO) is not bioavailable from foods as a source of Cu for dogs, cats, pigs, or poultry.
High S in the absence of Mo in the diet can cause Cu deficiency due to formation of insoluble CuS in the gastrointestinal tract. Cupric thiomolybdate also reduces Cu absorption. High Zn content can cause Cu deficiency.

Species	Clinical signs	Comments
Humans	Low plasma enkephalins, increased pancreatic leucine encephalin-containing peptides, reduced free leucine and methionine enkephalin-containing peptides *Infants:* anemia, impaired growth, neutropenia, osteoporosis, depigmentation, neurological disturbances, seizures *Adults*: anemia, anorexia, bone abnormalities, brain degeneration, cardiovascular disturbances, hypotonia, leukopenia, loss of muscle tone, sagging facial features, ? ischemic heart disease, ? atherosclerosis Menkes disease is a hereditary alters Cu distribution resulting in developmental problems, brain degeneration, intellectual disability, steely kinky hair, seizures (https://rarediseases.org/rare-diseases/menkes-disease/)	*See Menkes syndrome*
Birds (Captive)	aortic rupture (? ratites), convulsions, pale colored feathers, muscle weakness, polydipsia, weakness	
Buffaloes	Anemia, anestrus, ataxia, poor body condition score, leukoderma, increased malondialdehyde (MDA) increased nitric oxide (NO), low serum progesterone level, unthriftiness (calves), vitiligo; **decreased:** catalase (CAT), ascorbic acid (ASCA), superoxide dismutase (SOD), glutathione (GSH-R), total antioxidant capacity (TAC); zinc (Zn), iron (Fe), copper (Cu), ceruloplasmin	

Camels	*Young calves (4–6 months):* ataxia, angular deformity, bone demineralization, poor growth, limb incoordination(forelegs), *Adult camels:* anemia, loss of pigment in hair, rough haircoat, infertility (temporary), reduced milk production, weakness	
Cats	Achromotrichia, anemia, abortion, ataxia, congenital deformities (kinked tails, carpi), cannibalism, poor keratinization, poor reproductive performance (increased time to conception), poor weight gain	
Cattle	Anemia, ataxia (neonates), hair coat thin and sparse, diarrhea (severe—especially if high Mo), depressed growth, hair pigmentation loss (if around eyes looks like spectacles), infertility (delayed/depressed estrus, reduced conception rate, retained placenta), low hemoglobin, lameness, reduced milk production,), heart failure, myocardial fibrosis, neutrophil inhibition, congenital skeletal abnormalities (rickets, weak fragile bones, spontaneous fractures, bone swellings above joints),	
Deer	Ataxia, reduced conception rate, death, diarrhea, emaciation, impaired growth, hair depigmentation, hemosiderosis (liver), osteochondrosis, unthriftiness, weight loss	
Dogs	Achromotrichia, anemia, poor keratinization, hyperextension of the phalanges, *Unpublished study on sudden death in grayhounds: Death, major blood vessel rupture*	
Goats	Anemia, articular cartilage deformities, ataxia (neonates), cardiovascular defects, diarrhea (severe—especially if high Mo), fractures, hair depigmentation, impaired immune response, increased infection risk, infertility, osteoporosis, skeletal abnormalities, skin depigmentation	
Guinea pigs	Achromotrichia, cardiovascular defects, death, impaired endochondral ossification, growth retardation, hemorrhage due aortic aneurysms, CNS lesions (agenesis of cerebellar folia, cerebral edema, delayed myelination) fetal death, fetal resorption, low liver Cu content	
Horses	Osteochondrosis (Cu deficient grazing or secondary to high Zn) osteodysgenesis, hemorrhage at parturition (due to ruptured uterine artery)	Bridges et al. (1984), Bridges and Harris (1988), Bridges and Ghagan Moffitt (1990), Glade and Belling (1986), Knight et al. (1990), Pearce et al. (1998)
Marine mammals	*Dugong in Australia:*	
Mice	Anemia, impaired humoral immune response (antibody production)	
Pigs	(Uncommon) anemia (microcytic hypochromic),ataxia, blood vessel rupture, low mean corpuscular volume, low hemoglobin, lack of limb rigidity (hyperflexion hocks), crooked forelegs, reduced osteoblast activity, impaired endochondral ossification (ossification of cartilage), cardiomegaly,.	
Poultry	Anemia, poor growth rate, death (chicks), internal hemorrhage due ruptured vessels, thickened aorta walls, aneurisms (chicks), cardiomegaly, lameness (chicks), fragile bones, angular deformity metatarsals, thickened epiphyseal cartilage, failure of cartilage cross-linking, enlarged hocks, poor bone mineralization, reduced egg production, reduced hatchability, retarded growth, reduced monoamine oxidase activity, low plasma Cu, shell-less eggs, abnormal egg shape, abnormal egg texture, hypopigmentation	

Rabbits	Achromotrichia, alopecia, anemia, bone abnormalities, cartilage abnormalities, dermatitis, hypopigmentation (graying) of dark hairs,	
Rats	Achromotrichia, anemia, behavioral changes, cardiac function reduced, fetal death, resorption, high cholesterol, serum triglyceride and phospholipids,	
Red Deer	Osteochondrosis	
Sheep	Abnormal wool (lack crimp, steely, or stringy) (straight) hair, abortion, achromotrichia, anemia, diarrhea (severe—especially if high Mo), ataxia (swayback lambs), spastic paralysis, stiff gait, swaying of hindquarters, paralysis, death (lambs) infertility, reduced leukocyte phagocytosis, reduced resistance to infection	
Fish	Depressed growth	

Toxicity

Species	Clinical signs	Comments
Humans	Blood changes, cirrhosis of the liver, flu-like reaction, impaired neurological function *Wilsons disease*: Copper accumulation in brain, cornea, kidney, liver	"metal fume disease"
Cats	Interference with Fe and Zn	
Dogs	Death, hepatotoxicosis, vomiting; interference with Fe and Zn	*Hepatic accumulation of Cu is an autosomal recessive disease in Bedlington, Skye and West Highland White Terriers*
Fish	Death (increased mortality rate), reduced food utilization impaired growth, increased liver enzymes activity—SGOT and SGPT	
Goats	Death, hemoglobinuria, hemolysis, jaundice, liver damage, high liver enzymes (sorbitol dehydrogenase, gamma-glutamyl transferase, glutamyl oxaloacetic transaminase)	
Marine mammals	*Manatees:* Copper accumulation in liver, death	
Marsupials	*Wombats:* Copper accumulation in liver, death, hemoglobinuria, intravascular hemolysis, hepatic congestion, jaundice	
Pigs	High Cu can destroy natural tocopherols (vitamin E)	
Sheep	Death, hemoglobinuria, hemolysis, jaundice, liver damage, high liver enzymes (sorbitol dehydrogenase, gamma glutamyle transferase, and glutamic oxaloacetic transaminase)	

Carcinogenesis, mutagenicity, embryotoxicity

Cu deficiency is teratogenic causing early embryonic death and cardiac, skeletal, pulmonary, vascular, and neurological defects.

In animals, Cu deficiency during pregnancy results in heart, brain, and yolk sac vascular system being most affected.

Cu toxicity is also genotoxic and mutagenic in mice (Prá et al., 2008).

Diagnosis

Cu content of foods can be measured as can Cu in blood and tissue samples. As an example in Deer liver Cu <4 μg/g wet weight (ww) and serum concentrations <0.3 μg/mL are considered to represent Cu deficiency. These values will vary between species and diagnostic laboratory performing the analysis.

Prevention

Deficiency: Ensure adequate amounts of available Cu are in the ration, and avoid high intake of foods that may interfere with Cu bioavailability, e.g., fiber, phytate, Ca, Fe, S, Mo—high protein (ruminants), ascorbic acid.
Toxicity: Avoid high dietary intake of Cu especially in patients predisposed to develop Cu toxicosis and liver disease.

Treatment

Deficiency: Ensure adequate dietary intake of Cu with a balanced ration or Cu supplementation, and ensure low levels of factors that might reduce availability, e.g., fiber, phytate, Ca, Fe, S, Mo—high protein (ruminants), ascorbic acid.
Toxicity: Avoid high intake of Cu.
(*See Appendix 1 and 2*).

References

NORD (2020a), Handeland et al. (2008), Handeland and Bernhoft (2004), Handeland and Flåøyen, 2000, Prá et al. (2008), NRC (2006), Beckers-Trapp et al. (2006), Vikøren et al. (2005), Penland and Prohaska (2004), Hawk et al. (2003, 1998), Keen et al. (2003), Bernhoft et al. (2002), Wilson and Grace (2001), Klevay (2000), Waggoner et al. (1999), Pearce et al. (1998), Fyffe (1996), Thompson et al. (1994), McDowell (1992h), Bridges and Ghagan Moffitt (1990), Knight et al. (1990), Mackintosh et al. (1989), Wilson (1989), Bridges and Harris (1988), Davis and Mertz (1987), Knox et al. (1987), Clark and Hepburn (1986), Glade and Belling (1986), Bridges et al. (1984), Wilson et al. (1979), Barlow and Butler (1964).

Dysprosium (Dy)

Dy is a rare earth element. It is used to make control rods in nuclear reactors, neodymium-iron-boron high strength permanent magnets, laser materials and commercial lighting. It is also in alloys for neodymium-based magnets which are used in electric vehicles.

Clinical applications

The radioactive isotope dysprosium-165-FHMA has been used in the treatment of rheumatoid knee effusions. As Dy is poorly absorbed it has been used as a fecal marker to assess digestibility, e.g., Zn.

Essentiality

Human	No
Other species	No

Parameter	Location
Site of absorption	Poorly absorbed from the gastrointestinal tract
Sites of accumulation	Bone

Biological activity

Not reported.

Deficiency

Not reported.

Toxicity

Dy insoluble salts are considered not toxic, soluble salts are mildly toxic.

Species	Clinical signs	Comments
Guinea pigs	Epilation, nodule formation	Dy Injected intradermally
Mice	Tissue fibrosis, calcification ataxia, labored breathing, lacrimation stretching of limbs while walking, writhing	Dy injected intradermally
Rabbits	Conjunctivitis, dermatitis, reduced ileal contractility, severe irritation to abraded skin	Application of $DyCl_3$
Rat	Loss urine concentrating ability, renal vascular resistance increased	

Carcinogenesis, mutagenicity, embryotoxicity

Dy is not considered to be carcinogenic by the IARC https://www.espimetals.com/index.php/msds/551-dysprosium-oxide

Diagnosis

History of exposure.

Prevention

Avoid exposure.

Treatment

Treat symptomatically, avoid further exposure.
(*See Appendix 1 and 2*).

References

Damian (2014), Hirano and Suzuki (1996).

Erbium (Er)

Er is a rare earth element used as a coloring agent in glazes and glass, in alloys (often with V), in amplifiers and lasers.

Clinical applications

A large variety of medical applications (i.e., dermatology, dentistry) rely on the erbium ion's 2940 nm emission from Er-yttrium aluminum garnet (YAG) medical lasers: skin resurfacing, acne scars, mole or tattoo removal and warts.

Essentiality

Human	No
Other species	No
Parameter	**Location**
Site of absorption	NR
Site(s) of excretion	NR
Sites of accumulation	Bones mainly, liver, kidneys

Biological activity

None reported, however, Er does stimulate some metabolic processes.

Deficiency

None reported.

Toxicity

Er is regarded as having low toxicity.

Species	Clinical signs	Comments
Guinea pigs	Nodules	At intradermal injection site
Mice	Local calcification, fibrosis, multinucleated giant cell accumulation	At subcutaneous injection sites
	ataxia, labored breathing, stretching of limbs while walking, lachrymation, writhing	Acute toxicity study
Rabbits	Conjunctivitis, severe irritation on abraded skin	

Carcinogenesis, mutagenicity, embryotoxicity

Er is not considered to be carcinogenic, mutagenic, or embryotoxic.

Diagnosis

Known exposure to toxic levels.

Prevention

Avoid unnecessary exposure.

Treatment

Treat toxicity symptomatically, prevent further exposure.
(*See Appendix 1 and 2*).

References

Damian (2014), Hirano and Suzuki (1996).

Europium (Eu)

Eu is used in controlling rods in nuclear reactors and in superconducting alloys.

Clinical applications

Eu is tagged to complex biochemicals for live tracing living tissue research.

Essentiality

Human	No
Other species	No

Parameter	Location
Site of absorption	Poorly absorbed from gastrointestinal tract
Site(s) of excretion	NR
Sites of accumulation	Liver mainly, skeleton

Biological activity

None reported.

Deficiency

Not reported.

Toxicity

Species	Clinical signs	Comments
Humans	Abdominal pain, eye irritation, nausea, extensive scarring on abraded skin	
Guinea pigs	nodules	At intradermal injection site
Mice	Arched back, ataxia, death, epiphora, labored breathing, neurotoxicity, stretching of the hind limbs while walking, writhing *Neurological signs:* reduced motor activity, disorientation, loss of motor coordination, rapid breathing, tachycardia, and infrequent tremor	Acute toxicity study
Rabbit	Conjunctivitis, inflammatory and fibrotic responses in the skin, abraded skin irritation	
Rats	Reduced creatinine excretion, (damage to glomerular basement membrane), decreased glomerular filtration rate (GFR). *N*-acetyl-β-D-glucosaminidase (NAG) high (adverse effect on proximal convoluted tubule)	

Carcinogenesis, mutagenicity, embryotoxicity

No evidence for Eu carcinogenicity or mutagenicity.

Diagnosis

Toxicity: Evidence of exposure.

Prevention

Avoid exposure especially in the workplace.

Treatment

Treat symptomatically and avoid further exposure.

References

Destefani et al. (2016), Roth (2016), Ohnishi et al. (2011).

Fluorine (F)

F is a halogen gas that combines readily with metals to form fluorides. Its uses include: nuclear material for nuclear power plants, insulating electrical towers, to etch glass (as hydrogen fluoride), to make plastics (e.g., Teflon), and in dental health. It is also used in rodenticides and acaricides which can lead to accidental poisoning.

Clinical applications

Dental prophylaxis in human populations—water supply and dental products.
Fluorine products are widely used in the compilation of medicines to increasing bioavailability of lipophilic drugs, they are also a component of inhalation anesthetics (isoflurane, sevoflurane).
The radioactive isotope fluorine-18 emits positrons which are used in PET-scanners and Fluorine-19 is commonly used in NMR scans.

Essentiality

Human	No
Other species Goat	 ? Possibly

Parameter	Location
Site of absorption	Highly absorbed from the gastrointestinal tract (stomach and rumen)
Site(s) of excretion	Urine
Transportation	Combined with calcium in blood
Sites of storage/accumulation in the body	Bone teeth mainly, kidney
Interaction with other minerals	Combines with Ca in blood

Biological activity

Absorbed F binds with calcium and so interferes with many enzyme systems. Perfluorinated compounds (PFCs) can persist a long time in the environment.

Deficiency

Species	Clinical signs	Comments
Humans	Increased caries	On low F dietary intake—but F is not considered essential
Goats	Reduced food intake, impaired growth, joint deformity, increased mortality (kids), skeletal deformity, thymus hypoplasia	Following intrauterine F deficiency (Anke et al., 2005b)

Toxicity

Volcanic ash and industrial waste are sources of high mineral F in grazing pasture and cause fluorosis. Chronic toxicity (fluorosis) is common in humans and animals in some parts of the world, e.g., Africa, Australia, and India. In gas form F is highly toxic.
Human intoxication can be from air, food, or water.
If fluorine is absorbed too frequently, it can cause teeth decay, osteoporosis and harm to kidneys, bones, nerves, and muscles. Fluorine gas is released by industries which are very dangerous, as it can cause death. At low concentrations, it causes eye and nose irritations.

Species	Clinical signs
Humans	*Acute:* Death *Chronic:* Bone disease with exostosis, dental disease —mottling teeth and white patches on enamel (*children*), stained teeth (yellow/black), hypermineralization of skeleton, skeletal deformities, ligament and tendon calcification, spinal ankylosis *Called: "Darmous" in Africa* *HF acid:* hypocalcemia, cardiac arrhythmias, severe pain, skin burns, death *F Gas:* severe irritation to eyes, nose, lungs later damage to liver and kidneys, death
Amphibians	*Frogs:* Metabolic bone disease, poor bone mineralization
Cattle	Acetonemia (adults), anemia, chewing (constant), collapse, death (neonatal), dental mottling with pits and streaked with pigmentation (yellow, green, brown, black) feed intake decreased, fertility reduced, gastroenteritis, hyperesthesia, impaired growth, joint thickening and ankylosis, lameness, milk production decreased, muscle tremors, periosteal hyperostosis, pupillary dilation, ruminal statis, stiffness, renal tubular fibrosis and degeneration, renal mineralization, tetany, weakness, weight loss
Dogs	Reduced growth rate
Goat	Increased food intake, increased intrauterine and postnatal mortality, growth retarded, joint deformities (older), skeletal deformities (older). Thymus hypoplasia (fetus)
Horse	Bones thick, rough and irregular surfaced, dental abscesses, food intake reduced, hair coat dry and rough, hypersalivation, lameness, reluctant to move, shifting body weight, skin lost normal elasticity, swollen jaw (due dental abscesses), teeth discoloration, teeth wear excessively, teeth traumatize the cheek/gums, exposure of dental pulp cavity, unthriftiness
Pigs	Anorexia, bones soften and overgrow, death, dental erosion and excess wear, kidney damage, lactation reduced, thirst reduced due oral pain, poor weight gain
Poultry	Reduced growth rate and hatchability
Sheep	Death (neonatal). Fertility reduced, impaired growth, joint thickening and ankylosis, lameness, mandibular hyperostosis, long bone periosteal hyperostosis, stiffness
Fish	Reduced hatching time

Treatment

F poisoning: vigorous washing of skin, administer calcium gluconate gel and IV. In severe F burns cases surgical excision or amputation may be needed.

Carcinogenesis, mutagenicity, embryotoxicity

IARC: Fluoride and sodium fluoride: not classifiable as to its carcinogenicity to humans.

Diagnosis

Known exposure.

Prevention

Avoid exposure especially in the workplace.

Treatment

Treat symptomatically and avoid further exposure.
(*See Appendix 1 and 2*).

References

ATSDR Fluorides (2011), Ulemale et al. (2010), Anke et al. (2005b), McDonagh et al. (2000), Shupe et al. (1992), McDowell (1992n), Krishnamachari (1987).

Gadolinium (Gd)

Gd is a rare-earth metal used for making magnets, electronic components, and data storage disks.

Clinical applications

Gd compounds are components of magnetic resonance imaging (MRI) contrast agent. Serious disease has resulted in some patients following use of these agents.

Essentiality

Human	No
Other species	No

Parameter	Location
Site of absorption	Poorly absorbed from the gastrointestinal tract
Site(s) of excretion	Feces and urine (50% each)
Sites of accumulation	Liver

Biological activity

Not reported.

Deficiency

Not reported.

Toxicity

Species	Clinical signs	Comments
Humans	Allergic reactions, dizziness, dysgeusia, flushing, headaches, hypertension, inflammation, injection site pain, itchiness, nausea, nephrogenic systemic fibrosis (NSF*), parosmia, pruritus, respiratory signs, skin rash, vomiting *Contact exposure:* Conjunctivitis, skin ulcers on abraded skin	After injection for MRI *NSF, after Gd-based medical imaging contrast agents in patients with severe or end-stage kidney disease
Mice	Pneumonia was increased in most of the exposed groups but was not related to the duration of exposure Calcification of the lung	
Rabbits	Severe irritation to abraded skin	
Rats	Abdominal cramps, diarrhea,labored breathing, lethargy, increases in the white blood cell count (WBC), alanine aminotransferase (ALT), aspartate aminotransferase (AST), lactate dehydrogenase (LDH), prothrombin time (PTT), cholesterol and triglycerides, and decreases in platelet numbers, albumin and blood glucose *Post-mortem:* mineralization (especially lung and kidney, in mononuclear phagocytes), necrosis of the spleen and liver, and increases in liver and spleen weight	Acute toxicity study Spencer et al. (1997) Spencer, A.J., Wilson, S.A., Batchelor, J., Reid, A., Rees, J., Harpur, E., 1997. Gadolinium chloride toxicity in the rat. Toxicol. Pathol. 25 (3), 245–255. https://doi.org/10.1177/019262339702500301.

NSF* is a syndrome involving: thickening and hyperpigmentation of the skin, although it can also affect other areas of the body. Initial symptoms include pain, pruritis (itching) swelling, erythema (redness) of the skin, transient hair loss, nausea, vomiting, diarrhea, and abdominal pain. Subsequent symptoms include stiffness of the joints, deep bone pain, and muscle weakness. The condition may ultimately result in disability or death.

Carcinogenesis, mutagenicity, embryotoxicity

In vitro: Gd demonstrated mutagenicity and cytotoxicity using human peripheral blood lymphocytes—chromosomal effects are more sensitive than micronuclei induction.
However, the human carcinogenicity of gadolinium cannot be determined.
Based on animal and in vitro studies Gadolinium products are contraindicated for use in pregnancy because they may cause harm to the fetus.

Diagnosis

Confirmed exposure.

Prevention

Avoid exposure especially in the workplace.

Treatment

Treat symptomatically, avoid further exposure.
(*See Appendix 1 and 2*).

References

Damian (2014), EPA (2007a), Yongxing et al. (2000), Hirano and Suzuki (1996).

Gallium (Ga)

Ga is a metal used in electronics, microwave circuits, high-speed switching circuits, and infrared circuits, in blue and violet light-emitting diodes (LEDs) and diode lasers, and jewelry.

Clinical applications

Dentists use Ga alloys to replace mercury-containing amalgam. Radioactive gallium and stable gallium nitrate are used as diagnostic and therapeutic agents in cancer and disorders of calcium and bone metabolism. Radiogallium scanning is used as a diagnostic tool in medicine They have antiinflammatory and immunosuppressive activity in animal models of human disease.

Essentiality

Human	No
Other species	No

Parameter	Location
Site of absorption	Gastrointestinal tract (good humans, poor horses)
Site(s) of excretion	Urine
Sites of metabolism	NR
Interaction with other minerals	NR

Biological activity

Not reported.

Deficiency

Not reported.

Toxicity

Species	Clinical signs	Comments
Humans	Anemia (microcytic), azotemia, difficulty breathing, diarrhea, chest pain, creatinine clearance reduced, electrolyte abnormalities, low hemoglobin, hypercalciuria, hypocalcemia (if initially normocalcemic), hypokalemia, metallic taste in the mouth, nausea, pulmonary edema, paralysis, pulmonary toxicity (as GaAs), renal tubular acidosis, renal toxicity, throat irritation, blood and urine malondialdehyde (MDA), elevated, vomiting	Fumes exposure most serious Given as IV infusion to treat hypocalcemia or cancer
	Acute case: initially presented as dermatitis and appeared relatively not life-threatening, rapidly progressed to dangerous episodes of tachycardia, tremors, dyspnoea, vertigo, and unexpected black-outs, irreversible cardiomyopathy	Ivanoff (2012)
Mice	Anemia (microcytic), bone marrow hyperplasia, growth suppression, hemosiderosis (liver, spleen) reduced life span, plasma cell hyperplasia (lymph nodes), sperm count and motility low, testicular atrophy and weight loss, weight loss, high platelets, neutrophils, alanine aminotransferase, sorbitol dehydrogenase	
Rats	Growth suppression, hypocalcemia, life span reduced, nephrotoxicosis *In lungs:* inflammation, necrosis, neutrophil infiltration, and fibrosis (inhalation)	

Carcinogenesis, mutagenicity, embryotoxicity

Gallium is not considered mutagenic by the European Chemicals Agency (ECA) https://echa.europa.eu/registration-dossier/-/registered-dossier/23228/7/7/1.
In female rats significantly increased incidences of alveolar/bronchiolar neoplasms, benign pheochromocytoma and mononuclear-cell leukemia were observed with gallium arsenide.
There is inadequate evidence in humans for the carcinogenicity of gallium arsenide. There is limited evidence in experimental animals for the carcinogenicity of gallium arsenide.

Diagnosis

Confirmed exposure.

Prevention

Avoid exposure especially in the workplace.

Treatment

Treat symptomatically and avoid further exposure.
(*See Appendix 1 and 2*).

References

IARC (2020), Bomhard et al. (2013), Ivanoff et al. (2012), Chitambar (2010), Nielsen (1986), Krakoff and Newman (1979).

Germanium (Ge)

Ge is present in many foods and is marketed as dietary supplements, cosmetics, accessories, and warm bath services, especially in Japan. It is also used as a semiconductor in transistors, other electronic devices, fiber-optic systems, infrared optics, solar cell applications, light-emitting diodes (LEDs), and in the manufacture of nanowires.

Clinical applications

None.

Essentiality

Human	No
Other species	No

Parameter	Location
Site of absorption	Small intestine, almost completely absorbed as GeO_2 Transdermal
Site(s) of excretion	Urine
Sites of metabolism	Not reported
Interaction with other minerals	Reduced Cr uptake in tissues Increased Cu uptake in liver

Biological activity

Not reported.

Deficiency

Not reported.

Toxicity

Ge is present in many common foods. Supplements containing Ge (sold as having health benefits) have been a cause of toxic exposure as Ge accumulates in the body.
Natural insoluble forms are not toxic but synthetic soluble Ge compounds are nephrotoxic.

Species	Clinical signs	Comments
Humans	Anemia, death, dizziness, dyspnoea, fatty liver, hemolysis, headache, muscle weakness, nausea, peripheral neuropathy, pulmonary toxicity, renal failure, renal tubular degeneration, syncope, vomiting	Vacuolar degeneration in renal tubular epithelial cells
Mice	Reduced lifespan, fatty degeneration liver	*Experimental*
Rats	Anemia, reduced erythropoiesis, reduced lifespan, fatty degeneration liver, renal dysfunction (tubular disease, reduced creatinine clearance, high BUN and P), weight loss	*Experimental*

Carcinogenesis, mutagenicity, embryotoxicity

Ge is not considered to be carcinogenic or mutagenic (Gerber and Léonard, 1997).
In animals high doses cause embryonic resorption and teratogenic abnormalities with dimethyl germanium oxide administration (Gerber and Léonard, 1997).

Diagnosis

Known exposure.

Prevention

Avoid exposure.

Treatment

Treat symptomatically.
(*See Appendix 1 and 2*).

References

NIOSH Germanium (2019), Yokoi et al. (2008), Gerber and Léonard (1997); Tao and Bolger (1997), Anger et al. (1992), Sanai et al. (1991), Schauss (1991), Nielsen (1986).

Gold (Au)

Au is a transition metal regarded as precious is used in lieu of currency and widely used in the manufacture of jewelry and other decorative items, it also has unique properties as a soft relatively inert conductive metal and is used in dentistry and medicine, in mobile phones, laptops, computers, and other electronic devices as well as in some foods and drinks.

Clinical applications

Au is used to treat rheumatoid arthritis (e.g., aurothiomalate, aurothioglucose, and auranofin) and injectable gold has been shown to be beneficial in a Cochrane Library Systematic Review Clark et al. (1997). Also: https://wikem.org/wiki/Gold_toxicity.
Au is also used in dentistry for fillings and in medicine nanoparticles can be used to deliver chemotherapy and gene therapy.

Essentiality

Human	No
Other species	No

Parameter	Site
Site of absorption	From injection sites
Site(s) of excretion	Urine mainly (after injection)
Sites of accumulation	Liver, spleen. Blood
Interaction with other minerals	NR

Biological activity

Not reported.

Deficiency

Not reported.

Toxicity

Au is approved as a food additive in the EU (E175 in the Codex Alimentarius) and is a component of some alcoholic drinks. It is considered nontoxic when ingested as metallic gold which is quite inert but soluble gold salts, e.g., chloride, are toxic to the kidneys and liver.
Overdose of Au for treatment of rheumatoid arthritis can result in toxicity.

Species	Clinical signs	Comments
Humans	Alopecia, anemia (aplastic-rare) anaphylaxis (rare) bone marrow depression, dermatitis, diarrhea, encephalopathy (peripheral/cranial—rare), enterocolitis, eosinophilia (common), gastrointestinal bleeding, glomerulonephritis (common), headaches, jaundice, liver disease, metallic taste in the mouth, myokymia, interstitial pneumonitis (rare), nephrotic syndrome (rare), pancreatitis (rare), pruritus, skin rashes, stomatitis, trophic nails, proteinuria and hematological disorders, pseudocyanosis (nonblanching blue-gray skin discoloration), urticaria, vomiting	In the review cited below (Clark et al., 1997) 22% of patients withdrew mainly because of these side effects

Carcinogenesis, mutagenicity, embryotoxicity

Au is not considered to be carcinogenic—International Agency for Research on Cancer (IARC) and the US National Toxicology Program (NTP).
Au is not considered to be mutagenic or embryotoxic although in some studies DNA breaks have been linked to Au nanoparticles (Ávalos et al., 2018) and there is a justifiable concern based on a recent review (Wang et al., 2020). If Au nanoparticles are indeed causing cancer it could be associated with the particle size rather than the element itself.

Diagnosis

Known exposure.

Prevention

Avoid exposure.

Treatment

Gold toxicity can be treated by chelating the Au using Dimercaprol, and symptomatic treatment. Avoid further exposure.

References

Wang et al. (2020), NORD (2020a), Ávalos et al. (2018): Nielsen (1986).

Holmium (Ho)

Ho is a rare earth element.
Used to create the strongest magnetic fields and is used in lasers, optical isolators, and microwave equipment, spectrophotometers and for data storage.

Clinical applications

Ho is used in the construction of some medical lasers and it is also used in some radioisotopes for targeted treatment of some cancers.

Essentiality

Human	No
Other species	No

Parameter	Location
Site of absorption	NR
Site(s) of excretion	NR
Sites of accumulation	Bone

Biological activity

Not reported but Ho salts do stimulate metabolism.

Deficiency

Not reported.

Toxicity

Exposures mainly in the workplace include contact with eyes, skin, and gastrointestinal tract if ingested. Ho compounds are caustic and can cause severe injury.

Species	Clinical signs	Comments
Humans	Abdominal (stomach) pain, eye injury, skin injury (burns), skin blisters, gastrointestinal damage (corrosive)	Occupational exposure
Mice	*Acute toxicity:* arched back, ataxia (loss of muscular coordination), labored breathing, lachrymation stretching of limbs while walking, writhing	
Rabbits	Conjunctivitis, severe irritation on abraded skin	
Rats	inflammatory and fibrotic responses in the skin	

Carcinogenesis, mutagenicity, embryotoxicity

IARC: Not identified as carcinogenic.
Ho is regarded as being safe however it has been shown to have cytotoxic and genotoxic effects (Qu et al., 2004).

Diagnosis

Known exposure especially in the workplace.

Prevention

Avoid exposure.

Treatment

Treat symptomatically, avoid further exposure.

References

Leggett (2017), Damian (2014), Qu et al. (2004).

Indium (In)

The indium tin oxide (ITO) industry is massive and In is used to make cathode-ray tubes, light-emitting diodes, electronic devices, flat panel displays, gas sensors, injection lasers, photodiodes, plasma displays, photovoltaics, semiconductors, solar cells/panels, touch panels, transparent conductive films on glass or plastic panels, energy-efficient windows.

Clinical applications

Pentetate indium disodium In 111 (radioisotope) is a diagnostic drug for intrathecal use in brain imaging, and In is being examined for possible applications in cancer management.

Essentiality

Human	No
Other species	No

Parameter	Site(s)
Site of absorption	Gastrointestinal tract, inhalation
Site(s) of excretion	Urine—not fully reported
Sites of accumulation	Spleen, liver, lungs, brain, kidney
Interaction with other minerals	NR

Biological activity

None known. In mice (given In in water supply) had lower incidence of tumor formation (Schroeder and Mitchener, 1971).

Deficiency

Not reported.

Toxicity

Species	Clinical signs
Humans	*Serious lung disease following inhalation:* alveolar proteinosis, death, emphysema, interstitial fibrosis, pulmonary fibrosis increased DNA damage in blood cells and increased urinary 8-OHdG (leukocytes and urine)
Hamsters	Alveolar epithelial hyperplasia, severe inflammation in lungs, infiltration of alveolar macrophages and neutrophils
Mice	Alveolar epithelial hyperplasia, cancer (adenoma, bronchioalveolar carcinomas, thickened pleural wall, alveolar proteinosis and alveolar macrophage infiltration
Rats	Alveolar epithelial hyperplasia, rough coat, weight loss *Inhalation:* bronchioalveolar hyperplasia, alveolar wall fibrosis and thickened pleural wall, alveolar proteinosis and infiltrations of alveolar macrophages and inflammatory cells **Carcinogenic effect** seen bronchioalveolar adenomas and carcinomas

Carcinogenesis, mutagenicity, embryotoxicity

Indium phosphide and indium tin oxide (ITO) have both been shown to be carcinogenic causing lung tumors in mice and rats. (IARC, 2018).

There is also moderate evidence that ITO is mutagenic.

IARC: There is inadequate evidence in humans for the carcinogenicity of indium tin oxide. There is sufficient evidence in experimental animals for the carcinogenicity of indium tin oxide.

Indium tin oxide is possibly carcinogenic to humans (Group 2B).

Diagnosis

History of exposure. Measure In concentrations in air or tissue samples (plasma, serum, urine).

Prevention

Avoid exposure to In especially in the workplace.

Treatment

No specific treatment, so treat symptomatically and avoid further exposure.

References

Bomhard (2018), IARC (2017), Nagano et al. (2011), Nielsen (1986).

Iodine (I)

I occurs naturally in rocks and seawater and has a wide range of uses including catalysts, disinfectants, inks and dyes, photographic chemicals, polarizing filters for LCD displays.

Clinical applications

I is a component of disinfectants including surgical scrubs and dressings, and is used in the management of thyroid disease and as positive contrast media in radiography. Radioactive isotope I-131 is used to treat thyroid tumors (carcinomas and (especially in cats) adenomas).

Essentiality

Human	Yes
Other species	Yes

Parameter	Site(s)
Site of absorption	Gastrointestinal tract after conversion into iodide ion
Site(s) of excretion	Urine
Site(s) of accumulation in body	Thyroid, kidney

Biological activity

I is a constituent of thyroid hormones triiodothyronine (T3) and thyroxine (T4).

Deficiency

In all species a deficiency of I results in thyroid enlargement called goiter.
In dogs and cats I deficiency has been reported following the feeding of an all-meat ration.

Species	Clinical signs
Humans	abortions, brain development impaired, congenital anomalies, cretinism (mental deficiency, deaf mutism, spastic diplegia, neurological deficits), goiter, hypothyroidism, stillbirths
Birds (Captive)	Circulatory collapse (due compression), crop distension, dyspnoea, hypothyroidism, goiter, regurgitation, increased expiratory respiratory noise (chirp, wheeze),
Cats	alopecia, abnormal Ca metabolism, fetal resorption, goiter, death
Cattle	abortion, alopecia, blindness, death, estrus cycling impaired,fetal development arrested, fetal membrane retention, fetal resorption, food intake reduced, gestation period prolonged, goiter, infertility (male - lack of libido, poor semen quality), milk production reduced, stillbirths, weak offspring, weakness
Deer	Goiter
Dogs	Alopecia, apathy, coat dull and dry, cretinism (mental deficiency, deaf mutism, spastic diplegia, neurological deficits), drowsiness, dullness, goiter, hairlessness, myxedema of skin, skeletal deformities, low thyroid hormone levels (T3 and T4), timidity, weight gain
Elephants	Reduced reproductive performance
Fish	Goiter, mortality increased
Goats	Delayed brain development, low conception rate, hairless newborn, goiter, reduced milk production, skin disorders, stillbirths, sterility, susceptibility to cold stress
Hamsters	Goiter, growth retardation

Horses	Goiter, abnormal estrus cycles, male infertility (lack libido and poor semen quality), stillborn foals, weak neonates, neonatal death
Mice	Goiter
Pigs	Deaths (neonatal), dwarfism, hairlessness, lethargy, reproductive failure, shortened limb bones, skin dry rough and thickened, stillbirths, thyroid enlargement, weakness
Poultry	Rare. *Experimental* poor growth, poor feather condition poor egg production, poor egg size, thyroid enlargement in embryos, small testes, lack of sperm, decreased size of comb in males, weight gain
Sheep	Goiter, poor brain development, reduced wool production, reproductive problems (abnormal estrus cycling, arrested fetal development, resorption, abortion, stillbirths, prolonged gestation, weak offspring, fetal membrane retention, male infertilitypoor semen quality)

Toxicity-

I excess causes hyperthyroidism.
Seaweed is a natural source that is very high in I content (1850 mg I/kg). In dogs feeding raw carcass material containing thyroid tissue has caused clinical hyperthyroidism.
Maximum tolerance levels for I vary between species which highlights the large difference that sometimes exist between species:

Species	Maximum tolerance
Humans	2 mg/day
Dogs	Not reported
Cats	Not reported
Other species	
Cattle	50 ppm
Sheep	50 ppm
Pigs	400 ppm
Poultry	300 ppm
Horses	5 ppm

Species	Clinical signs
Humans	Goiter, hypersalivation, skin ulcers (kelp acne), brassy taste in mouth
Amphibians	Excess iodine added to water may cause species such as the axolotl to shed its external gills
Birds	Central nervous system signs, poor growth
Cats	Hypersalivation, lachrymation, nasal discharge,
Cattle	Anorexia, cough (hacking), depression, dullness, listlessness, excessive tears, scales and sloughing of skin, difficulty swallowing, reduced humeral and cell-mediated immunity
Dogs	hyperthyroidism:, aggressiveness, panting and restlessness, tachycardia, weight loss
Horses	Alopecia, goiter (foals)
Mink	Reduced number of females whelping, reduced number of offspring, reduced birth weight, reduced viability
Pigs	Feed intake reduced, reduced hemoglobin levels, growth impaired, reduced liver Fe content
Poultry	Reduced egg production, reduced egg size, reduced hatchability
Rabbits	Death (neonates)
Sheep	Anorexia, depression, hypothermia, poor weight gain

Carcinogenesis, mutagenicity, embryotoxicity

Chronic dietary I deficiency causes thyroid follicular adenomas and carcinomas in rats and thyroid tumors have been reported in people living in low I areas.
In mice on high I intake their fetuses developed reduced body weight, decreased number of live fetuses, resorptions, skeletal abnormalities (Yang et al., 2006).
Radioactive I was one of the main pollutants resulting from the Chernobyl and Fukushima nuclear accidents and is carcinogenic (IARC Monograph, 2006). Guidelines have been published for treating patients exposed to radioactive iodine Aaseth et al. (2019).

Diagnosis

Deficiency: characteristic clinical signs and low blood thyroid hormone levels (T3 and T4). Check dietary intake levels of I.
Toxicity: characteristic clinical signs and high blood I concentrations. Measure thyroid hormone levels in blood. Check dietary intake (e.g., I content of pet foods, pastures or raw meat diets).

Prevention

Ensure adequate but not excessive dietary intake.

Treatment

Deficiency: Increase iodine intake in diet.
Toxicity: Remove source of excess intake from dietary intake, e.g., supplements, seaweed.

References

Aaseth et al. (2019), NRC (2006), Yang et al. (2006), McDowell (1992j), Hetzel and Maberly (1986), Ward and Ohshima (1986).

Iron (Fe)

Fe is widely used in the manufacture of steel for the construction industry.

Clinical applications

Oral iron tablets are available and fatalities have occurred in humans after overdose.
Iron dextran may be injected into infants and is routinely administered to piglets as they have very low Fe at birth and are prone to develop Fe-deficiency anemia.
Various Fe salts—ferrous fumarate, ferrous gluconate, ferrous sulphate—are used to treat Fe-deficiency anemia.

Essentiality

Human	Yes
Other species	Yes—All

Parameter	Site(s)
Site of absorption	Gastrointestinal tract after conversion into iodide ion
Site(s) of excretion	Faeces, urine
Site(s) of accumulation in body	Liver mainly, but also other tissues blood, brain, pancreas, etc.

Biological activity

Fe is essential in:

- Enzyme systems, e.g., cytochrome activity—electron transport; oxygenases and oxidases—activation of oxygen
- Heme compounds, e.g., hemoglobin in erythrocytes– oxygen carrying capacity; transferrin a protein in serum, erythrocytes, leukocytes (monocytes)—involved in defence mechanisms in the body

Deficiency

According to the World Health Organisation (WHO, 2008) 28.4% of the global human population (1.62 billion people) suffer from anemia, 50% of which are caused by Fe deficiency.
High copper content of food can interfere with Fe availability and cause signs of deficiency. Other minerals known to interfere with Fe absorption when present in high amounts, probably due to competition for receptor sites include: Cd, Ca, Co, Mn, Zn. Ni deficiency also decreases Fe absorption.
Aluminum can blunt the effect of erythropoietin by interfering with iron bioavailability.
Cottonseed meal contains a toxin gossypol which chelates with Fe reducing availability and causing deficiency (pigs). High parasite burdens in young animals can cause Fe-deficiency anemia. Some fish impair Fe absorption which can lead to Fe deficiency in fish-dependent species, e.g., mink.
In people and pets important causes of deficiency include chronic hemorrhage due to neoplasia, infections, gastrointestinal ulcers or parasites. Urinary losses may also be a factor.

Species	Clinical signs	Comments
Humans	Abortion, anemia (microcytic-hypochromic), listlessness, fatigue, palpitations on exertion, sore tongue, angular stomatitis, cheilitis (erythema at corners of the mouth), dysphagia, koilonychia (spoon nail) prolonged cardiorespiratory recovery after exercise (women), irritability, lethargy, malabsorption, pallor, pica *Children:* anorexia, enteropathy, cognitive deficits (attention span, intelligence, sensory perception functions, abnormal emotions, and behavior), gastric achlorhydria, gastric atrophy, gastritis, gastrointestinal hemorrhage, reduced growth rate, increased risk of infections, reduced serum myoglobin concentrations.	Abortion linked to Zn status

Buffaloes	Unthriftiness	
Cats	Anemia, diarrhea, lethargy, low hemoglobin, low hematocrit, hematuria,melena, pallor, weakness, weight loss	
Cattle	*Rare*: Calves low hemoglobin, anemia, low liver nonhem Fe content, low serum Fe, low weight gain, listlessness, labored breathing, reduced appetite, atrophy of tongue papillae, pallor, low performance, increased susceptibility to infections	
Dogs	Anemia (microcytic, hypochromic), diarrhea, poor growth, lethargy, low hematocrit, low hemoglobin, hematochezia, melena, reduced muscle myoglobin concentrations, pallor, low transferrin, weakness	
Elephants	Anemia	
Fish	Anemia, low hematocrit, low hemoglobin, low plasma Fe, low transferrin	
Foxes	anemia, depigmentation of underfur	
Horses	anemia, labored breathing,diarrhoea, exercise intolerance (tires easily), reduced growth rate, haircoat rough, lethargy, pneumonia, pallor, weakness,	
Marine mammals	Anemia	
Mice	Anemia, reduced birth weights, reduced litter sizes	
Mink	Anemia, death, emaciation, retarded growth, rough fur, lack of fur pigmentation (called "cotton fur" or "cotton pelt")	
Nonhuman primates	Anemia	diets based on soybean protein or cows milk
Pigs	*Common*. Anemia (hypochromic-microcytic),ascites, dilated cardiomegaly, death, ears hang limp, eyelids droop, food intake reduced, hair dull rough and coarse, hepatomegaly, immune response reduced, infection risk increased (enteritis eg E.coli; respiratory infections) listlessness, fatty liver, lung collapse, mortality increased, mucus membranes pale, muscle myoglobin low, palor, pulmonary edema, skin wrinkled, subcutaneous edema, tail hangs limp, thoracic fluid accumulation, ventroflexion of neck, weight gain reduced	
Poultry	*Rare*: anemia, abnormal erythrocytes, reduced growth rate, low hemoglobin, low hematocrit, lack of pigment in feathers (achromatrichia—seen in chickens and turkeys), poor feathers, embryonic mortality, *Chicks*: high serum triglycerides, low muscle myoglobin	
Rabbits	*Rare:* anemia	
Rats	Achromotrichia, anemia, white incisor teeth, cardiomegaly, splenomegaly, enlarged cecum, increased risk of infections, high serum triglycerides, low serum folic acid, low muscle myoglobin, low tissue cytochrome c concentrations	
Reptiles	Anemia (hypochromic), anorexia, depression, lethargy, pallor	
Sheep	*Experimental Lambs*: anemia	
Fish	Anemia	

Toxicity

Fe toxicosis is either corrosive or cellular and can be corrosive to in-contact tissues such as the gastrointestinal tract. A common cause is children eating Fe supplements.

High Fe intake can interfere with other minerals, e.g., Mn. In humans in South Africa food, iron cooking pots and Kaffir beer can provide toxic levels of Fe.

Dogs may ingest a variety of Fe containing products around the home including contraceptive pills, fertilizers, hand/foot warmers, heat patches, mineral supplements, oxygen absorber sachets, slug/snail bait.

Excessive burden of iron (iron overload), or iron storage disease (ISD) has been reported in a large variety of captive mammal species, including browsing rhinoceroses; tapirs; fruit bats; lemurs; marmosets and some other primates; sugar gliders; hyraxes; some rodents and lagomorphs; dolphins; and some carnivores; including procyonids and pinnipeds.

Species	Clinical signs	Comments
Humans	Abdominal pain, high aminotransferase, anuria, impaired cardiac and central nervous system metabolism, cardiomyopathy, coagulopathy, coma, convulsions, death (accidental intake children); diarrhea, free radicle production, hematemesis, hematochezia, hemochromatosis, hemorrhagic necrosis of gastrointestinal tract, hepatic failure, hepatic necrosis, hypoglycemia, hypotension, hypovolemia, jaundice, liver damage, acute liver failure, metabolic acidosis, multiple organ system failure, nausea, perforation of the gastrointestinal tract, peritonitis, renal failure, seizures, shock (cardiogenic), vomiting	Iron overload occurs. (genetic heritable condition, or high dietary intake; excess Fe supplementation; postblood transfusion) Lethal dose in humans: 200–250 mg iron/kg body weight
Amphibians	Excess iodine added to water may cause species such as the axolotl to shed its external gills	
Bats (fruit)	Iron overload with hemosiderosis/hemochromatosis	
Birds	Iron overload with hemosiderosis/hemochromatosis	Frugivores and insectivores
Cats	Diarrhea, vomiting	
Cattle	Abdominal pain, anorexia, death, depression, dyspnoea, hepatic degeneration, jaundice (buccal mucosa, lips), renal congestion	
Coati	Iron overload with hemosiderosis/hemochromatosis	
Dogs	Coagulopathy, death, depression, diarrhea, gastrointestinal damage, gastric obstruction, hematemesis, hematochezia, hypotension, lethargy, liver failure, metabolic acidosis, renal failure (acute), shock (hypovolemia), skin discoloration and edema (at injection site), tachycardia	
Fish	Death, diarrhea, impaired growth, hepatocellular damage	
Goats	Causes Cu deficiency (*see Cu deficiency*)	
Horses	*Foals:* coma, death, dehydration, diarrhea, jaundice, jejunal villi erosion, pulmonary hemorrhage, hepatic accumulation of Fe, liver degeneration, reduced serum, and liver Zn *Acute*: anorexia, coagulopathy, depression, jaundice, neurological signs, *Chronic:* hemosiderosis: asymptomatic, hemochromatosis: liver failure	
Marine mammals	Hemochromatosis, high postprandial insulin, hyperglobulinemia, hyperglycemia, leucocytosis, fatty liver disease	
Nonhuman primates	Iron overload with hemosiderosis/hemochromatosis (*Lemurs*)	
Pigs	Convulsions, death, diarrhea, dyspnoea, incoordination, hyperpnea, jaundice, pallor, tetanic posterior paralysis, shivering	
Racoon	Iron overload with hemosiderosis/hemochromatosis	
Rhinoceros	Iron overload with hemosiderosis/hemochromatosis	
Sheep	High alanine aminotransferase (ALT, high serum alkaline phosphatase (ALP) anorexia, high blood urea nitrogen (BUN), circulatory failure, low serum copper, high creatinine, death, diarrhea, depression, hemosiderosis, hepatic degeneration, high serum iron, myocardial degeneration, high serum phosphorus, pulmonary edema, renal degeneration, respiratory failure, splenic degeneration, high total iron binding capacity (TIBC), weight loss	
Tapir	Iron overload with hemosiderosis/hemochromatosis	

Supplementary information of clinical relevance

Measurement of low hemoglobin concentrations in blood with microcytic, hypochromic erythrocytes on blood smears are characteristic of Fe deficiency. Normal hemoglobin concentrations vary between species:

Species	Normal reference range (adults) *
Human	13–17 g/dL 8.07–10.55 mmol/L
Dog	13–14 g/dL 8.07–8.69 mmol/L
Cat	9–18 g/dL 5.59–11.17 mmol/L
Pig	10–11 g/dL 6.21–6.83 mmol/L
Cattle	11–12 g/dL 6.83–7.45 mmol/L
Sheep	10–11 g/dL 6.21–6.83 mmol/L
Goat	10–11 g/dL 6.21–6.83 mmol/L
Horse	10–11 g/dL 6.21–6.83 mmol/L
Rat	13–17 g/dL 8.07–10.55 mmol/L
Rabbit	13–14 g/dL 8.07–10.55 mmol/L

*Note: Different diagnostic laboratories may have their own reference ranges

Diseases and serum Fe/transferrin

The hormone hepcidin regulates Fe metabolism binding it to the main Fe export protein ferroportin. Increased hepcidin leads to iron deficiency and/or iron restricted erythropoiesis. Insufficient hepcidin causes Fe overload and occurs in multiple diseases.
Some disease states and serum Fe/transferrin concentrations are:

Disease	Fe/Transferrin concentrations	Comments
Iron deficiency anemia	Low serum Fe	Low intake. Loss of body stores of Fe and less hemoglobin breakdown
Aplastic and pernicious anemia	High serum Fe	Increased Fe absorption; reduced hemoglobin synthesis in bone marrow.
Hemochromatosis	High serum Fe	Excessive absorption and retention of Fe in body
Hepatitis	High serum Fe	Release from liver stores
Infections	Low serum Fe	Leukocyte endogenous mediator (LEM) increases which increases hepatic uptake of Fe from serum
Malignant cancer	Low serum Fe	
Nephrotic syndrome	Low serum Fe	Urinary loss of protein-bound Fe due to proteinuria

Reduce absorption	Increase absorption
Phytate—from cereals	Vitamin C (ascorbic acid) and other organic acids increase absorption of nonhem Fe.
High Phosphate	Meat-based foods compared to vegetable-based foods.
Dairy products—milk, cheese, egg	Simple sugars (e.g., fructose, sorbitol)
Deferoxamine: inhibits nonhem iron absorption—used to treat iron overload	Some amino acids, e.g., cysteine, histidine, lysine which chelate with Fe
Soybean protein	

Wheat Bran (not due to phytate alone)	
Clays (associated with pica)	
Tea, Coffee (tannins)	
Achlorhydria (lack of gastric acid)	
In monogastric species—duodenal disease	

Carcinogenesis, mutagenicity, embryotoxicity

According to the IARC occupational exposures during iron and steel founding are carcinogenic to humans (Group 1). Iron is also genotoxic and mutagenic in mice (Czarnek et al., 2018; Prá et al., 2008)

Diagnosis

Measure tissue Fe content or plasma ferritin.
Plasma ferritin <30 ng/mL for men, <15 ng/mL for women.

Prevention

Ensure adequate intake to avoid deficiency. Avoid excess intake of minerals that might interfere with Fe availability (Ca, Zn).
Stop Fe supplements.

Treatment

Deficiency—Iron supplementation—oral or injection, avoid excessive supplementation.
Toxicity—Supportive treatment. IV fluid replacement, chelator (Deferoxamine 15 mg/kg/h for up to 24 h).
(*See Appendix 1 and 2*).

References

Yuen and Becker (2020), Czarnek et al. (2018), Ahmadi et al. (2017), Abhilash et al. (2013), Oruç and Uzunoğlu (2009), Toyokuni (2008), Prá et al. (2008), WHO (2008), Chua et al. (2010), IARC Monograph (2006), Albretsen (2006), Velásquez and Aranzazu (2004), Baranwal and Singhi (2003), Singhi et al. (2003), Casteel (2001), Greentree and Hall (1995), Pearson et al. (1994), Lavoie and Teuscher (1993), Rallis et al. (1989), Mullaney and Brown (1988), Morris (1987), Acland et al. (1984), McDowell (1992g), Donnelly and Smith (1990).

Lanthanum (La)

La is a rare earth metal and a transition metal.
La has many uses as catalysts, additives in glass, carbon arc lamps for studio lights and projectors, GTAW electrodes, ignition elements in lighters and torches, electron cathodes, scintillators.

Clinical applications

La is used as a phosphate binder for patients in end stage renal failure.

Essentiality

Human	No
Other species	No

Parameter	Location
Site of absorption	Poorly absorbed after ingestion
Site(s) of excretion	NR
Sites of metabolism	NR

Biological activity

Not reported.
Lanthanum acts at the same modulatory site on the GABA receptor as zinc, a known negative allosteric modulator.

Deficiency

Not reported.

Toxicity

La has low to moderate toxicity.
In general, in rodents La poisoning causes immediate defecation, writhing, incoordination, labored breathing, and inactivity. Respiratory and heart failure may follow causing death. (Santa Cruz, 2020 online).

Species	Clinical signs	Comments
Humans	Appetite reduced, headache, nausea, pneumoconiosis, lowers cholesterol, liver damage (high ALP, ALT, Gamma-GT, low albumin), lowers blood pressure, increased risk of blood coagulation	Exposure to arc light
Mice	Hypoglycemia, low blood pressure, degeneration of the spleen and hepatic damage, low uric acid, high creatinine, low Ca, low P, low BUN	
Poultry	*Newborn Chicks:* reduced activity of Ca2+-ATPase,Mg2+-ATPase, and cholinesterase, inhibition of calcium binding to brain synaptosomal membrane	
Rats	Behavioral changes (impaired spatial learning and memory), morphological and histological changes in hippocampus, neural degeneration, lung lesions (granulation, giant cells), stomach hyperkeratosis, gastric eosinocyte infiltration, reduced sense of smell, reduced body weight, Increased enzymes alkaline phosphate (ALP) and alanine aminotransferase (ALT), low blood urea (BUN) lowered creatinine	
Fish	Embryotoxic and developmental effects reduced survival and hatching rates, tail malformations.	

Carcinogenesis, mutagenicity, embryotoxicity

In vitro: cytotoxic to lung tissue alveolar macrophages, chromosome aberrations.
No evidence for carcinogenicity in vivo.

Diagnosis

Toxicity: Known exposure. Body or environmental sample analysis.

Prevention

Prevent exposure.

Treatment

Treat symptomatically and avoid further exposure.

References

EPA (2018b), Damian (2014), Fei et al. (2011), Namagondlu et al. (2009), Feng et al. (2006), Wells and Wells (2001), Yongxing et al. (2000).

Lead (Pb)

Pb is a naturally occurring metal. It is widely used in batteries, by welders and solders and in the building trade. It is in insecticides, leaded paint, linoleum, motor oil, petrol, putty, old toys (lead figures), vehicle exhaust fumes, and water from lead pipes and tanks.

Clinical applications

Lead gloves, gowns, boxes and linings to walls are in widespread use on medical premises to protect against exposure to X-rays and other radiation.
Historically medics used lead to treat a myriad of conditions without success and precipitated toxic side effects instead!

Essentiality

Human	No
Other species:	No Rats—possibly

Site of absorption	Gastrointestinal tract, respiratory tract, skin
Site(s) of excretion	Feces (via bile), urine, milk
Site(s) of accumulation	Mainly in bone, also liver, kidney and blood
Interaction with other minerals	Several minerals can interfere with Pb absorption including Ca, Fe, Mg, P, S, Zn Lead interferes with Se

Biological activity

Not reported.
Pb inhibits many enzymes and it interferes with heme formation.

Deficiency

Species	Clinical signs	Comments
Rats	Anemia (hypochromic, microcytic), growth depression, reduced hepatic Fe stores	Insufficient evidence to support a claim for essentiality

Toxicity

Lead is a common cause of illness in Veterinary patients, for example it accounted for most (43% of 1341) cases of poisoning recorded between 1998 and 2013 at the Western College of Veterinary Medicine, Saskatoon, Saskatchewan.
Exposure to lead in automotive repairs, battery manufacturing, brass or copper foundries, old car batteries (licking - common in, ivestock), construction industry, foreign "digestive remedies". firearms, glass, hair dyes, kajal, kohl, mining, paints, pasture contamination (near lead mines and smelters), petrol, printing, smelting, surma, water from lead pipes

Species	Clinical signs	Comments
Humans	*Children:* abdominal pain, anemia (low Fe in red blood cells), anorexia, ataxia, basophil stippling, behavioral problems (e.g., mental retardation, selective deficits in language, cognitive function, balance, behavior, school performance), impaired consciousness, constipation, convulsions, dysarthria (slurred speech), clumsier, encephalopathy, headaches, hyperproteinemia, learning difficulties, less playful, pallor, papilledema (swelling optic disc), sluggish (lethargic), seizures, slurred speech (dysarthria), changes in kidney function, vomiting *Adults*: abdominal pain, anemia, anorexia, behavioral changes (e.g., hostility, cataracts (*increased risk*), depression, anxiety), basophil stippling, altered consciousness, constipation, death, encephalopathy, endurance loss, fatigue, fever, hallucinations, headaches, hypertension, incoordination, insomnia, irritability, joint pain, kidney disease, lethargy, listlessness, muscle strength loss, neuropathy (peripheral—demyelination, slow conductance), optical atrophy, skill loss (recent), damage to reproductive organs, seizures, sluggishness, vomiting, wrist drop	Lead poisoning can trigger Fanconi syndrome
Amphibians	Death, erythrocytopenia, excitation, hypersalivation, leukocytopenia, metamorphosis delayed (tadpoles), monocytopenia, loss of muscle tone, muscle twitches, neutrocytopenia, skin sloughing, sluggishness	Frogs
Birds	Decreased ALAD activity (blood, brain, liver), anemia, anorexia, balance loss, death, dehydration, depression, loss of depth perception, diarrhea (green), dropped wings, emaciation, bright green feces, gastrointestinal signs, hemoglobinuria (Amazon parrots), lethargy, locomotion impaired, neurological signs (seizures), edema (cephalic), esophageal impaction, renal, elevated protoporphyrin IX levels (blood, brain, liver), skeletal abnormalities, urinary signs (polydipsia, polyuria), tremors, weakness (neck and limb), weight loss (gradual)	
Buffaloes	Blindness, head pressing, salivation, violent movements	
Cats	Anorexia, excitation, seizures	
Cattle	*Common.* *Acute:* blindness, convulsions, death (can be sudden), depression, hyperirritability, hypersalivation, muscle twitches, teeth grinding, depressed reflexes *Chronic:* Anemia, growth retardation, low hemoglobin	
Dogs	Abdominal pain, anxiety, ataxia, blindness, encephalopathy, lethargy, seizures, tremors, vomiting	
Fish	Death (increased mortality rate), reduced food utilization, impaired growth, increased liver enzymes activity—SGOT and SGPT	
Horses	Erythrocyte stippling, loud respiratory noises ("roaring") due pharyngeal and laryngeal paralysis	
Marine mammals	Death, liver damage	Bottle nose dolphins
Mink	Anorexia, convulsions, dehydration, ocular discharge (mucopurulent), muscular incoordination,stiffness, trembling	
Poultry	Death, erythrocyte changes, liver damage, renal intranuclear inclusions	Lead shot in wild birds
Reptiles	Anemia (hemolytic), anorexia, depression, lethargy, pallor	
Sheep	Abortions, death, hearing loss, hydronephrosis, osteoporosis, sterility, stiffness, unthriftiness	

Carcinogenesis, mutagenicity, embryotoxicity

Not all authorities agree about Pb carcinogenicity: EPA: Probable human carcinogen. IARC: Inorganic lead—probably carcinogenic to humans. IARC: Not classifiable as to carcinogenicity to humans. IARC: Possibly carcinogenic to humans. NTP: Reasonably anticipated to be a human carcinogen (ATSDR Lead, 2011).
Evidence for genotoxicity in animals are conflicting (Winder and Bonin, 1993) but in humans Pb damages the developing baby's nervous system which can affect later behavior and intelligence. Lead also causes miscarriage, stillbirths, and infertility (in both men and women). (Centers for Disease Control and Prevention https://www.cdc.gov/niosh/topics/lead/health.html.

Diagnosis

Toxicity: History of exposure. Blood tests for Pb. Blood Pb is not reliable in cattle. Pb content in hair, urine, deciduous teeth, and other body tissues.

Prevention

Avoid exposure.

Treatment

Avoid further exposure, treat symptomatically possibly use chelation.
(*See Appendix 1 and 2*).

References

Cowan and Blakley (2016), Nigra et al. (2016), ATSDR Lead (2011), Schaumberg et al. (2004), Winder and Bonin (1993), McDowell (1992o), Quarterman (1986).

Lithium (Li)

Li has many uses including to make batteries, glasses, ceramics and pharmaceuticals. It is used as a flux for iron, steel, and aluminum production.

Clinical applications

Li is used to treat manic-depressive psychoses in humans. At high doses it suppresses thyroid function.

Essentiality

Human	No
Other species	
Goats	Yes
Rats	Possibly

Site of absorption	Gastrointestinal tract
Site(s) of excretion	Kidneys
Sites of metabolism	Li is not metabolized
Site(s) of accumulation	Brain and kidney mainly, also liver, bone, muscle, thyroid
Interaction with other minerals	NR

Biological activity

Not reported.

Deficiency

Species	Clinical signs
Goats	Abortion, low birth weight, low serum dehydrogenases (aldolase and liver monoamine oxidase) impaired growth,lifespan reduced, decreased milk production, reduced fertility, weight gain reduced

Toxicity

Accidental or deliberate intake of excessive lithium-containing medications is a source of toxicity as well as industrial exposure.

Species	Clinical signs
Humans	Abdominal pain, acne, agitation, ataxia, cardiac ECG changes (flattened T waves, sinus node dysfunction, prolongation QT, intraventricular conduction defects, U waves), coma, confusion, consciousness changes, constipation, delirium, allergic dermatitis (rash), death, diarrhea, disorientation, drowsiness, dry mouth, dysarthria, fatigue, flatus, heart failure, hyperreflexia, hyperthermia, hyperthyroidism (rare), hypertonia, hypotension, hypothyroidism, kidney failure, lung cancer (bronchogenic carcinoma), muscle weakness, uncontrolled muscle movements, myoclonus, nausea, nephrogenic diabetes insipidus, nephrotic syndrome, neurological signs (pyramidal, extrapyramidal, and cerebellar signs) nystagmus, polydipsia, restlessness, rigidity, seizures, skin ulcers, slurred speech, sodium-losing nephritis, stupor, tachycardia, tremors, urine concentrating ability lost, urination frequency increased, vomiting, weakness, weight change (gain or loss)
Cattle	Inflammation and ulceration of abomasum, rumen and stomach
Dogs	Body mass loss, dehydration, diarrhea, polydipsia polyuria, seizures, weakness

Goats	Ataxia, death, depression, diarrhea
Mice	Reduced growth, kidney damage, liver damage
Rats	Reduced growth, kidney damage, liver damage *Teratogenic changes*: reduced numbers and weight of the litter, increase in resorptions, with skeletal deformities including wavy ribs, short, deformed limb bones, incomplete ossification of sternebre, and wide bone separation in the skull.
Various	Dermatitis, respiratory tract inflammation, ulceration, and perforation of nasal septum

Carcinogenesis, mutagenicity, embryotoxicity

Li is not considered to be carcinogenic.
Embryotoxicity has been demonstrated in rats with teratogenic effects—see above (Marathe and Thomas, 1986).

Diagnosis

A blood or urine test to determine lithium levels.

Prevention

Avoid exposure

Treatment

Stop lithium intake and medications that may reduce lithium elimination. Treat symptomatically.

References

Hedya et al. (2019), Healthline (2020), Altschul et al. (2016), Harari et al. (2015), Canan et al. (2008), Davis et al. (2018), Ott et al. (2016), Kibirige et al. (2013), McDowell (1992p), Davies (1991), Mertz (1986), Marathe and Thomas (1986).

Lutetium (Lu)

Lu is a rare earth element used in carbon arc lamps for movie projection, in lasers, in cancer radiotherapy, as a catalyst in petroleum cracking, and as a phosphor in LED-light bulbs.

Clinical applications

Lutetium texaphyrin has been used to treat macular degeneration. Lutetium therapy (lutetium-177-DOTA-oxodotreotide) is a type of treatment known as a targeted radionuclide therapy or peptide receptor radionuclide therapy (PRRT).

Essentiality

Human	No
Other species	No

Parameter	Location
Site of absorption	Poorly absorbed from the gastrointestinal tract
Site(s) of excretion	Feces
Sites of accumulation	Bones, liver, kidneys

Biological activity

Not reported.
When administered artificially to rats Lu can increase soft tissue Ca concentrations, potentiate GABA-induced chloride currents (possible neurological effects) and in vitro stimulate fibroblast proliferation.

Deficiency

Not reported.

Toxicity

Species	Clinical signs	Comments
Humans	Alopecia, nausea, pneumoconiosis, extensive scarring on abraded skin, tiredness	
Cats	Death, cardiovascular collapse with respiratory paralysis.	IV: Acute toxicity study
Dogs	Increased clotting and prothrombin times, increased blood pressure and heart rate, death	
Guinea pigs	Mineralization, subcutaneous nodules	Intradermal injection sites
Mice	Arched back, ataxia (loss of muscular coordination), death, labored breathing, sedation, stretching of limbs while walking, and excessive tearing, writhing. *IV:* Impaired blood clotting	Acute toxicity
Rabbits	Conjunctivitis, severe irritation to abraded skin	

Carcinogenesis, mutagenicity, embryotoxicity

No evidence for carcinogenicity mutagenicity or embryotoxicity.

Diagnosis

History of exposure.

Prevention

Avoid exposure.

Treatment

Treat symptomatically and avoid further exposure (*See Appendix 1 and 2*).

References

EPA (2018a,b), Damian (2014).

Magnesium (Mg)

Mg is an alkaline earth metal used in alloys and the manufacture of many items including cameras, car seats, laptops, luggage, power tools.

Clinical applications

Magnesium sulfategiven intramuscularly or by infusion is used in the management of preeclampsia and eclampsia, to protect against cerebral palsy in babies delivered preterm. It is also being investigated for use in other conditions including asthma, diabetes mellitus type II, cardiovascular diseases, hypertension, and pain.
Oral mg promotes defecation via osmotic retention of fluids, it is a natural calcium channel blocker, a cofactor of the Na-K-ATP pump, controls atrioventricular node conduction so can cause myocardial excitability and arrhythmias including ventricular tachycardia and torsades de pointes.

Essentiality

Humans	Yes
Other species All	Yes

Parameter	Location
Site of absorption	Small intestine
Site(s) of excretion	Kidney
Sites of accumulation	Bone (50%), muscle (25%), red cells, serum, soft tissues
Interference with absorption	Zn, Fiber, Phytate
Active form	IonizedMg

Biological activity

A major intracellular cation involved in over 300 metabolic processes, and is essential for

- protein synthesis
- ATP activity and DNA and RNA metabolism
- muscle and nerve cell membrane stability
- secretion of hormones
- platelet activation
- lymphocyte proliferation
- structure of bones and teeth
- calcium channel regulation in the heart
- cell to cell adhesion in tissues
- extracellular matrix communications
- Mg depresses the central nervous system (CNS) with anticonvulsant effects
- it inhibits acetylcholine release, blocking peripheral neuromuscular transmission.
- is a tocolytic during preterm labor stimulating calcium reuptake by sarcoplasmic reticulum, causing muscle relaxation and vasodilation
- is a cofactor of parathyroid hormone (PTH) synthesis, so in hypomagnesemia hypoparathyroidism results
- causes bronchial smooth muscle relaxation

Deficiency

High dietary P may exacerbate the effects ofMg deficiency. Low Mg may interfere with K, Na and Ca metabolism.
In humans Mg deficiency is usually associated with: alcoholism, hyperaldosteronism, hyperparathyroidism (e.g., postparathyroidectomy), malabsorption, malnutrition, prolonged IV infusions, renal disease.
Low Mg intake has been associated with increased risk of development of type II diabetes.

Species	Clinical Signs
Humans	Anorexia, anxiety, appetite loss, arrhythmia, cardiac arrest, Chvostek sign, dysphagia, fatigue, hyperreflexia hypocalcemia, lethargy, memory loss, muscle fasciculations, hyperreflexia, hyporeflexia, nausea, muscle weakness, incoordination, irritability, mental derangement, muscle cramps, muscle fasciculations, myographic changes, personality change, tonic-clonic seizures, tetany, tremors, Trousseau sign, vertigo, vomiting, weakness
Birds	Angular deformity long bones, convulsions, death (chicks), poor growth, lethargy
Buffaloes	Death, depression, head shaking, opisthotonos, muscle fasciculations, internal organ congestion and hemorrhages, over-reaction to stimuli,
Cats	Anorexia, aortic calcification, anorexia, arrhythmias, ataxia, behavioral changes, convulsions, depression, poor growth, hyperextension carpi, hyperirritability, hyperreflexia, hypomagnesemia, muscle twitches, muscle weakness, tachycardia, tetany, torsades de pointes, soft tissue calcification
Cattle	Convulsions, hypocalcemia (calves) hypomagnesemia, ketosis, reduced feed intake, lethargy, ataxia, stiffness, excitability, hypersalivation, muscular tremors, uncontrolled limb movements, collapse, death-called grass tetany
Dogs	Anorexia, aortic mineralization, ataxia, hyperextension of carpal joints, hypocalcemia, hypomagnesemia, lameness, hindleg paralysis, muscle necrosis, poor weight gain, rhabdomyolysis, seizures, vomiting, weight loss ? mitral valve prolapse in Cavalier King Charles Spaniels (Pederson and Mow, 1998)
Fish	Anorexia, cataracts, convulsions, death, gill filament degeneration, poor growth, lethargy, reduced tissue Mg content, muscle fiber degeneration, nephrocalcinosis, pyloric ceca epithelium degeneration, spinal curvature
Gerbils	Alopecia, death, seizures, weight loss
Goats	Convulsions, tetany
Guinea pigs	Activity decreased, alopecia, anemia, ataxia, poor growth, hair loss, hyperphosphatemia, loss muscle coordination, soft tissue mineralization, stiffness, teeth discoloration and erosion, tetany, poor weight gain
Horses	Ataxia, aortic mineralization, collapse, convulsions, death. hyperpnea, hypomagnesemia, muscle tremors, nervousness, sweating, with limb movements, soft tissue mineralization and tissue degeneration at postmortem examination, tetany
Mice	Convulsions, death
Nonhuman primates	*Rhesus monkeys*: cardiovascular (ECG changes due low K or low Ca), gastrointestinal signs, hypocalcemia, irritability (Rhesus), neuromuscular signs
Pigs	Ataxia, collapse, death, growth impairment, hyperirritability, hypocalcaemia, muscle twitches, stepping syndrome, tetany, weakness
Poultry	*Rare.* ataxia, convulsions, coma, death, egg production reduced, egg quality poor, feathering poor, gasping, growth poor, hypocalcemia, hypomagnesemia, loss of muscle tone, panting, poor survivability,
Rabbits	Alopecia, blanching of ears, cardiac changes (endocardial, subendocardial, interstitial, and perivascular fibrosis), convulsions, fur chewing, poor growth, hyperexcitability
Rats	Convulsions, death, hyperirritability, soft tissue mineralisation, nephrocalcinosis, failure to suckle young, vasodilation,
Sheep	Collapse, convulsions, death, hypersalivation, hypocalcemia, hypomagnesemia, limb movements when collapsed, tetany

Toxicity

Human cases of Mg toxicity are often associated with excessive intake through the use of antacids or laxatives which can supply over 5000mg Mg/day. Excessive Mg should be excreted through the kidneys so individuals with impaired renal function are at higher risk to develop hypermagnesemia.

Species	Clinical signs
Humans	Abdominal cramps, arrhythmias, breathing difficulty, cardiac arrest, death, depression, diarrhea, ECG changes (prolonged PR, wide QRS, peaked T waves) flushing (face), hypermagnesemia, hypocalcemia, hyporeflexia, hypotension, ileus, lethargy, muscle weakness, nausea, respiratory depression, urinary retention, vasodilation, vomiting
Birds	Reduced egg production, thin egg shells, diarrhea, dermatitis, feather abnormalities (brittle, frayed), irritability
Cats	Struvite urolithiasis
Cattle	Degeneration of rumen papille, diarrhea, nephroliths (calves—calcium appetite), weakness
Goats	Death, diarrhea, reduced food intake, lethargy, disturbed locomotion, reduced performance
Pigs	Depressed weight gain/growth
Poultry	Diarrhea, decrease egg production, thin egg shells, depressed growth, increased mortality
Sheep	Diarrhea, reduced nutrient utilization

Carcinogenesis, mutagenicity, embryotoxicity

There is evidence that Mg might protect against carcinogenicity as cancer rates increase in humans with Mg deficiency (Kasprzak and Waalkes, 1986).

Diagnosis

Deficiency: Check dietary intake.
Toxicity: Check dietary intake. Measure blood/tissue and food/supplement content.

Prevention

Deficiency: Ensure adequate dietary intake and a balanced mineral ration.
Toxicity: Avoid exposure to excess Mg in food or supplements.

Treatment

Deficiency: oral Mg salts or intravenous (IV) or intramuscular (IM) magnesium sulphate.
Toxicity: Avoid further excessive intake in diet or supplements.

References

Linus Pauling Institute (2020), Allen and Sharma (2020), Rodriguez-Moran et al. (2011), Dong et al. (2011), German (2010), Simmons et al. (2010), Larsson and Wolk (2007), Schulze et al. (2007), NRC (2006), Shils, 1999, Pederson and Mow (1998), McDowell (1992e), NRC (1989), Kasprzak and Waalkes (1986), Daniel et al. (1986), Cronin et al. (1982).

Manganese (Mn)

Mn is used in the production of several metals.

Clinical applications

In pallid mice, administration of a high dose Mn supplement to mothers with the mutant gene coat totally prevented the development of congenital ataxia in the offspring (Erway et al., 1971).
In pastel mink screwneck (caused by a mutant gene) and abnormal otolith and postural reflex development can also be prevented by Mn supplementation during pregnancy.

Essentiality

Humans	Yes—with Vitamin K deficiency
Other species Various: Pigs, mice, rats, cattle, goats, sheep, rabbits, guinea pigs, poultry	Yes

Parameter	Location
Site of absorption	Small intestine
Site(s) of excretion	Bile mainly, pancreatic juice, small intestine
Transportation	Bound to alpha2-macroglobin or transferrin
Factors blocking absorption	Co, P, Fe
Factors increasing absorption	Citrate, L-histidine
Nutritional sources	Fiber, menhaden fish meal, dicalcium phosphate

Biological activity

Mn activates enzyme systems and is a metalloenzyme. It is important for bone and cartilage development (through activation of glycotransferases), reproduction, cell membrane integrity (mitochondria), and lipid metabolism (synthesis of choline and cholesterol). Mn is high in the corpus luteum in cattle.
Mn is essential for:

- Brain function
- Skeletal development
- Chondroitin sulphate synthesis needed for cartilage synthesis
- Pancreatic activity, insulin synthesis, glucose tolerance
- Mitochondrial structure
- Cell membrane integrity
- Lectin synthesis
- Immune function
- Carbohydrate metabolism
- Lipid metabolism

A component of several metalloenzymes of importance in biochemical pathways, for some Mg can replace Mn:

Enzyme system	Biological Activity
Arginase	Responsible for urea formation
Pyruvate carboxylase	Involved in gluconeogenesis, lipogenesis, and neurotransmitter synthesis
Manganese superoxide-dismutase	Protects mitochondria from oxidative damage

Hydrolases	Important in hydrolysis to breakdown large molecules
Kinases	Responsible for transfer of phosphates from high energy molecules (ATP) important for many metabolic processes including cell signaling, protein regulation, cellular transport, secretory processes
Decarboxylases	Various metabolic pathways, nonoxidative decarboxylation of α- and β-keto acids and carbohydrate synthesis
Transferases	Transfer functional groups (e.g., a methyl or glycosyl) from one molecule to another

Deficiency

Species	Clinical signs
Human	*Children*: bone demineralization, impaired growth, skeletal abnormalities *Adults*: ataxia, high alkaline phosphatase, high blood calcium, low blood cholesterol, impaired glucose tolerance, hair depigmentation, high blood phosphorus, skin rash
Birds	Ataxia, bone deformities (chondrodystrophy), death (embryos), dermatitis, reduced egg production, feather abnormalities (brittle, frayed), growth impairment, reduced hatchability, perosis, slipped hock tendon, tetanic spasms (chicks), tibiometatarsal joint enlargement, tibial deformity (twisting), slipping of gastrocnemius muscle from condyle (especially young cranes, gallinaceous and ratites, also cockatiels, raptors, rosellas).
Cattle	*Calves:* malformed crooked limbs, joint enlargement, short forelimbs, difficulty standing, *Adults:* abortion, birth weight low, conception rate low, estrus delayed or depressed, stillbirths
Dogs	Angular deformities of the limbs, fatty liver, reduced growth, reproduction impaired
Fish	Short body dwarfism, impaired growth, reduced hatchability, low Mn and Ca in vertebrae, low Mn in eggs, low superoxide dismutase (Cu-Zn and Mn in liver and heart), skeletal abnormalities
Goats	Abortion, ataxia, angular deformities, birth weight low, conception rate low, joint swelling (tarsus), estrus delayed or depressed, stillbirths
Guinea pigs	Angular deformity (bowing) forelegs, ataxia (neonatal), balance loss, body righting reflexes delayed development, bone growth retarded, forelimb shortening, glucose tolerance impaired, head retraction, head tremors, hyperreactivity to stimuli, litter size reduced, neonatal mortality, otolith abnormalities, pancreatic aplasia or hypoplasia, premature birth, skull deformities, vestibular syndrome
Mice	Ataxia (congenital), fat deposition in the liver, otolith deformity, impaired melanin synthesis (in pallid mice)
Mink	A condition called "screwneck" in pastel mink, otolith abnormalities, abnormal postural reflexes
Pigs	Angular deformity limbs, fat deposition in the liver, joint enlargement (hocks), lameness, shortened limbs
Poultry	*Chickens:* Embryo: chondrodystrophy chicks and poults: ataxia (neonatal), death, perosis (slipped tendon—enlarged tibiometatarsal joint; distortion of the tibia, thickening and shortening long bones, gastrocnemious tendon slips off condyles, reluctant to walk, walk on hocks Chondrodystrophy: abnormal skull development (globular) short thick limbs and wings, short lower beak "parrot beak," *adults:* Reduced egg production, poor eggshells, poor hatchability *Ducks:* death, perosis (slipped tendon)—enlarged tibiometatarsal joint; distortion of the tibia, thickening and shortening long bones, gastrocnemius tendon slips off condyles, reluctant to walk, walk on hocks, die. Chondrodystrophy: abnormal skull development (globular) short thick limbs and wings, short lower beak "parrot beak".
Rabbits	Angular deformity (bowing) forelegs, bone poorly mineralized, bone growth retarded, libido lack, otolith abnormalities, shortened forelimbs, skull deformity, sperm lack, sterility (male), tubular degeneration testes
Rats	Angular deformity (bowing) forelegs, ataxia (neonatal), balance loss, body-righting reflexes delayed development, bone growth retarded, convulsions (epileptiform), epiphyseal dysplasia, forelimb shortening, glucose tolerance impaired, head retraction, head tremors, hyperreactivity to stimuli, insulin synthesis impaired, libido lack, male sterility, neonatal mortality, otolith abnormalities, ovulation defective, poor survivability, reproductive abnormalities, skull deformity (doming), skeletal abnormalities (short radius, ulna, tibia, and fibula), sperm lack, sterility (male), testicular degeneration, vaginal orifice delayed development, vestibular syndrome
Sheep	Reduced appetite, balance difficulty, death, impaired growth, joint pain, locomotion impaired, reproductive failure

Toxicity

Species	Clinical signs
Humans	Ataxia, central nervous system damage, confusion, fatigue, abnormal gait, hallucinations, hyperirritability, muscle spasms (dystonia), pneumonia, psychiatric abnormalities, rigidity of trunk, stiffness, tremors (hands), violent acts, weakness
Rats	Hyperglycemia, hypoinsulinemia
Sheep	Reduced appetite, impaired growth, impaired Fe metabolism

Carcinogenesis, mutagenicity, embryotoxicity

Mn deficiency can affect fertility and be teratogenic.
High Mn exposure affects fertility and is toxic to mammalian embryo and foetus. The fungicide MANEB and the contrasting agent MnDPDP are also embryotoxic at higher than normal exposures.
Inorganic Mn is not carcinogenic.

Diagnosis

Blood Mn levels (Human adults 4.7 and 18.3 nanograms per milliliter (ng/mL)).

Prevention

Ensure adequate dietary intake and avoid high Fe, Ca, or P intake.
Avoid excess Mn intake.

Treatment

Deficiency: Mn supplementation.
Toxicity—avoid excess intake.

References

Gerber et al. (2002), McDowell (1992k), Johnson and Lykken (1991), Norose et al. (1992), Friedman et al. (1987), Hurley and Keen (1987).

Mercury (Hg)

Hg is used by dental assistants and hygienists, chemical workers, goldsmiths, mirror makers, electrical industry, and in many instruments, such as thermometers and barometers. It is also in some topical creams and ointments.

Clinical applications

Dental amalgams, skin preparations.

Essentiality

Humans	No
Other species	No

Parameter	Location
Site of absorption	Metallic mercury is poorly absorbed from gastrointestinal tract. However, salts, e.g., methylmercury is highly available. Lungs, transdermal, and across the conjunctiva
Site(s) of excretion	Feces (via bile), Urine
Sites of accumulation	Brain, kidneys, muscle

Biological activity

None reported.

Deficiency

Not reported in any species.

Toxicity

Metallic mercury is relatively nontoxic and it is not readily absorbed from the gastrointestinal tract. Inorganic Hg salts and especially the alkyl derivative are more toxic. Common sources of human toxicity are dental amalgam, eating contaminated fish and seafood and occupational exposure to mercury vapor.
Mercury toxicity has been recognized for many years in humans as Erethism or mad hatter syndrome (felt hat makers).

Species	Clinical signs
Humans	Abdominal cramps, abdominal pain, abnormal thoughts, abortion (wives of men chronically exposed), acrodynia (painful swelling and pink color to fingers and toes), anorexia, anuria, anxiety, ataxia (progressive), basophil stippling, behavioral changes (excessive anger, deficits in mental concentration), brain damage (permanent), breathlessness, cerebellar ataxia, cerebral palsy (congenital), chest burning pain or tightness, choreoathetosis (uncontrolled jerky movements combined with slow, writhing movements), coma, concentration loss, confusion, changes in consciousness, cough (dry), death, dehydration, depression, dermatitis (rare except contact rash), diarrhea (bloody), dysarthria (slurred speech), dysmenorrhea, dysphagia, dyspnoea, emotional lability, erythema, abnormal excitability, eye irritation, facial expression changes, fatigue, forgetfulness, gastrointestinal disturbances including necrosis, gingivitis, glomerulonephritis, headaches, hearing loss, hyperesthesia (extreme sensitivity), hypersalivation (ptyalism), hypertension, impaired immune function, insomnia, interstitial nephritis, irritability, kidney damage, kidney failure, lethargy, locomotor disturbances, lymphocyte aneuploidy, memory loss, metallic taste in mouth, mood swings, nausea, nerve conduction speeds slowed, neuropsychological changes, paraesthesia, peritonitis, severe pharyngitis, pneumonia, pneumonitis, polyarthritis, polyneuropathy, proteinuria, pulmonary fibrosis, pulmonary edema, quick temperedness, renal failure (acute), renal tubular necrosis, respiratory distress or failure, restrictive lung disease, chronic respiratory insufficiency, seizures, shock, shortness of breath, skin irritation, sluggishness, sensory disturbances, spermatogenesis reduced, severe stomatitis, excessive shyness, tachycardia, tingling, tremors (legs, arms, tongue, lips), uremia (severe), vertigo, visual problems (e.g., loss of color vision, constriction of visual fields, tunnel vision, and blindness) vomiting, weakness

Cats	Anorexia, ataxia (cerebellar), loss of balance, blindness, death, fetal anomalies, motor incoordination, loss of nerve cells with replacement by reactive and fibrillary gliosis, hypermetria, neuronal necrosis, nystagmus (vertical), impaired reflexes, e.g., righting and diminished sensitivity to pain, proprioceptive deficits, rigidity (hindlegs) difficulty walking, tonic-clonic seizures, tremors, uncontrolled vocalization (howling), vomiting
Cattle	Ataxia, convulsions, dyspnoea, muscle tremors, neuromuscular incoordination, oral ulcers, paresis, proteinuria, renal failure
Dogs	Abdominal pain, anuria, anxiety, ataxia, blindness, depression, bloody diarrhea, death, extensor rigidity, glycosuria, head tilt, hypocalcemia, hypokalemia, hypoproteinemia, incoordination, lymphocytopenia, metabolic acidosis, motor disabilities, nervousness, nystagmus (horizontal) opisthotonos, proteinuria, loss of pupillary reflexes, kidney damage (inability to urinate, abdominal swelling), acute renal tubular necrosis, tetraparesis, uremia (severe), visual disturbances, vomiting, loss of withdrawal reflexes
Ferrets	Anorexia, paroxysmal convulsions, death, incoordination, tremors, weight loss
Guinea pigs	Death, neurotoxicity, ototoxicity
Horses	*Acute:* colic, death, dermatitis (ulcerative), edema (ventral), stomatitis, acute renal failure, tubular necrosis, shock, *Chronic*: ataxia, depression, dermatitis, gingival swelling and necrosis, head bobbing, hypermetria, masseter muscle atrophy, trembling, neurologic signs. interstitial renal fibrosis, chronic renal failure, weakness, weight loss
Marine mammals	*Seals:* Anorexia, death, hepatitis, lethargy, renal failure, uremia, weight loss *Bottle-nose dolphins*: lipofuscin accumulation in liver *Harp seals:* Damage to organ of Corti sensory cells
Mice	Behavioral changes in neonates, hind leg weakness and degenerative changes in the corpus striatum, cerebral cortex, thalamus, and hypothalamus. Renal tumors
Mink	Anorexia, ataxia, blindness, chewing, circling, convulsions, death, dysphonia, head tremors, incoordination, paralysis, falling to recumbency, salivation, tremors, vocalization (high pitched), vomiting, weight loss
Nonhuman primates	Abortions, low conception rate, incoordination, reduced visual sensitivity, restricted visual field, impairment of spatial vision, intention tremors, somasethic impairment,
Pigs	Anorexia, ataxia, blindness, coma, crystalluria (white), bloody diarrhea, death, gastrointestinal tract necrosis, paralysis (partial), kidney damage, uremia, wandering, weight loss
Rats	degeneration of the cerebellum and dorsal route ganglia, and clinical signs of neurotoxicity
Sheep	Alopecia, anorexia, blindness, death, depression, diarrhea, convulsions, acute hemorrhagic diarrhea, incoordination, stiff gait, paresis, polio encephalomalacia, pruritus, acute renal failure, scabs, tooth loss, weight loss

Carcinogenesis, mutagenicity, embryotoxicity

There is evidence that methylmercury is mutagenic, carcinogenic, embryotoxic, and highly teratogenic.
However, the International Agency for Research on Cancer (IARC, 1993) declared that methylmercury compounds are possibly carcinogenic to humans (Group 2B). and metallic mercury and inorganic mercury compounds are not classifiable as to their carcinogenicity to humans (group 3). There is sufficient evidence for the carcinogenicity of methylmercury chloride in animals.
Based on in vitro testing using the embryonic stem cell test (EST) Hg demonstrated strong embryotoxicity (Imai and Nakamura, 2006).

Diagnosis

Methylmercury content of red blood cells, whole blood ideal <10 mcg/L, or hair are considered good indicators of Hg toxicity. Urinalysis can be done following inhalation of Hg ideal <20 mcg/L

Treatment

a. Charcoal has been used but does not bind to mercury
b. Oral sodium thiosulfate at a dose of 0·5–1·0 g/kg bodyweight) binds mercury

c. Dimercaptosuccinic acid (10 mg/kg administered orally three times a day for 10 days) may increase urinary excretion of inorganic mercurial. Contraindicated for organic Hg
d. DMSA is widely used in humans for both chronic and acute Hg toxicity

(*See Appendix 1 and 2*).

References

EC (2020), MSD Veterinary (2020), RAIS (2020), PigSite (2020), Bernhoft (2012), CDC Mercury (2011), Imai and Nakamura (2006), Farrar et al. (1994), McDowell (1992o), Clarkson (1987), Irving and Butler (1975), Wobeser et al. (1976), Wobeser and Swift (1976), Aulerich et al. (1974), Falk et al. (1974), Hanko et al. (1970).

Molybdenum (Mo)

Mo trioxide is used in metallurgical processes, as a pigment in cosmetics, and in contact lens solution. Mo is present in many vitamin and mineral supplements. Exposure of humans to Mo is almost exclusively through food.

Clinical applications

Magnesium molybdate and molybdenized iron (II) sulphate have been used to treat anemia. Mo has a cariostatic effect.

Essentiality

Species	Essential nutrient?	Daily Requirement (adults)
Human	Yes	45 μg/day (0.64 μg/kg/day) (NAS, 2000)
Dog	Probable	
Cat	Probable	
Other species Goats Poultry	 Yes Yes	

Parameter	Location
Site of absorption	Small intestine
Site(s) of excretion	Mainly urine and some in bile and milk.
Sites of metabolism	Low amounts in all tissues—mainly bone and liver
Factors blocking absorption	Fe, Pb, S, Zn, W, vitamin C, vitamin D, protein
Nutritional sources	Soybean meal, cereal grains

Biological activity

Mo is an important component of several metalloenzymes, notably aldehyde oxidase, sulfite oxidase and xanthine oxidase, the latter is involved in the conversion of nucleic acid metabolites (purines) to uric acid. These enzyme activities are thought involve a common cofactor (molybdopterin). Mo is also a component of mitochondrial amidoxime reducing component (mARC).

Mo with S reduces ceruloplasmin synthesis by reducing Cu in tissues, and Cu excretion is mainly urine, not bile. High Cu reduces Mo in liver. High S reduces Mo in tissues and increases Mo excretion in urine.

Enzyme system	Biological Activity
Aldehyde oxidase	Electron transport chain- with cytochrome C Niacin metabolism
Sulfite oxidase	Sulfate excretion in urine
Xanthine oxidase	Electron transport chain—with cytochrome C

Deficiency

Sulphur in particular may compete with Mo absorption.

Species	Clinical signs	Comments
Humans	*Genetic:* brain atrophy/lesions, early postnatal death, intellectual disability, lens dislocation, mental retardation, opisthotonos, seizures High plasma and urine sulfite, sulfate, thiosulfate, S-sulfocysteine, and taurine. *Nutritional:* coma, headache, irritability, nausea, night blindness, tachycardia, tachypnoea, vomiting. High sulfite and xanthine, low sulfate, and uric acid in the blood and urine, Caries incidence is higher in the UK where Mo levels are low	Genetic defect: Type A—mutations in MOCS1 gene, Type B deficiency mutations in MOCS2 result in inability to form molybdopterin and the molybdenum coenzyme sulfite oxidase. Long term parenteral nutrition
Cattle	Not reported	Excess Cu can induce Mo deficiency
Goats	Low conception rates, abortion, poor growth, increased mortality (kids)	
Poultry	abnormal growth and development (impaired ossification of long bones; changes in articular cartilage with the inability to move), feather defects and loss, high embryonic mortality, reduced viability of chicks	
Sheep	Xanthine calculi	

Toxicity

Human toxicity rare, but possible for workers in mining and metalworking. High Mo intake may contribute to Cu deficiency (ruminants).
Toxicity depends on other dietary factors: Ag, Cu, Cd, Zn, and S-containing amino acids. High Cu and S would protect against Mo toxicity.

Species	Clinical signs	Comments
Humans	*Inhalation:* cough, dyspnoea, reduced lung function Arthralgia, expiratory capacity reduced (dust inhalation), gout-like signs, hallucinations, hepatotoxicity (Mendy et al., 2012), hyperuricemia, hyperuricuria, joint aches, multiple sclerosis (Pitt, 1976), neurological signs, psychosis (acute), seizures, reduced testosterone (Lewis and Meeker, 2015), high tissue xanthine oxidase activity *Associations:* reduced sperm count, reduced serum testosterone concentrations; urate and cystine calculi (Abboud, 2008)	Mo concentration in urate and cystine calculi was 10x that in other uroliths (Abboud, 2008)
Buffaloes	Cu deficiency. Neutropenia	
Cattle	anorexia, anemia, poor condition, death, depigmentation hair, diarrhea (severe—especially if low Cu), reduced growth, reduced fertility (conception rate), lameness, loss of libido, mandibular exostoses, osteoporosis, delayed puberty, skeletal deformities, spermatogenesis impaired, staring coat, high serum phosphatase activity, testicular damage, weight loss	Depigmentation due low Cu Excess Mo causes fatal Cu deficiency. Molybdate and sulfide in the rumen, resulting in the formation of thiomolybdates (Gould and Kendall, 2011)
Deer	Impaired growth rate, low Cu (blood and liver)	
Fish	Death of fertilized eggs	Rainbow trout
Guinea pigs	Death, growth retardation, contact sensitization (topical or intradermal exposure), weight loss	
Horses	Rickets in foals	

Mice	*Inhalation:* hyaline degeneration of nasal epithelium, squamous metaplasia of nasal planum, increased cancer (alveolar/ bronchiolar adenomas or carcinomas)	
Poultry	*Chicks:* Anemia, growth retardation	
Rabbits	*Oral:* Alopecia, anemia, anorexia, death, dermatosis, epiphyseal cartilage widening, fractures, growth impairment, limb deformity (weakness) of front legs, metaphyseal cartilage irregularly calcified, muscle degeneration, thyroxine T4 decline, weight loss *Instilled into eye:* conjunctivitis (mild)	Due induced Cu deficiency
Rats	*Oral ingestion:* Alkaline phosphatase in liver decreased activity, alkaline phosphatase in small intestine and kidney increased activity, increases in serum alanine aminotransferase (ALT) and aspartate aminotransferase (AST) activities, albuminuria, anemia, anorexia, high blood urea, bone strength reduction, caries, cartilaginous dysplasia, creatinuria, death, dental enamel lesions, impaired endochondral ossification, diarrhea, diuresis (polyuria), impaired endochondral ossification, fetal death rate increased, fetal resorption rate increased, fetal size decreased, growth retardation, growth plates widened, hemorrhages and tears in tendons and ligaments, increases in urinary kallikrein (distal tubule enzyme), kidney damage (degeneration, increased renal lipid, hyperplasia renal proximal tubules), limb shortening, liver glucose-6-phosphatase activity decreased, neurocyte degeneration (cerebral cortex and hippocampus), impaired estrus cycles, abnormal oocysts, male sterility, mandibular/ maxillary exostoses, milk production reduced, estrus prolonged, spermatogenesis impaired, seminiferous tubule degeneration, survivability decreased, testicular damage, weight loss *Inhalation:* alveolar epithelial metaplasia, histiocytic cellular infiltration, hyaline degeneration of nasal epithelium, pulmonary inflammation, squamous metaplasia of nasal planum	*Note:* low confidence in the hepatic and renal changes reported in these studies, treat findings with caution. There are studies showing conflicting results for the effect of Mo toxicity on reproductive function. See ATSDR (2020a)
Sheep	Alkaline phosphatase increased, connective tissue changes (subperiosteal hemorrhages on the humerus, spontaneous fractures), central nervous system degeneration, motor neurone disease, reduced weight gain, depigmentation of skin and hair, delayed puberty, loss of crimp in wool	*Oral intake*

Carcinogenesis, mutagenicity, embryotoxicity

Mo inhalation is associated with lung cancer development in mice and rats. Serum Mo (and Mn) concentrations were significantly higher in breast cancer patients in one study from Korea (Choi et al., 2019).
The International Agency for Research on Cancer (IARC, 2018) categorized molybdenum trioxide as possibly carcinogenic to humans (Group 2B).
Mo salts demonstrated genotoxicity at relatively high doses in vitro (human cells) and *in vivo* in mice (Titenko-Holland et al., 1998) and hamsters (Campbell et al., 2006).
In vitro Mo compounds have exhibited anticancer effects (Mitsui et al., 2006).

Diagnosis

Deficiency: Analyse dietary intake.
Toxicity: Analyze dietary intake and other exposures.

Prevention

Ensure adequate Mo intake in foods to avoid deficiency, and also avoid high intake of substances that can interfere with availability. Avoid grazing ruminants on low or high Mo pastures, especially with high or low S or Cu.
Avoid excessive dietary or supplemental intake.

Treatment

Humans: with copper toxicosis (Wilson's disease) can give oral tetrathiomolybdate.
Cattle: Mo toxicity can be treated by moving stock off high Mo pasture, and treating with Cu supplementation.
Mice: molybdenum toxicity in mice chelators: ethylenediaminetetraacetate, diethylenetriaminepentaacetate, unithiol and deoxycholate.

References

ATSDR (2020a), Toxicology (2020), Johnson (2020), NIH Molybdenum (2020), IMOA (2020), Choi et al. (2019), IARC (2017), Bourke (2018, 2016, 2015, 2012), Lewis and Meeker (2015), Smith et al. (2014), Higdon et al. (2013), Mendy et al. (2012), Abboud (2008), Campbell et al. (2006), Mitsui et al. (2006), Grace et al. (2005), NAS (2000), Titenko-Holland et al. (1998), McDowell (1992h), Mills and Davis (1987), Pitt (1976), USEPA (1975), Friberg et al. (1975).

Neodymium (Nd)

Nd is a rare earth element. Used in carbon arc lamps for movie projection, metal alloys, high performance magnets, lasers, as a catalyst for cracking crude petroleum and in glass coloring.

Clinical applications

Neodymium is an anticoagulant when given by injection.

Essentiality

Human	No
Other species	No

Parameter	Location
Site of absorption	Inhalation
Site(s) of excretion	–
Sites of metabolism	–

Biological activity

None reported.

Deficiency

None reported.

Toxicity

Nd soluble salts have low to moderate toxicity insoluble Nd salts are nontoxic.

Species	Clinical signs	Comments
Humans	Conjunctivitis (severe), dermatitis, mucus membrane inflammation (severe), liver damage, pneumoconiosis, pulmonary embolism abdominal cramps, increased blood clotting time, chills, fever, muscle aches, hemoglobinuria *In vitro:* toxic to pulmonary alveolar macrophages	Industrial exposure: inhalation and accumulation in the body
Mice	*Acute toxicity* ataxia (loss of muscular coordination), labored breathing, sedation, and stretching of limbs while walking, writhing	Acute toxicity
Rabbits	Conjunctivitis, severe irritation to abraded skin	
In vitro *studies*	Toxic to pulmonary alveolar macrophages in vitro	Suggests Nd is cytotoxic to lung tissue and potentially fibrogenic

Carcinogenesis, mutagenicity, embryotoxicity

No evidence to show Nd is carcinogenic, mutagenic or embryotoxic.

Diagnosis

Confirmation of exposure and analysis of body or environmental samples.

Prevention

Avoid exposure.

Treatment

Treat symptomatically.

References

Lenntech, 2013, Damian (2014), Beaser et al. (1942).
Beaser, S.B., et al., 1942. The anticoagulant effects in rabbits and man of intravenous injection of salts of the rare earths. J. Clin. Invest. 21 (4), 447–454. https://doi.org/10.1172/JCI101321.

Nickel (Ni)

Ni is used in many industries including the production of alkaline (nickel-cadmium) batteries, alloys (nonferrous and super-alloys), catalysts, ceramics, coins in electronic products, electroplating, magnets, pigment, stainless, and heat-resistant steels, and welding products (coated electrodes, filter wire).

Clinical applications

Ni is used in prostheses and implants (joint prostheses, plates, and screws) and is found in contaminated intravenous medications (albumin; dialysis solutions, X-ray contrast).

Essentiality

Human	No
Other species Cattle	Yes
Goat	Yes
Minipig	Yes
Poultry	Yes
Rat	Yes
Sheep	Yes

Parameter	Location
Site of absorption	Small intestine (poorly absorbed)
Site(s) of excretion	Urine, feces (bile), sweat
Sites of accumulation	Lung, bone, thyroid and adrenals; Humans: lungs kidneys, liver and other tissues
Interferes with	Ca, Fe, Zn

Biological activity

May interfere with Zn activity in skin and Ca in skeletal tissue, also can be synergistic with Fe metabolism and facilitate Fe absorption from the intestine.
High Ni in plants interferes with Fe and Mn availability.

Deficiency

Species	Clinical signs
Goats	Impaired growth rate, abortions, reduced conception rates, low hemoglobin, lower milk production, skin lesions (scales and crusts—parakeratosis), skeletal lesions,survivability reduced lower testicular weight
Pigs	*Minipigs*: Impaired growth rate, skin scales and crusts, delayed estrus, poor mineralization of skeleton
Poultry	Low hematocrit, liver lesions
Rats	Impaired growth rate

Toxicity

Human occupational exposure is usually due to skin contact or inhalation or ingestion of dusts, fumes or mists containing nickel or by inhalation of gaseous nickel carbonyl.

Species	Clinical signs
Humans	Abdominal pain, adrenal insufficiency, asthma, allergic contact dermatitis, cancer, (lung and nasal) cough, cyanosis, death, dermatitis, diarrhea, dyspnoea, ECG changes (myocarditis) fever, giddiness, headache, hemianopsia, hyperglycemia, lassitude, leukocytosis, nausea, nephrotoxicity, neurasthenic signs, palpitations, pneumonitis (interstitial), prostration, pulmonary edema, shortness of breath, respiratory disease substernal pain, upper airway irritation, vertigo, vomiting, weakness *Industrial exposure mainly*
Amphibians	*Salamander*: melanoma (after intraocular injection)
Cattle	Reduced feed intake
Dogs	Anemia (mild), growth depressed, hypersalivation, vomiting 2500 μg/g food (Ambrose et al., 1976)
Ducks	Reduced feed intake
Fish	Death (increased mortality rate), reduced food utilization impaired growth, increased liver enzymes activity—SGOT and SGPT
Guinea pigs	Adenomatous alveolar lesions (inhalation) *Subcutaneous injection:* hem oxygenase activity increased
Hamsters	*Inhalation:* alveolar septal fibrosis, atelectasia, bronchial hyperplasia, death, emphysema, granulomatous pneumonia, interstitial pneumonia, squamous metaplasia, decrease in tracheal ciliary activity, tracheal mucosal degeneration *Intratracheal injection* Lung tumors (adenoma, adenocarcinoma) *Subcutaneous injection:* hem oxygenase activity increased *Intramuscular injection:* sarcomas
Marine mammals	*Ringed seals Finland:* Stillbirths
Mice	Reduced feed intake *Subcutaneous injection:* heme oxygenase activity increased *Intramuscular or subcutaneous injection*: local fibrosarcomas, sarcomas, immunosuppression *Inhalation:* degeneration of bronchial epithelium, lymph node hyperplasia (bronchial and mediastinal), atrophy of olfactory epithelium, thymus atrophy, inflammation, fibrosis, hyperplasia alveolar macrophages, necrotizing pneumonia *Intratracheal:* pulmonary hemorrhage
Nonhuman primates	Reduced feed intake
Pigs	Diarrhea, growth impairment, poor haircoat, melena
Poultry	Reduced feed intake
Rabbits	Reduced feed intake *Inhalation:* pneumonia *Intramuscular implants*: rhabdomyosarcomas
Rats	Reduced feed intake *Inhalation*: alveolar macrophage hyperplasia, bronchial epithelium degeneration, fibrosis, lung adenomas (inhalation), squamous metaplasia, inflammation, necrotizing pneumonia,lymoh node hyperplasia (bronchial and mediastinal), thymus atrophy *Intratracheal injection:* alveolitis, lung tumors (mainly squamous cell carcinomas and adenocarcinomas) *Intrapleural injection*: spindle-cell sarcomas, mesothelioma, rhabdomyosarcomas *Subcutaneous injection* (*dental materials*)*:* sarcomas heme oxygenase activity increased (many organs), lipid peroxidation increased (liver kidney lung)

	Intramuscular injection: injection site fibrosarcomas, rhabdomyosarcomas, sarcomas, suppression of natural killer cell activity *Intraperitoneal injections*: carcinomas, mesothelioma, rhabdomyosarcomas, sarcomas, myocardial, degeneration liver necrosis, renal failure (proximal tubular degeneration), seminiferous tubular degeneration *Intravenous injection:* sarcomas *Subperiosteal or intramedullary injection:* bone tumors *Intrarenal injection:* erythrocytosis, renal-cell carcinomas *Intratesticular injection*: testicular sarcomas *Intraocular injection:* Various tumors, e.g., melanoma, gliomas, retinoblastomas *Dermal application:* epidermal acanthosis, epidermal atrophy, disordered epidermal cells, hyperkeratinization. liver necrosis,
Salamander	*Intraocular injection:* melanoma

Carcinogenesis, mutagenicity, embryotoxicity

Some nickel compounds are extremely potent carcinogens after inhalation usually following occupational exposure.
No evidence that nickel is mutagenic, however, nickel can cross the placenta and has embryotoxic and teratogenic properties.
There is sufficient evidence in humans for the carcinogenicity of nickel sulphate, and of the combinations of nickel sulfides and oxides encountered in the nickel refining industry. Nickel compounds are carcinogenic to humans (Group 1). Metallic nickel is possibly carcinogenic to humans (Group 2B).
There is sufficient evidence for the carcinogenicity of metallic nickel, nickel monoxides, nickel hydroxides and crystalline nickel sulfides in animals.

Diagnosis

Measure air and blood serum, plasma, or urine Ni.

Prevention

Ensure adequate PPE in the workplace.

Treatment

Treat symptomatically, avoid further exposure.
(*See Appendix 1 and 2*).

References

NORD (2020a), McDowell (1992p), Nielsen (1987a), Léonard et al. (1981).

Niobium (Nb)

Nb is a hard transition metal with many uses including in alloys, jewelry, jet, and rocket engines, and superconducting materials in magnets and MRI scanners.

Clinical applications

Nb is present in MRI scanners and as it is inert it is used in various prosthetics and implants including pacemakers.

Essentiality

Human	No
Other species	No

Parameter	Location
Site of absorption	Small intestine
Site(s) of excretion	Urine

Biological activity

Nb has no biological role.

Deficiency

Not reported.

Toxicity

Exposure is most likely to occur in the workplace through inhalation or direct contact.

Species	Clinical signs	Comments
Humans	Eye inflammation, skin inflammation	Contact with Nb dust
Cats	Cardiovascular collapse, respiratory paralysis	
Dogs	*IV injection:* glomerular filtration rate (GFR) decreased, the maximal rate of tubular secretion of *p*-aminohippurate decreased, maximal rate of tubular reabsorption of glucose decreased, injuries in both proximal and distal tubules, granular pigment in tubules (cortex and medulla), casts, epithelial cells flattened, cellular necrosis.	Potassium niobite given IV single dose
Mice	Fatty liver, growth suppressed, reduced lifespan	
Rats	*IV injection*: increased thirst, increased urine volume, decreased urinary specific gravity and osmolality. Urinary sodium excretion increased initially then decreased. Lost urine-concentrating ability, failed to respond to injection of antidiuretic hormone, delayed peak urine excretion rate after a water load, proximal and distal tubules showed tubular dilation, hyaline casts, deposits of granular pigment. Occasional necrosis, regeneration, and early fibrosis. Increased Cu, Mn, and Zn in organs	Heart and liver especially

Carcinogenesis, mutagenicity, embryotoxicity

Nb is not considered to be carcinogenic, mutagenic, or embryotoxic.

Diagnosis

History of exposure.

Prevention

Avoid exposure.

Treatment

No specific treatment. Treat symptomatically and avoid further exposure.
(*See Appendix 1 and 2*).

References

Wong and Downs (1966), Haley et al. (1962).

Palladium (Pd)

Pd is one of the platinum group metals used in catalytic converters, jewelry, photography, watch making, glucose test strips, aircraft spark plugs, electrical contacts, and concert flutes.

Medical Uses

In dental amalgams, surgical instruments, and palladium chloride used to be used in the treatment of tuberculosis.

Essentiality

Human	No
Other species	No

Parameter	Location
Site of absorption	Gastrointestinal tract (poorly absorbed), inhalation, skin
Site(s) of excretion	Feces and urine
Sites of metabolism	Not reported
Site(s) of accumulation	Kidney, liver, spleen, lymph nodes, lung, bone

Biological activity

None reported.

Biological effects

Palladium compounds have an affinity for nucleic acids, form bonds with amino acids including cysteine, cystine, and methionine, to proteins and vitamin B_6.

Interferences

Not reported.

Deficiency

Not reported.

Toxicity

Species	Clinical signs
Humans	Contact dermatitis, erythema (topical), eye irritation, fever, hemolysis, subcutaneous injection site reaction (discoloration, necrosis) mucositis, edema (topical), oral lichen planus, respiratory tract inflammation skin sensitivity reaction, stomatitis
Cats	Bronchospasm (IV administration), increased serum histamine
Mice	Amyloidosis, carcinogenic, death, longer life span (males), reduced testes weight
Rabbits	Anorexia, death, reduced defecation, dermatitis: eschar, erythema, and edema in intact and abraded skin. Eye irritation, bone damage, heart damage, kidney damage, liver damage, sluggishness, reduced urine output, water intake reduced
Rats	Anemia, anorexia, ataxia, cardiovascular effects, conjunctivitis, convulsions (clonic-tonic), death, emaciation, exudates, gastrointestinal hemorrhage, gastric ulcers, glomerulosclerosis, hematomas, reduced hepatic enzyme activity, keratoconjunctivitis, ketonuria, liver damage, peritonitis (intraperitoneal administration), proteinuria, pulmonary hemorrhage, pulmonary inflammation, renal damage, sluggishness, testicular damage, tiptoe gait, urination reduced, water intake reduced, weight loss

Clinical applications

None reported.

Carcinogenesis, mutagenicity, embryotoxicity

Pd is carcinogenic in mice.

Diagnosis

Tissue analysis.

Prevention

Avoid exposure.

Treatment

Symptomatic.
(*See Appendix 1 and 2*).

References

Hosseini et al. (2016), Melber and Mangelsdorf (2006), Melber et al. (2002), Emsley (2011).

Phosphorus (P)

P is important in fertilizers, flame retardants, matches, nerve agents, plasticizers, pesticides, steel production, and water softeners.

Clinical applications

As a supplement in patients with low P.

Essentiality

Human	Yes
Other species All	Yes

Biological activity

P has many roles in the body:

1. Structural component of bone (with Ca andmg) and teeth
2. P is in high amounts in red blood cells, skeletal muscle and nerve tissue.
3. P is involved in most metabolic reactions as it is a component of intracellular ATP a major high energy store
4. P is a component of DNA and RNA. Cell growth differentiation and replication
5. Is a buffer in blood and other body fluids
6. Controls appetite
7. Energy metabolism—transfer and use
8. Muscle function
9. Nerve function
10. metabolism (energy metabolism
11. muscle contractions
12. nerve function
13. metabolite transport
14. nucleic acid structure
15. carbohydrate (gluconeogenesis), fat (fatty acid transport), and amino acid (protein synthesis) metabolism
16. production of ATP
17. assists the kidney function
18. helps maintain regularity in heartbeat
19. Cell membrane function and integrity
20. Na:K pump activity

Phytate P is not available to many species, but in ruminants microorganisms produce phytase which makes the bound P available for absorption.

Parameter	Location
Site of absorption	*In simple-stomached:* predominantly as inorganic phosphate in the proximal small intestine, with Na. Facilitated by vitamin D
Site(s) of excretion	*Simple-stomached:* Urine *Ruminants:* Feces and saliva (reabsorbed) and urine
Sites of metabolism	Bone, kidney
Site(s) of accumulation	Bone

Deficiency

Pastures may be deficient in phosphorus for example in a survey of 1106 New Zealand pastures (Knowles and Grace, 2014) 13% would not provide the P requirements of cattle.
P bioavailability reduced by phytate, dietary fiber, and high amounts of other minerals.
Reduced appetite and skeletal abnormalities due to impaired endochondral ossification include poor stunted growth, rickets with poor mineralization of bone, developmental abnormalities, swollen joints, angular deformities lameness and spontaneous fractures.

Species	Clinical signs
Humans	Anorexia, anxiety, bone pain, bone fragility, poor bone mineralization, breathing irregularly, poor growth, fractures, irritability, joint stiffness, numbness, poor tooth development, weakness
Birds	Angular deformity of limbs, beak malformations, bone demineralization, egg abnormalities (shape), egg binding, fractures, poor growth, metabolic bone disease (hyperparathyroidism), osteomalacia, polydipsia, polyuria, reproductive problems, rickets, skeletal deformities, (especially African gray parrots; Timneh parrots)
Buffaloes	Anorexia, death, emaciation, growth retardation, reduced fertility, fractures, hemoglobinuria, hypophosphatemia, joint stiffness, limb abduction, reduced meat production, reduced milk production, neck extension, reluctance to move, paralysis (hindleg), pica, prostration, stiffness weakness
Cats	*Rare due to carnivorous diet.* Anemia (hemolytic), ataxia, metabolic acidosis, dental enamel hypoplasia, rickets, weight loss (*and Ca deficiency*)
Cattle	Poor growth, hypophosphatemia, inappetence, poor milk production, poor fertility, osteomalacia, pica, rickets (calves), unthriftiness
Deer	Antler problems: deformity, delayed velvet shedding
Dogs	angular deformities in long bones, anorexia, emaciation, poor growth, depressed myocardial performance, pica, seizures, rhabdomyolysis, rickets, swelling of joints, tremors
Fish	Alkaline phosphatase increases, reduced feed utilization, poor growth, poor bone mineralization, low hematocrit, skeletal abnormalities *Atlantic salmon*: bone weakness, spinal curvature; *Carp*: cephalic deformities frontal bones *Halibut*: scoliosis *Atlantic halibut and haddock juveniles*: twisted spines)
Foxes	Angular deformity of legs, bones undermineralized, joint enlargement, lameness, undershot jaw
Goats	Bone undermineralized, reduced conception rate, reduced food intake, reduced milk production, suppressed estrus, pica, skeletal deformities, poor weight gain
Horses	*Foals:* Rickets, nutritional secondary hyperparathyroidism *Adults:* lameness, osteomalacia, osteofibrosis, fibrous osteodystrophy "big head", jaw enlargement
Reptiles	Metabolic bone disease, poor bone calcification, renal damage, urolithiasis
Sheep	Bone undermineralized, low conception rate, reduced food intake, reduced milk production, suppressed estrus, pica, skeletal deformities, reduced weight gain

Toxicity

P toxicity results in skeletal disorders especially in young animals, reduced food intake, and poor growth. The maximum tolerance varies with species:

Species	Maximum tolerance
Dogs	Not reported
Cats	Not reported
Cattle, Horses, Rabbits	1% of ration
Sheep	0.6% of ration
Pigs	1.5% of ration
Poultry (most) Laying hens	1% of ration 0.8% of ration

High P intake will reduce the efficacy of utilization of Ca causing signs of deficiency.

Species	Clinical signs
Humans	Abdominal cramps, bronchial spasms, eyelid drooping (upper-ptosis), headaches, hypersalivation, hypocalcemia, muscular weakness, neurological damage (atrophy), pupil constriction (miosis), sensorimotor polyneuropathy, ptosis, respiratory paralysis, soft tissue calcification, sweating, twitching, vomiting, weakness, wheezing
Birds	Bone fragility, poor eggshell
Cats	Dehydration, depression, hyperphosphatemia, metabolic acidosis
Cattle	Hypocalcemia, hypophosphatemia, bone resorption (osteomalacia), urinary calculi, diarrhea, abdominal discomfort (mild, inhibits vitamin D synthesis, reduced reproductive performance
Fish	Impaired growth, skeletal deformities
Giraffes	Urolithiasis
Goats	Poor bone mineralization
Guinea pigs	Nephrocalcinosis
Horses	Poor bone calcification, nutritional secondary hyperparathyroidism, enlarged facial bones (called "big head" or "bran disease")
Marsupials	*Opossums:* Nutritional secondary hyperparathyroidism: bone underminerized, fractures *Sugar-gliders*: osteodystrophy
Rats	Calcinosis, nephrocalcinosis
Sheep	Undermineralized bone, (metabolic bone disease) *Lambs:* urolithiasis

Carcinogenesis, mutagenicity, embryotoxicity

P and its compounds are not considered to be carcinogenic, embryotoxic, or mutagenic.

Diagnosis

P status can be measured in blood and tissue samples.

Prevention

Ensure adequate intake to avoid deficiency and avoid excess intake and exposure to toxic salts to avoid adverse effects.

Treatment

Supplement deficient patients, and avoid further exposure in patients with toxic signs.
(*See Appendix 1 and 2*).

References

NRC (2006), Knochel (1999), Knochel et al. (1978), McDowell (1992b), Fuller et al. (1978).

Potassium (K)

P is used in some soaps and fertilizers.

Clinical applications

Humans: Increasing dietary potassium chloride intake decreases blood pressure in some hypertensive patients.

Essentiality

Human	Yes
Other species All	Yes

Parameter	Location
Site of absorption	Passive diffusion across the small intestine *In ruminants* K is absorbed across the ruminoreticulum, omasum, and intestines
Site(s) of excretion	Urine mainly, feces, very small amount in sweat
Site(s) of accumulation	Plasma, blood, and sweat, bones, muscle

Biological activity

Potassium is the major intracellular electrolyte of importance in the regulation of acid-base balance. It is important in the transportation of carbon dioxide and oxygen in the blood. Like sodium, potassium is important in nerve conductivity and muscle contractions. Other functions include energy transfer, protein synthesis, amino acid uptake into cells, and carbohydrate metabolism.
High K in forage can reduce Mg availability influencing the occurrence of grass tetany.

Deficiency

Deficiency can occur as a result of losses in chronic diarrhea. In poultry, as a response to severe stress and urinary potassium loss under the influence of aldosterone. In ruminants stress is also an important factor in K requirements Ruminants are more prone to develop K deficiency than monogastric species.
K deficiency has been reported in kittens within 2 weeks of being fed a vegetarian diet (Leon et al., 1992).

Species	Clinical signs	Comments
Humans	Arrhythmias, intestinal muscle tone loss, intestinal distension, impaired neuromuscular function, paralysis, loss of reflexes, mental confusion, soft flabby muscles, muscle weakness, respiratory failure, weakness	K deficiency often seen with severe dehydration due diarrhea, surgery or burns
Cats	anorexia, ataxia, cardiac failure, poor growth, hypercreatininemia, immune function depressed, muscle weakness, depressed reflexes, ventroflexion of neck	Usually due to K losses in urine or feeding a vegetarian diet.
Cattle	*Experimental*: Anorexia, death, degeneration of vital organs, emaciation, food utilisation reduced, growth impairment, hide loses pliability, hypokalaemia, intracellular acidosis, licking objects, milk yield reduced, muscle wekness, nervus disorders, pica, stiffness, water intake reduced, weakness	
Dogs	Bradycardia, cardiac output reduced, dehydration, growth impaired, hypotension, muscle potassium reduced, paralysis (hindlegs), renal blood flow reduced, stroke volume reduced, ventroflexion of the neck, weakness	
Fish	Anorexia, convulsions, death, tetany	

Goats	Arched back, emaciation, reduced food intake, impaired growth, muscle paralysis, rhabdomyolysis, stiffness	
Horses	Reduced appetite, growth impaired, hypokalemia, unthriftiness	
Mice	Coat dull, emaciation, tail dry and scaly, weight loss	
Pigs	*Experimental*:appetite reduced, ataxia, bradycardia, emaciation, growth rate reduced, haircoat rough, lethargy, increased PQRS intervals	
Poultry	*Experimental:* Cardiac failure, muscle weakness, ileus, respiratory failure. *Chicks*: Bone mineralisation poor, death, reduced feed intake, impaired growth, high mortality,paralysis, polyuria, protein metabolism reduced, renal and ureteral lesions, tetanic seizures *Laying hens*: collapse, death, reduced egg production, egg weight, shell thickness, albumin content, weakness,	
Rabbits	Muscular dystrophy	
Rats	Coat rough, death, edema, food intake reduced, poor growth rate, Lesions in many organ tissues	
Sheep	*Experimental*:Death, emaciation, food intake reduced, poor growth, listlessness, hair rough, hypoglycaemia, liver glycogen increased, stiffness, responses impaired, urinary calculi, weight loss, wool loss. Histological changes in many organs.	

Toxicity

High dietary Mg intake may protect against K toxicity.
Hyperkalemia has been documented in deer with postcapture myopathy and during the rut.

Species	Clinical signs
Humans	Confusion, death, ECG changes (tall T waves, absent P waves, wide QRS), hyperkalemia, lethargy, muscle weakness, palpitations, flaccid paralysis/paresthesia of hands and feet
Cattle	*Experimental:* death, diarrhea with mucus, excess salivation, muscle tremors, excitability, hyperkalemia, high PCV food intake and milk yield reduced, polydipsia, polyuria, hyperkalemia *Calves*: death
Goats	Hypocalcemia, hypomagnesemia
Pigs	Bradycardia
Sheep	Low Ca (milk fever), low Mg (grass tetany)
Various species	Cardiac insufficiency (bradycardia), death, edema, muscle weakness

Carcinogenesis, mutagenicity, embryotoxicity

Potassium and its salts are not considered mutagenic or embryotoxic.
Potassium bromate is carcinogenic in rats and nephrotoxic in both man and animals when given orally.

Diagnosis

Measure blood K concentrations.

Prevention

Ensure adequate dietary intake and avoid excessive intake.

Treatment

Deficiency: Supplements include potassium salts—acetate, bicarbonate, carbonate, phosphate, (dibasic or monohydrate), or sulfate.
Toxicity: treat symptomatically.
(*See Appendix 1 and 2*).

References

Tao et al. (2016), NRC (2006), McDowell (1992d), Leon et al. (1992), Kurokawa et al. (1990), Patterson et al. (1983), Hove and Herndon (1955).

Praseodymium (Pr)

Pr is a rare earth element used in carbon arc lamps for movie projection, metal alloys for the aerospace industry, as a catalyst for cracking crude petroleum, and glass coloring.

Clinical applications

None reported.

Essentiality

Human	No
Other species	No

Parameter	Location
Site of absorption	Gastrointestinal tract (poor)
Site(s) of excretion	Feces
Sites of accumulation	Liver and bone

Biological activity

Not reported.

Deficiency

Not reported.

Toxicity

Pr has low to moderate toxicity. Case reports of respiratory disease following environmental exposure are confounded by concomitant exposure to more than one rare earth element and other substances such as silica dust, so the actual contribution of Pr to the signs is unknown. Nevertheless, it is accepted that environmental exposure and inhalation of Pr could account for the signs seen (Haley, 1991).
The reliability of the reported studies on toxicity in laboratory animals is considered low

Species	Clinical signs	Comments
Humans	Eye irritant, lung embolisms, liver damage, pneumoconiosis, pulmonary fibrosis, skin irritant	
Dogs	Deaths, drop in blood pressure and heart rate, increase in respiratory rate, increased prothrombin time	*Intravenous injection* (Graca et al., 1964) Variable results between different salts
Mice	*Acute toxicity* arched back, ataxia, labored breathing, sedation, stretching of limbs while walking, walking on toes, writhing	
Rabbits	Conjunctivitis, severe irritation on abraded skin	
Rats	Hepatotoxicity (fatty degeneration, altered hepatic microsomal lipid content, decreased RNA polymerase activity, reduced gluconeogenesis, reduced drug metabolizing enzymes)	*Intravenous injection*

Carcinogenesis, mutagenicity, embryotoxicity

In vitro praseodymium chloride increased the frequency of micronuclei in human lymphocytes (Hui et al., 1998) and in mice praseodymium oxide (single intraperitoneal injection) induced chromosomal aberrations in bone marrow cells (Jha and Singh, 1995).
In accordance with the 2005 Guidelines for Cancer Risk Assessment (USPA, 2009) data for stable (nonradioactive) praseodymium provided "Inadequate Information to Assess [the] Carcinogenic Potential" of praseodymium or its compounds.

Diagnosis

Known exposure, tissue levels.

Prevention

Avoid exposure.

Treatment

Treat symptomatically and avoid further exposure.

References

ATSDR (2020), Lenntech (2013), Damian (2014), USPA (2009), Hui et al. (1998), Jha and Singh (1995), Haley (1991), Haley et al. (1964), Graca et al. (1964).

Promethium (Pm)

A rare-earth metal in the lanthanide group. The radioisotope Promethium-147, with a half-life of 2.6 years is a beta radiation source used in in nuclear batteries for guided missiles, luminous paint, radios, pacemakers, and watches.

Clinical applications

Not reported.

Essentiality

Human	No
Other species	No

Parameter	Location
Site of absorption	Poorly absorbed from gastrointestinal tract
Site(s) of excretion	Feces, urine
Site(s) of metabolism	Liver
Sites of accumulation	Gastrointestinal tract, liver, bone (very small amounts)

Biological activity

Not reported.

Deficiency

Not reported.

Toxicity

Bone tissue may be affected.

Species	Clinical signs	Comments
Humans	Potentially dangerous	Protective clothing should be worn to prevent skin/ocular contact or inspiration

Carcinogenesis, mutagenicity, embryotoxicity

No evidence for carcinogenesis, mutagenicity or embryotoxicity.

Diagnosis

History of exposure, tissue analysis.

Prevention

Avoid exposure.

Treatment

Treat symptomatically and avoid further exposure
(*See Appendices 1, 2, and 3*)

References

Lenntech (2020b), EPA (2007b).

Rubidium (Rb)

Rb is an alkali metal used in the manufacture of special glasses, photocells, and vacuum tubes.

Clinical applications

Rubidium chloride is used in medicine as an antidepressant to treat patients with neurological disease such as depression and bipolar disease. These studies have generated most of the data on Rb clinical signs.

Essentiality

Human	No
Other species	No

Parameter	Location
Site of absorption	Gastrointestinal tract
Site(s) of excretion	Gastrointestinal tract and urine
Sites of metabolism	Not reported

Biological activity

Rb has interchangeability with K in lower forms of life than mammals.

Interferences

Not reported.

Deficiency

Species	Clinical Signs	Comments
Goats	Abortion, high mortality, low birth weight, low weaning weight	*Need corroborating—not accepted as an essential nutrient yet*

Toxicity

Species	Clinical Signs	Comments
Humans	Agitation, confusion, dermatitis, diarrhea, excitement, nausea, polyuria, skin rash, vomiting, weight gain	
Dogs	Colonic mucosal congestion (at PME) vomiting	*PME—post mortem examination*
Mice	Seizures (audiogenic)	
Nonhuman primates	Aggression, increased low frequency changes on EEG, hyperactivity, increased locomotor activity	*EEG—electroencephalograph*
Rats	Aggression (after intraperitoneal injection), death, degeneration of internal organs, (myocardium, liver, kidney on PME), low red cell count, low hemoglobin (females), motor activity increased, pulmonary edema and pneumonitis (at PME), reduced saliva production, seizures (audiogenic), reduced survival time	

Carcinogenesis, mutagenicity, embryotoxicity

Rb does not appear to be carcinogenic, mutagenic or embryotoxic in animals or humans.

Diagnosis

Confirmed exposure.

Prevention

Avoid exposure.

Treatment

Treat symptomatically, avoid further exposure.

References

EPA (2016), Nielsen (1986).

Samarium (Sm)

Sm is a rare earth element. Used in carbon-arc lamps for movie projection, permanent magnets, organic reagents, lasers, and alloys.

Clinical applications

Radioactive samarium Sm 153 lexidronam has been used to provide pain relief in some cancers (e.g., bone).

Essentiality

Human	No
Other species	No

Parameter	Location
Site of absorption	Poorly absorbed after ingestion. Inhalation
Site(s) of excretion	Feces mainly, urine
Sites of accumulation	Liver mainly, bone

Biological activity

Not reported.

Deficiency

Not reported.

Toxicity

Sm insoluble salts are nontoxic, soluble salts are slightly toxic.

Species	Clinical signs	Comments
Humans	Cyanosis, dyspnoea, pneumoconiosis, progressive pulmonary fibrosis, ulcers on abraded skin	Confounded as other rare earth elements or silicon often present at same time as exposure to Sm
Dogs	*Acute toxicity study*: circulatory failure, death, reduced heart rate, hypotension	
Mice	*Acute toxicity IP injection* abdominal cramps, diarrhea, labored breathing, lethargy, muscular spasms Growth and fertilization rates decreased, LDH enzyme activity decreased in testes	
Rabbits	Increased blinking but no conjunctivitis, severe irritation to abraded skin	
Rats	decreased growth, liver damage, higher liver weight/ body weight increased ratio, a substantial reduction in the antioxidant enzyme superoxide dismutase (in liver and brain), increase in liver malondialdehyde (MDA)	
In vitro studies	Antineoplastic and cytotoxic effects against some cancers	Reported

Carcinogenesis, mutagenicity, embryotoxicity

No evidence for Sm cytotoxicity, carcinogenicity, or mutagenicity.

Diagnosis

Known exposure, measurement of body, or environmental samples.

Prevention

Avoid exposure especially in the workplace.

Treatment

Treat symptomatically.

References

Rim et al. (2013), EPA (2009b), Damian, 2014, Graca et al. (1964).

Scandium (Sc)

Sc is a rare earth element and a transition metal that is naturally occurring in the environment.
Sc alloys with Al are used in the aerospace industry, for sports equipment (baseball bats, bicycle frames, fishing rods, golf irons), in mercury vapor lamps for film/television industry.

Clinical applications

Sc is being used as a marker in food studies.

Essentiality

Human	No
Other species	No

Parameter	Location
Site of absorption	Not reported
Transportation	Transferrin, protein-bound
Site(s) of excretion	Urine
Sites of accumulation	Liver

Biological activity

Sc is found in higher amounts in heart and conductive tissues. No biological role determined.

Interferences

Not reported.

Deficiency

Not reported.

Toxicity

Sc is generally regarded as being safe but scandium chloride is toxic by ingestion and intraperitoneal injection. Inhalation through industrial exposure can cause lung disease (Rim et al., 2013).

Species	Clinical signs	Comments
Humans	Severe burns (skin), severe eye damage, respiratory irritation	Corrosive
Fish	Reproductive failure	
Rats	Death	Acute study

Carcinogenesis, mutagenicity, embryotoxicity

Sc may be mutagenic, teratogenic (Gebhart and Rossman, 1991) and carcinogenic (Horovitz, 2000) However, according to the IARC: Not identified as carcinogenic.

Diagnosis

History of exposure. Environmental analysis for Sc.

Prevention

Avoid exposure in the workplace.

Treatment

Treat symptomatically and avoid further exposure. Caustic so advised not to induce emesis or perform gastric lavage due to risk of perforation.

References

ThermoFisher Scientific (2020), Bilčíková et al. (2016), Horovitz (2000), Gebhart and Rossman (1991).

Selenium (Se)

Uses of Se include: antidandruff shampoos, copper refining, dietary supplements, gun bluing solution, glass manufacturing, insecticides, and rubber vulcanization.

Clinical applications

There is some good evidence that supplemental Se may be beneficial in the management of cancer (Dennert et al., 2011 and HIV (Burbano et al., 2002; Hurwitz et al., 2007), but more studies are required to confirm these findings and also whether it is beneficial in preventing cardiovascular disease, cognitive decline or thyroid disease.
Radioisotope Se is used in diagnostic medicine.

Essentiality

Human	No
Other species Horses	No Yes—0.1 mg/kg diet

Parameter	Location
Site of absorption	Absorbed efficiently in nonruminants.
Site(s) of excretion	Fecal (camels)
Sites of metabolism	Not reported
Site(s) of storage	Skeletal muscle

Biological activity

Se is a component of the enzyme glutathione peroxidase involved in detoxification of lipo- and hydrogen peroxides which are toxic to tissues. Linked to vitamin E activity, Se has a critical role in DNA synthesis, as an antioxidant, in immunity, in reproduction and metabolism of thyroid hormone.

Interferences

Not reported.

Deficiency

Pastures may be deficient in Se for example 24% in a survey of 1106 pastures in New Zealand (Knowles and Grace, 2014). In humans Se deficiency is most likely to occur in vegan/vegetarian people where soil is low in Se. Se levels are low in patients on long term renal dialysis, and HIV patients.

Species	Clinical signs
Humans	Cardiomyopathy (Keshan disease China), low glutathione peroxidase, infertility (males), muscle pain, osteoarthritis (Kashin-Beck disease -China, Siberia, Tibet) weakness
Birds	Dermatitis, encephalomalacia, exudative diathesis, feather abnormalities (brittle, frayed), malabsorption, maldigestion, muscular dystrophy, subcutaneous edema, oral paralysis (cockatiels), ventricular muscle degeneration
Buffaloes	Unthriftiness
Cats	*Kittens:* Reduced GSH-px concentrations in blood

Cattle	Abortions, cardiac changes (frequency, rhythm), conception rate reduced, fetal heart lesions, glutathione peroxidase increased in granulosa cells, mastitis, metritis, muscular dystrophy, myopathy, ovarian cysts, phagocytosis (white cells) impaired, recumbency, impaired reproduction, retained placenta, stiffness, stillbirths, weakness, white muscle disease (*calves*)
Deer	Low conception rate, myopathy (white muscle disease), unthriftiness
Dogs	Anorexia, coma, depression, dyspnoea, edema (ventral), muscle pallor, muscle degeneration, skeletal muscle discoloration (white streaks), intestinal muscle discoloration (brownish-yellow) due to lipofuscinosis, myopathy, renal white mineral deposits at corticomedullary junction, subendocardial necrosis in cardiac ventricular muscles
Fish	Exudative diathesis, high glutathione peroxidase (liver, plasma), impaired growth, muscular dystrophy
Goats	Cardiac changes (frequency, rhythm), muscular dystrophy, recumbency, stiffness, weakness, white muscle disease
Horses (*and Donkeys*)	Anorexia, aspartate aminotransferase activity increased, increased aspartic-pyruvic transaminase, blood urea nitrogen (BUN) increased, cardiac function impairment, creatine kinase activity increased, dysphagia, fever, gamma-glutamyl transferase increased, hematocrit low, hyperkalemia, lactate dehydrogenase high locomotor difficulties myopathy, neck pain, edema (ventral), recumbency, respiratory distress, steatitis (yellow fat disease), stiffness, suckling difficulty, weakness, weight loss, white muscle disease
Pigs	Death, hepatosis, myopathy, mulberry heart disease (*and Vitamin E deficiency*)
Reptiles	Myopathy, impaired immune function, impaired vision, neurological signs (*and Vitamin E deficiency*)
Sheep	Arrhythmia – cardiac changes (frequency, rhythm), muscular dystrophy, early embryonic loss/failure to implant, reduced growth rate recumbency, stiffness, tachycardia, high P wave, shorter PR, ST and QT interval, short T wave weakness, white muscle disease, muscular dystrophy (white muscle disease), reluctance to move,

Toxicity

Ingestion of excessive Se in foods (e.g., Brazil nuts = 68–91 mcg per nut) or nutritional supplements (MacFarquhar et al., 2010) can result in toxicosis. Chronic exposures to nutritional supplements and in rural farmland that has high Se content. In production animals selenium toxicity is called "alkali disease syndrome."

In humans serum Se of 400–30,000 μg/L associated with acute toxicity, 500–1400 μg/L associated with chronic toxicity, and <1400 μg/L free of toxicity (Nuttall, 2006).

Species	Clinical signs
Humans	Abdominal pain, alliaceous breath, alopecia, anorexia, anesthesia, ataxia, bone erosion (long bones) cardiac failure, cardiomyopathy (toxic), chest pain, prolonged clotting time, coma, confusion, congestion (internal organs), high serum creatinine, high serum creatine kinase (CK) activity, cyanosis (extremities), death, delirium, dermatitis (chronic), diarrhea, dizziness, fingernail changes (brittleness, deformity, discoloration, white spots and longitudinal streaks, loss), ECG changes (T-wave flat and inverted, high ST, prolonged QT interval), fasciculations, fatigue, flushing (face), gastroenteritis (severe), garlic smell on breath, glycosuria, hemiplegia, hepatic degeneration, hyperreflexia, hyporeflexia, hypotension, irritability, joint pain, kidney failure, lassitude, light headedness, liver dysfunction, mesenteric infarction, metabolic acidosis, metallic taste in mouth, mucosal damage (mouth, esophagus, stomach), muscle spasms, muscle tenderness, myocardial depression, myocardial infarction, myoclonus, nausea, neurological damage, neurological dysfunction (impaired vision, ataxia, disorientation), pain in extremities, paraesthesia (peripheral), ptyalism, pulmonary edema, renal insufficiency, acute respiratory distress syndrome, restlessness, skin rashes, smooth muscle degeneration (bladder, gall bladder, gastrointestinal tract), splenomegaly, tachycardia, teeth discoloration (mottling), tremors, ventricular dysrhythmias, vomiting, weakness *Inhalation* (*Hydrogen selenide*): eye pain, pneumomediastinum, restrictive and obstructive pulmonary disease, rhinorrhea, throat pain, wheezing. *Selenium dioxide and selenium oxide:* forms selenius acid in presence of water in respiratory tract: bronchospasm, diarrhea, fever, hypotension, pneumonitis (chemical), tachycardia, tachypnoea, vomiting *Dermal/eye contact:* caustic burns, conjunctival edema, corneal ulcers, lacrimation, skin erythema, heat, rashes, swelling, and pain *Chronic selenosis*: pruritic scalp rash Erythematous skin blisters, anesthesia, hemiplegia, paraesthesia

Birds	Abdominal distension, ascites, coughing, death, depression, dyspnoea, hemochromatosis (iron storage disease) hemosiderosis (liver) hemochromatosis (iron storage disease—most commonly passerines, e.g., toucans (often asymptomatic) Mynah birds, birds of paradise) weakness.
Buffaloes	Emaciation, death, hoof and horn deformities, detached hooves, gangrenous of extremities, skin cracks, skin sloughing, recumbency
Cattle	Abortion, anorexia ataxia, body condition loss, cardiac atrophy, coma, constipation, coronary band inflammation, death, decubital ulcers, depression, diarrhea, dyspnoea, emaciation, excitability, garlic smell on breath, gastrointestinal irritation, hypersalivation, incoordination, lameness (joint erosion); horizontal grooves or cracks in the hooves; hoof .horn malformation, occasional sloughing of hooves; incoordination, inappetence, infertility, polioencephalomalacia (on histology), pulmonary edema, paralysis, reluctance to move, long periods of recumbency; reproductive failure, tachycardia, severe unthriftiness, defective vision *India:* gangrene of distal extremities (below the tarsus or carpus, tips of ears, muzzle, tip of tail, or tongue) occurs occasionally.
Deer	Anorexia, death, lameness, myocardial fibroplasia, myocardial mineralization, myocardial necrosis, weight loss
Dogs	Anemia (microcytic, hypochromic), ascites, reduced food intake, reduced growth rate, low hemoglobin, liver damage (atrophy, cirrhosis, necrosis)
Fish	Reduced activity, impaired growth, reduced feed efficiency, high mortality, nephrocalcinosis, teratogenic malformations
Horses	*Acute:* Abdominal pain, blindness, colic, diarrhea, head pressing, lameness, lethargy, sweating, tachycardia, tachypnoea. *Deaths reported following injection of Se Vitamin E preparations* *Chronic:* alopecia (mane and tail especially), cracking of hooves around coronary band, emaciation
Pigs	Alopecia, anorexia, cachexia, central nervous system depression, conception rate low, coma, death, depression, emaciation, emesis, increased glutathione peroxidase levels, hoof lesions (cracks, deformities), lethargy, postnatal mortality high, focal symmetrical poliomyelomalacia, respiratory distress, rough skin, reproductive failure, reduced conception rate, paralysis (posterior), high SGOT, high SGPT, poor weight gain, weight loss
Poultry	Failure of eggs to hatch due embryo deformities, increased mortality rate, reduced food intake, weight loss
Rabbits	Death
Reptiles	Abnormalities to the nails and skin and softening of the bones (selenosis).
Rats	High bilirubin, cataracts, cirrhosis, reduced food intake, increased glutathione peroxidase levels, reduced growth, low hemoglobin, hepatic atrophy, impaired reproduction
Sheep	Death, kidney damage, congenitally deformed eyes in offspring of pregnant ewes

Carcinogenesis, mutagenicity, embryotoxicity

Se compounds cause teratogenic deformities in hamster and rat fetuses and bird embryos. (Usami and Ohno, 1996).
According to IRIS (1991) "The evidence for various selenium compounds in animal and mutagenicity studies is conflicting and difficult to interpret; however, evidence for selenium sulfide is sufficient for a B2 (probable human carcinogen) classification."
According to the ATSDR website (accessed 25th August 2020): NTP: Reasonably anticipated to be a human carcinogen. EPA: Not classifiable as to human carcinogenicity. EPA: Probable human carcinogen. IARC: Not classifiable as to carcinogenicity to humans.

Diagnosis

Tissue concentrations of Se can be used: blood, milk, plasma, serum, erythrocytes, urine, hair, and nails.
Glutathione peroxidase activity reflects Se status
Horses: Serum, plasma or whole blood concentrations. Serum or erythrocyte glutathione peroxidase (GSH-px) can also be used.
Urinary Se to creatinine ratio, serum Se, and serum and whole blood glutathione peroxidase can be used as biomarkers of selenium status in dogs (van Zelst et al., 2016).

Prevention

Avoid exposure especially in the workplace.

Treatment

Symptomatic treatment and avoid further exposure, chelation is not recommended nor is the induction of emesis. (*See Appendix 1 and 2*).

References

NIH (2020), WikEM (2020), Hosnedlova et al. (2017), van Zelst et al. (2016), Stoffaneller and Morse (2015), ATSDR Selenium (2011), Al-Dissi et al. (2011), Dennert et al. (2011), Calellor (2011), MacFarquhar et al. (2010), Faye and Seboussi (2009), Hurwitz et al. (2007), Sunde (2006), See et al. (2006), Nuttall (2006), Burbano et al. (2002,), Usami and Ohno (1996), McDowell (1992m), IRIS (1991), Rogers et al. (1990), NRC (1989), Olson (1986), Levander (1986), Van Vleet (1975)

Silicon (Si)

Si is widely used in manufacturing industries for alloys (Al-Si; Fe-Si) to make for example dynamo and transformer plates, engine blocks, cylinder heads, machine tools.

Clinical applications

In the form silicone (Si, Oxygen, and other molecules) is used to make breast implants, medical tubing, and other devices. Over-the-counter products containing Si are used by humans for perceived health benefits but adequate efficacy and safety data are not available. There is evidence that Si supplementation may improve bone mineral density (Jugdaohsingh, 2007).

Essentiality

Humans	No
Other species	
Chicks	Yes
Rats	Yes

Parameter	Location
Site of absorption	Small intestine
Site(s) of excretion	Urine
Sites of accumulation	Lungs (inhaled), soft tissues (kidney, liver, heart)

Biological activity

Si is important in calcification and maturation of bone. It is also involved in collagen synthesis probably mediated through prolyl hydroxylase enzyme activity (Carlisle, 1986).

Deficiency

Si deficiency in chicks reduced glycosaminoglycans and collagen concentrations in tibia and pathological changes in epiphyseal cartilage (Carlisle, 1986).
Si deficiency may be associated with atherosclerosis, osteoarthritis, and hypertension in humans (Nielsen, 1988)

Species	Clinical signs	Comments
Humans	Atherosclerosis, hypertension, osteoarthritis,	Nielsen, 1988 – Not confirmed. Si is not considered an essential nutrient for humans
Poultry	*Chicks:* cartilage damage, reduced collagen and glycosaminoglycans	Carlisle, 1986
Rats	Impaired growth	Schwarz and Milne, 1972

Toxicity

Intake of Si in soil is associated with Silicate uroliths in dogs and uroliths/nephroliths in ruminants.
Occupational exposure to Silica (Si and oxygen) is the main cause of toxicity in humans.

Species	Clinical signs	Comments
Humans	Increased incidence of autoimmune diseases (e.g., rheumatoid arthritis, systemic lupus erythematosus), death (silicosis), decreased lung function, chronic obstructive pulmonary disease (COPD), kidney disease, lung cancer, silicosis	Following long term exposure in the workplace
Dogs	Silica urolithiasis	Usually following ingestion of soil
Rats	Pulmonary fibrosis, pulmonary inflammations	Silicosis not reported

Carcinogenesis, mutagenicity, embryotoxicity

According to the IARC (2017) Fibrous silicon carbide is possibly carcinogenic to humans (Group 2B) and Silicon carbide whiskers are probably carcinogenic to humans (Group 2A).
Silica is carcinogenic to humans.
There is sufficient evidence in animals for the carcinogenicity of silicon carbide whiskers.
Silicon nanocrystals caused DNA damage in mice and rats after intraperitoneal injection (Durnev et al., 2010).

Diagnosis

Confirmed exposure.

Prevention

Avoid exposure especially in the workplace.

Treatment

Treat symptomatically, avoid further exposure.

References

ATSDR Silica (2020), Durnev et al. (2010), Jugdaohsingh (2007), McDowell (1992p), Carlisle (1986), Schwarz (1977), Schwarz and Milne (1972).

Silver (Ag)

Ag is a precious metal widely used in metallic form, as salts and as nanosilver for many purposes including for air quality management (indoor), contraceptive devices decorative and functional items, cosmetic products, wound dressings, jewellery, textiles, water purification systems,

Clinical application

Ag has been used to treat Se toxicity and is used in bone prostheses, surgical instruments, and Ag-impregnated dressings.

Essentiality

Human	No
Other species	No

Parameter	Location
Site of absorption	Poorly absorbed
Site(s) of excretion	Feces
Interferes with other mineral biovailability	Cu, Se

Biological activity

Not reported.

Deficiency

Not reported.

Toxicity

Exposure from skin contact, inhalation, and ingestion especially from occupational exposure.

Species	Clinical signs
Humans	Allergy, basophil stippling, diarrhea, nausea, organ discoloration (gray), skin discoloration (gray- Agyria), vomiting
Ducks	Signs of Vit E/Se deficiency—see Se
Pigs	Signs of Vit E/Se deficiency—see Se
Poultry	*Chicks*: Reduced aortic elastin, cardiomegaly, death, growth depression, low hemoglobin
Rats	Signs of Vit E/Se deficiency—see Se
Turkeys	Signs of Vit E/Se deficiency—see Se

Carcinogenesis, mutagenicity, embryotoxicity

Ag is not considered to be carcinogenic, embryotoxic, or mutagenic for humans.
However, in rats tumors developed at subcutaneous injection sites, and silver nanoparticles (Ag-NPs) may have the potential to induce oxidative stress mediated genotoxicity (Patlolla et al., 2015).
Low embryotoxicity has been demonstrated linked to Ag dental alloys.

Diagnosis

Analysis of blood, urine, hair, skin can be used.

Prevention

Avoid exposure especially in the workplace.

Treatment

Treat symptomatically, avoid further exposure.

References

NORD (2020a), Imai et al. (2016), Patlolla et al. (2015), ATSDR Silver (1990), Nielsen, 1986.

Sodium (Na)

Na is widespread in the environment, being a component of mineral deposits on land and salt water. It is used in many industries including the manufacture of bleaches, cleaners, cotton fabrics, dyes, explosives, paper, petroleum products, soaps and in electroplating, and electrolytic extraction.

Clinical applications

In sodium chloride intravenous fluid solutions.

Essentiality

Human	Yes
Other species	Yes—All

Parameter	Site(s)
Site(s) of absorption	Co-transported with glucose and amino acids, and by Na^+/H^+ exchange across the small intestine. Gills in fish
Site(s) of excretion	Kidneys, sweat
Transportation	In blood and extracellular fluids
Accumulation	30% in bones, 70% in bodily fluids (plasma, blood, sweat)

Biological activity

Na is inextricably linked to chlorine (Cl) in body systems as well as in nature and it is sometimes difficult to determine whether clinical signs are due to Na or Cl levels.
Na is important in:

- Maintaining acid-base balance
- Controlling extracellular volume through osmotic pressure
- Electrical potential in neurological and muscular tissue
- Nutrient transfer across the intestinal tract and into cells

Deficiency

Pastures may be deficient in Na, for example, in a survey of 1106 New Zealand pastures (Knowles and Grace, 2014) 22% would not meet the requirements of cattle.

Species	Clinical signs	Comments
Humans	Collapse, death, dizziness, exhaustion, fatigue, headaches, muscle cramps, vomiting	Due to excessive sweating or prolonged vomiting or diarrhea
Birds	Cannibalism, convulsions, death, reduced egg production, feather plucking due dry pruritic skin, gout (visceral), kidney damage, paralysis, pecking behavior, seizures	
Cats	Anorexia, impaired growth, polydipsia, polyuria, high hematocrit, low urine specific gravity, high plasma aldosterone	
Cattle	Loss of appetite, cardiac arrhythmias, death, dehydration, rough haircoat, incoordination, reduced milk production, pica, shivering, unthriftiness, weakness, weight loss	

Dogs	Alopecia, dehydration, exhaustion, increase hematocrit, increased hemoglobin concentration, fatigue, mucous membranes dry and tacky, polydipsia, polyuria, restlessness, retarded growth, dry skin, tachycardia	
Gerbils	Alopecia	
Goats	Loss of appetite, emaciation, reduced food intake, reduced efficacy of food utilization, poor growth shaggy dull haircoat, pica, poor weight gain, wobbly gait	
Horses	Anorexia, chewing uncoordinated, fatigue, reduced rate of feed intake, gait unsteady, decreased milk production, muscle spasms, pica (licking materials, e.g., mangers, rocks, skin turgor reduced, reduced thirst	Na deficiency can be induced by excessive sweating.
Marine mammals	Anorexia, collapse (Addisonian crisis), convulsions, death, hypochloridemia, hyponatremia, incoordination, lethargy, tremor, weakness	
Mink	Nursing sickness	Low Na may have a role in this
Pigs	Reduced appetite, poor growth rate, reduced efficacy of food utilization, licking metal cages, performance reduced, low body weight in neonates	
Poultry	Adrenal hypertrophy, appetite reduced corneal keratinization, gonadal inactivity, impaired feed utilization, decreased plasma volume,increased water consumption, impaired protein and energy metabolism. *Laying hens:* cannibalism, reduced egg production and egg size, reduced hatchability, weight loss	
Rabbits	Poor weight gain	
Rats	Reduced appetite, corneal lesions, death, undermineralized (soft) bones, infertility (males), growth impaired, delayed sexual maturity (females)	
Sheep	Reduced food intake, impaired growth, pica	

Toxicity

High Na intake can be compensated for by increasing water intake to maintain balance.
Many Na containing household products such as bleach (sodium hypochlorite) or sodium hydroxide can be very corrosive.

Species	Clinical signs	Comments
Humans	Confusion, chemical burns (to the oral cavity, pharynx, esophagus, stomach, small intestine), coma, death, hypernatremia, hypertension, paralysis respiratory muscles, seizures, vomiting *Teratogenic effects (sodium valproate)*: neural tube defects, cleft lip and palate, cardiovascular abnormalities, genitourinary defects, developmental delay, endocrinological disorders, limb defects, and autism	Normal plasma Na is 136–145 mM, over 152 mM can cause seizures and death Death can result from accidental intake of salt instead of sugar, baking soda
Birds	Diarrhea, muscle weakness, polydipsia, polyuria	
Cattle	Anhydremia, anorexia, collapse, udder edema, reduced milk yield, increased water intake, reduced water intake (if very saline water), weight loss	
Dogs	Abdominal pain, reduced food intake, gastritis, hematemesis, hypersalivation (ptyalism), pharyngeal edema, vomiting	
Ferrets	Brain edema, cerebellar coning, death, depression, meningitis (nonsuppurative), periodic choreiform spasms	
Fish	Impaired growth, reduced feed conversion (*Salmon and Trout*)	
Marsupials	*Wombat: Sodium fluoroacetate*: Heart failure	
Mink	? Impaired reproductive performance, reduced water intake	

Carcinogenesis, mutagenicity, embryotoxicity

Sodium saccharin is carcinogenic causing bladder cancer in rats and under test conditions NaCl is also genotoxic (Ashby, 1985) Na salts (e.g., sodium valproate) can be teratogenic (Alsdorf and Wyszynski, 2005).
However, the common form of Na ingested (NaCl) and others authorized for use in foods are considered safe.

Diagnosis

Calculated intake from content in foods and water, and analysis of Na in blood.
Measure blood Na concentration.

Prevention

Avoid ingestion of high Na foods or drinking salt water. Ensure adequate intake in food and water.

Treatment

Restore normal Na and electrolyte balance by the use of appropriate intravenous fluids. Ensure adequate nutritional supply.

References

NRC (2006), Alsdorf and Wyszynski (2005), Zentec and Meyer (1995), McDowell (1992c), Ashby (1985).

Strontium (Sr)

Sr is naturally occurring element used in manufacture of ceramics, fluorescent lights, glass, paint pigments, and pyrotechnics
Radioactive isotopes (e.g., Sr^{90} are artificially formed in nuclear reactors or after nuclear weapon use.

Clinical applications

Radioactive Sr rods have been used to treat cancer of eye and skin, and it is used in the manufacture of medicines.

Essentiality

Human	No
Other species	No

Parameter	Location
Site of absorption	Small intestine, transdermal, inhalation
Site(s) of excretion	Kidneys (mainly) and bile, sweat
Interferes with other mineral bioavailability	NR
Accumulates in body	Bone, teeth
Sr absorption interfered with by	Ca, P

Biological activity

None reported.

Deficiency

Not reported in any species.

Toxicity

Exposure is through inhalation, ingestion (food or water).
Stable Sr signs only likely if animal has low Ca intake, strontium chromate may be carcinogenic due Cr component.

Species	Clinical signs
Humans	*Children:* Bone growth impairment *Radioactive isotope:* Anemia, reduced red and white blood cells and platelets, hemorrhages, leukemia, susceptibility to infections, tiredness
Mice	Growth depression, rickets, death (at very high doses) *Radioactive isotope:* Anemia, cancers (bone, leukemia, lung, nose) reduced red and white blood cell counts and platelets, hemorrhages, susceptibility to infections, thinning of the cornea, thinning of the skin
Pigs	Incoordination, posterior paralysis, weakness
Rats	Growth depression, rickets, death (at very high doses) *Radioactive isotope*: Anemia, cancers (bone, leukemia, lung, nose) reduced red and white blood cell counts and platelets, hemorrhages, susceptibility to infections, thinning of the cornea, thinning of the skin

Carcinogenesis, mutagenicity, embryotoxicity

Cancers of bone, nose, lung, and skin have been seen in animals, and leukemia in humans and animals exposed to radioactive strontium.
The International Agency for Research on Cancer (IARC) has determined that radioactive strontium is a human carcinogen.

Diagnosis

Confirmed exposure to Sr, measurement of Sr in body tissues or environmental samples.

Prevention

Avoid exposure, especially to radioactive isotopes in the workplace or contaminated environments.

Treatment

Treat symptomatically and avoid further exposure.
(*See Appendices* 1, 2, *and* 3).

References

ATSDR Strontium (2014), Nielsen (1986).

Sulphur (S)

S is used in concrete, batteries, detergents, fertilizers, fireworks, fungicides, matches, gun power, solvents. Concrete and various chemical and pharmaceutical industries.

Clinical applications

S containing cream, lotion, ointment, and soaps are used in the management of many skin conditions.

Essentiality

Human	Yes
Other species	Yes—All

Parameter	Site(s)
Site(s) of absorption	Intestine
Site(s) of excretion	Urine
Accumulation	Stored mainly in proteins containing sulphur-containing amino acids cystine and methionine

Biological activity

Sulphur-containing amino acids, e.g., methionine, cysteine, and cystine, are important for many metabolic processes in all living cells and sulfur is an essential component of:

- Chondroitin sulphate—structural component of cartilage, bone, tendons, and blood vessel walls
- Coenzyme A—involved in enzyme reactions
- Cytochromes—electron transport in energy metabolism
- Glutathione
- Hemoglobin—oxygen carrier
- Lipoic acid—a coenzyme
- S-adenosylmethionine

Functions of sulfur include:

- metal transport
- free radical scavenging
- regulation of gene expression
- protein stabilization and synthesis
- tissue integrity and protection
- enzyme functionality
- DNA methylation and repair
- regulation of gene expression
- remodeling of extracellular matrix components
- lipid metabolism
- detoxification of xenobiotics/signaling molecule

(Hewlings and Kalman, 2019).

Deficiency

Species	Clinical signs
Humans	NR
Cattle	Reduced food intake, reduced milk production
Goats	Death, emaciation, reduced feed intake, reduced dietary digestibility and utilization, hypersalivation, lacrimation, poor production, high blood urea
Pigs	Reduced weight gain
Poultry	Reduced growth rate
Rats	Reduced collagen levels, reduced growth rate
Sheep	Low blood urea, death, reduced food intake, reduced food utilization, hypersalivation, lacrimation, poor production, reduced weight gain, emaciation, reduced wool production, weight loss, weakness

Toxicity

High S intake can inhibit Cu and Se absorption.

Species	Clinical signs
Humans	Bronchiectasis, bronchitis (chronic), cough, dyspnoea, emphysema, expectoration (with blood streaks), eye irritation, mucosal irritation, nasal irritation, nasal hyperplasia, nasal secretion, reduced pulmonary function, skin irritation, rhinitis, tracheobronchitis
Cattle	Anorexia, blindness, coma, enteritis (severe) halitosis (hydrogen sulfide) muscle twitches, reduced rumen activity, nervousness, polioencephalomalacia-like Syndrome, respiratory distress. Depressed milk production, peritonitis, petechial hemorrhages (organs, e.g., renal). Recumbency
Dogs	*Experimental:* Death Sodium bisulfite LD50 244 mg/kg body weight
Goats	Death, diarrhea, dyspnoea, halitosis (hydrogen sulfide smell on breath), muscle tremors
Horses	*Accidental consumption:* convulsions, cyanosis, death, dullness, jaundice, labored breathing, lethargy, nasal discharge (yellow), respiratory failure
Rabbits	*Experimental:* Carbon disulfide exposure in atmosphere: reduced weight gain, loss of muscle control, neurological damage, increased Cu in tissues (thyroid, pancreas, spinal cord). Death (sodium tetrathionate)
Rats	*Experimental:* Sodium bisufite LD50 650–740 mg/kg body weight
Sheep	Death, diarrhea, dyspnoea, halitosis (hydrogen sulfide), muscle tremors

Carcinogenesis, mutagenicity, embryotoxicity

None reported.

Diagnosis

History of exposure.

Prevention

Avoid contact especially in the workplace.

Treatment

Treat symptomatically.

References

Hewlings and Kalman (2019), ATSDR Sulphur (2014), Nimni and Han (2007), ETN (1995), OHS Database (1993), McDowell (1992f).

Tellurium (Te)

Te is a mildly toxic metalloid used as alloys with Cu, Cd, and steel in semiconductors and solar panels and as a coloring agent in chinaware, porcelain, and glass.

Clinical applications

None reported.

Essentiality

Human	No
Other species	No

Parameter	Location
Site of absorption	Colon (sheep and pigs)
Site(s) of excretion	Fecal after ingestion; Urine after parenteral administration
Sites of metabolism	Not reported

Biological activity

None reported.

Interferences

Not reported.

Deficiency

Not reported.

Toxicity

Exposures by inhalation, ingestion, contact.

Species	Clinical signs
Humans	Anorexia, dermatitis, dizziness, drowsiness, eye irritation, garlic odor on breath, headache, irritability fatigue, metallic taste, dry mouth, nausea, pulmonary edema, no sweating, vomiting, weakness
Rats	Alopecia, garlic odor on breath, erythema and edema of feet, muscle necrosis (cardiac and skeletal muscle) paralysis (hind legs—temporary) *Teratogenic effects*: exophthalmia, hydrocephalus, ocular hemorrhage, small kidneys, edema, umbilical hernia, retained testicles

Carcinogenesis, mutagenicity, embryotoxicity

Te is not regarded as being carcinogenic or mutagenic, however, it is teratogenic (NJDHH, 2009; Perez-D'Gregorio et al., 1988).

Diagnosis

Confirmed exposure, analysis of body tissues urine or environmental samples.

Prevention

Avoid exposure.

Treatment

Treat symptomatically, avoid further exposure
(*See Appendices 1, 2, and 3*).

References

NIOSH Tellurium (2020), NJDHH (2009), Perez-D'Gregorio et al. (1988)

Terbium (Tb)

Tb is a rare-earth metal used mainly in fluorescent phosphors in visual displays (e.g., televisions and computer screens).

Clinical applications

Terbium oxide is in monitor cathode Ray tubes (CRTs).

Essentiality

Human	No
Other species	No

Parameter	Location
Site of absorption	Gastrointestinal tract
Site(s) of excretion	Feces, urine
Sites of accumulation	*After injection:* Liver, lung, spleen

Biological activity

None reported.

Deficiency

None reported.

Toxicity

Species	Clinical signs	Comments
Human	SConjunctivitis, skin irritation *Chronic exposure:*Anemia, changes in blood cell distribution lung scarring, pneumoconiosis	Contact with powder Inhalation of dust
Mice	Lung gamma-glutamyl transpeptidase (gamma-GTP) activity was increased; alkaline phosphatase (ALP) was increased; lipid peroxidation increased, pulmonary superoxide dismutase (SOD), catalase (CAT), and glutathione peroxidase (GSH-Px) activities were all decreased	*Tb Administered IV*
Rats	Cardiovascular collapse, conjunctivitis, corneal damage, death, growth suppression, respiratory paralysis, skin irritation, abraded skin ulceration	

Carcinogenesis, mutagenicity, embryotoxicity

No evidence for Tb carcinogenicity or mutagenicity.
There is some evidence that other rare earth elements (lanthanides) are embryotoxic (Moreira et al., 2020).

Diagnosis

History of exposure. Analysis of body tissues or environment.

Prevention

Avoid exposure, especially in the workplace.

Treatment

Treat symptomatically, avoid further exposure.
(*See Appendices 1 and 2*).

References

Moreira et al. (2020), Santa Cruz (2020), Damian, 2014, Rim et al. (2013), Shimada et al. (1996).

Thallium (Tl)

Tl is a toxic metal, used in infrared optical equipment, low melting glasses, and photoresistors.

Clinical applications

Used to be used as a rodenticide.

Essentiality

Human	No
Other species	No

Parameter	Location
Site of absorption	Rapid from the gastrointestinal tract. Absorbed transdermally.
Site(s) of excretion	Urine, feces
Sites of metabolism	Hair, nails

Biological activity

Can cause autonomic nervous system dysfunction and activates enzyme systems including pyruvate kinase.

Interferences

Not reported.

Deficiency

Not reported.

Toxicity

In the body Tl replaces K in body systems which is responsible for some of the clinical signs it causes, such as color blindness and baldness (alopecia).

Species	Clinical signs	Comments
Humans	Abdominal pain, acne (severe), alopecia, anorexia, arrhythmia, cardiac failure, cerebellar ataxia, color blindness, coma, confusion, cranial nerve palsies, death, delirium, diarrhea, drowsiness (can be extreme), eye movement impaired, headaches, hemolysis red blood cells, hematemesis, hypersalivation, hypertension, inflammation of mouth, lips, gums, mental retardation, motor neuropathy, muscle aches, nausea, nerve pain, numbness in fingers and toes, ophthalmoplegia (impaired function of ocular muscles), optic atrophy, organic brain syndrome, peripheral nerve damage (like walking on hot coals), psychoses, renal failure, respiratory failure, retrobulbar neuritis, seizures, sensory polyneuropathy (painful) skin lesions (dry, crusty), swallowing difficulty, tachycardia, vomiting (bloody) weakness	Accumulates in body with age
Mice	Death	Used as rodenticide
Rats	Death, adverse reproductive effects. male reproductive system damage, teratogenic effects	

Carcinogenesis, mutagenicity, embryotoxicity

There is no evidence thallium is carcinogenic or mutagenic. Thallium has some teratogenic properties on cartilage and bone formation (chicks)—Léonard and Gerber (1997) and developmental defects in rats (ATSDR, 2020b).

Diagnosis

Evidence of exposure, especially occupational exposure. Measure high Tl in body tissues, environment (air).

Prevention

Avoid exposure.

Treatment

No specific treatment, treat symptomatically.

References

ATSDR Thallium (1992), CDC (2020), Léonard and Gerber (1997).
ATSDR Thallium (1992) Accessed online at https://www.atsdr.cdc.gov/toxprofiles/tp54.pdf

Thorium (Th)

Th is a weakly radioactive metal. It is used widely in aircraft engines, heat-resistant ceramics, optics for cameras and scientific instruments, light bulbs, in vacuum tubes, as light source as a coating for tungsten wire in electrical equipment, in high-temperature crucibles, in glasses and missile guidance systems.

Clinical applications

None.

Essentiality

Human	No
Other species	No

Parameter	Location
Site of absorption	Inhalation, ingestion, transdermal
Site(s) of excretion	Bile and feces. Exhalation.
Sites of storage	Liver, bones, lungs
Transport	With transferrin

Biological activity

None reported.

Deficiency

Not reported.

Toxicity

Exposure mainly by inhalation.

Species	Clinical signs	Comments
Humans	Bone cancer, blood disease (acute myeloid leukemia), dermatitis, liver cancer (hemangiosarcoma cholangiosarcoma), lung disease, lung cancer, pancreatic cancer	Inhalation of Th dust Following administration by injection
Hamsters	Carcinogenic	By injection
Mice	Oxidative stress (in liver, spleen, bone) increased lipid peroxidation, decreased activity antioxidant enzymes (superoxide dismutase, catalase), death	

Carcinogenesis, mutagenicity, embryotoxicity

No evidence, but Th radioactivity is of concern.

Diagnosis

History of exposure such as occupational or environmental contamination. Chemical analysis of body tissues.

Prevention

Avoid exposure.

Treatment

No specific treatment. Treat symptomatically.

References

ATSDR Thorium (2019) Thorium, Kumar et al. (2013).

Thulium (Tm)

Tm is a rare-earth metal used as a radiation source.

Clinical applications

Tm is used in surgical lasers and some portable X-ray machines.

Essentiality

Human	No
Other species	No

Parameter	Location
Site of absorption	Poorly absorbed from the gastrointestinal tract
Site(s) of excretion	Urine
Sites of accumulation	Skeleton, liver, kidney

Biological activity

Not reported

Deficiency

Not reported.

Toxicity

Species	Clinical signs	Comments
Mice	Degeneration of the liver and spleen	After injection
Rats	Conjunctival ulcers, growth suppression, skin ulcers on abraded skin death, cardiovascular collapse, respiratory paralysis	

Carcinogenesis, mutagenicity, embryotoxicity

No evidence for Tm carcinogenicity or mutagenicity (Rucki et al., 2020).
There is some evidence that other rare earth elements (lanthanides) are embryotoxic. (Moreira et al., 2020).

Diagnosis

Analysis of tissue sample Tm content.

Prevention

Avoid exposure especially in the workplace.

Treatment

No specific antidote. Treat symptomatically.
(*See Appendix 1 and 2*).

References

Rucki et al. (2020), Moreira et al. (2020), Haley et al. (1963).

Tin (Sn)

Sn is a metal that is widely used in tin cans and tin alloys in soft solder, pewter, bronze phosphor bronze. A niobium-tin alloy is used for superconducting magnets, and as indium tin oxide (ITO) for producing transparent conductive films on glass or plastic panels used in electronic devices, touch panels, plasma displays, flat panel displays, solar panels, cathode-ray tubes, energy efficient windows, gas sensors, and photovoltaics.

Clinical applications

None.

Essentiality

Human	No
Other species Rats	Possibly

Parameter	Location
Site of absorption	Poorly absorbed
Site(s) of excretion	Feces
Interferes with other mineral biovailability	Ca, Cu, Fe, Se, Zn

Biological activity

Tin can interfere with porphyrin synthesis and increase heme breakdown.

Deficiency

Low Sn diet was associated with impaired growth—however, the rats were also riboflavin-deficient so the effect cannot be attributed to Sn alone.

Toxicity

Organotin compounds are increasingly being used in industry and reports of toxicity in humans due to exposure in the workplace (e.g.leather or plastic factories) are increasing. Toxicity results in altered calcium metabolism, and Cu in blood, brain, liver, and kidneys.

Species	Clinical signs
Humans	Abdominal pain (stomach-aches), amblyacousia, anemia, coma, confusion, conjunctivitis, convulsions, dermatitis, diarrhea, dizziness, cholangitis, hallucinations, headache, hepatotoxicity; hypersomnia, hypodynamia, inappetence, nausea, neurotoxicity (CNS edema), psychomotor disturbances, psychotic behavior, renal damage, restlessness, somnipathy, tinnitus, tremor, twitches, vomiting
Rats	Brain spongiosis, growth depression, food efficiency reduced, pancreatic atrophy, renal calcification, testicular degeneration, reduced alkaline phosphatase activity in bone and liver, low hemoglobin
Various	Anorexia, impaired growth, impaired hematopoiesis due reduced Fe absorption, impaired porphyrin synthesis, increase breakdown of hem, liver, pancreatic and kidney lesions (inorganic Sn), and bile duct inflammation and CNS edema (organic Sn)

Carcinogenesis, mutagenicity, embryotoxicity

Based on *in vitro* testing using the embryonic stem cell test (EST) Sn demonstrated weak embryotoxicity (Imai and Nakamura, 2006).
Indium tin oxide (ITO) which is mainly Indium, is carcinogenic causing lung tumors in mice and rats. (IARC, 2017).
There is also moderate evidence that ITO is mutagenic.
IARC: There is inadequate evidence in humans for the carcinogenicity of indium tin oxide. There is sufficient evidence in animals for the carcinogenicity of indium tin oxide.
Indium tin oxide is possibly carcinogenic to humans (Group 2B).
Tin and tin compounds are not considered carcinogenic to humans by the ATSDR https://www.atsdr.cdc.gov/toxfaqs/tfacts55.pdf

Diagnosis

History of exposure, especially occupational. Sn concentrations in blood, urine, and tissues.

Prevention

Avoid exposure especially in the workplace.

Treatment

No specific antidote so treat symptomatically.
(*See Appendix 1 and 2*).

References

NORD (2020a), IARC (2017), Guo et al. (2010), Blunden and Wallace (2003), Imai and Nakamura (2006), McDowell (1992p), Winship (1988), Nielsen (1986)

Titanium (Ti)

Ti is present in many cosmetic powders and creams, in ceramics, inks, lacquers, paints, paper, plastics, rubber, and varnishes. Some foodstuffs including sweets contain titanium dioxide as a coating E-number: E171.

Clinical applications

Ti alloys are used for dental implants, endoprostheses, eyeglass frames, orthodontal brackets, pacemakers, and stents.

Essentiality

Human	No
Other species	No

Parameter	Location
Site of absorption	Poorly absorbed, inhalation
Site(s) of excretion	Urine and feces
Interferes with other mineral biovailability	Not reported

Biological activity

No biological functions reported.

Deficiency

None reported.

Toxicity

Exposures are through inhalation, ingestion, medical implants, or dermal contact.

Species	Clinical signs	Comments
Humans	Bone loss around dental implants, hypersensitivity (chronic fatigue syndrome, muscle pain, skin rashes), yellow nail syndrome includes: discolored thick nails, bronchial obstruction, bronchiectasis, cough, lymphedema, pleural effusion, rhinosinusitis	Bone loss due to inflammatory reaction
Fish	Death (increased mortality rate), reduced food utilization impaired growth, increased liver enzymes activity (SGOT and SGPT)	
Mice	Inflammation	Inhalation
Rats	Alveolar septal fibrosis, metaplasia, lung tumors (adenomas, squamous cell carcinomas)	Inhalation

Carcinogenesis, mutagenicity, embryotoxicity

Ti causes lung cancer in rats exposed to inhalation.
According to the IARC Titanium dioxide is a possible carcinogen to humans (Group 2B) based on sufficient evidence in animals and inadequate evidence from epidemiological human studies.

Diagnosis

Energy-dispersive X-ray fluorescence (EDXRF) can be used to measure the titanium content in nails.

Prevention

Avoid unnecessary exposure especially occupational exposure.

Treatment

In event of toxic effects remove Ti implants if appropriate and feasible and avoid unnecessary exposure through foods and occupational risk.
(*See Appendix 1 and 2*).

References

Melisa Org (2020), Kim et al. (2019), IARC (2018), NIOSH Titanium (2011), Bermudez et al. (2004), Heinrich et al. (1995), NCI National Cancer Institute (1979), Nielsen (1986).

Tungsten (W)

W is a transition metal used widely for the production of darts, fishing weights, golf clubs, gyroscope wheels, light bulb filaments, phonographic needles, X-ray tubes, welding electrodes, and munitions manufacture. W is present in fire retardant fabrics and pigments for ceramics.

Clinical applications

None reported.

Essentiality

Human	No
Other species	No

Parameter	Location
Site of absorption	Inhalation, ingestion, skin contact
Site(s) of excretion	Urine mainly
Sites of accumulation	Bone, thyroid—small amounts in lung, kidney, adrenal glands, spleen, femur, lymph nodes, and brain

Biological activity

None reported. W may interfere with Cu and Mo metabolism.

Interferences

Not reported.

Deficiency

Not reported.

Toxicity

Exposure is by inhalation or ingestion. Very poor transdermal transfer.

Species	Clinical signs	Comments
Humans	Dermatitis, lung cancer, memory loss, increased mortality (due cancer), pulmonary fibrosis, sensory deficits	Inhalation or contact—may be due cobalt exposure rather than Tungsten itself
	Acute: Rare anuria, nausea, seizures, clouded consciousness, coma, encephalopathy, moderate renal failure, acute tubular necrosis, hypocalcemia	Soldiers drinking alcohol from discharged gun barrels
Mice	Reduced longevity	
Rats	Reduced longevity, nephrotoxicity, loss of body weight	

Carcinogenesis, mutagenicity, embryotoxicity

W may be carcinogenic but early studies gave other minerals concurrently. However, in vitro W induces carcinogenic related endpoints including cell transformation, increased migration, xenograft growth in nude mice, and the activation of multiple cancer related pathways suggesting carcinogenic potential (Laulicht et al., 2015).
Lung cancer and increased mortality has been attributed to occupational exposure to W dusts. (ATSDR Tungsten, 2020).
Decreased sperm motility, increased embryotoxicity, and delayed fetal skeletal ossification in animals. (ATSDR Tungsten, 2020).

Diagnosis

Blood and urine analysis.

Prevention

Avoid unnecessary exposure to W.

Treatment

No specific treatment, treat symptomatically. Avoid further exposure.
(See Appendix 1 and 2).

References

ATSDR Tungsten (2020), VanderSchee et al. (2018), Laulicht et al. (2015), Lemus and Venezia (2015), Van der Voet et al. (2007), Nielsen (1986).

Uranium (U)

A naturally occurring radioactive metallic element as isotopes ^{234}U, ^{235}U, and ^{238}U. It emits alpha radiation when it transforms to Thorium.
U is used in nuclear power stations, helicopter, plane, and munitions construction.

Clinical applications

None.

Essentiality

Human	No
Other species	No

Parameter	Location
Site of absorption	Inhaled—lungs, ingested—gastrointestinal tract, transdermal (small amount)
Site(s) of excretion	Kidneys, Feces
Sites of accumulation	Lungs (inhaled), bones (60%), liver, kidneys

Biological activity

None reported.

Deficiency

None reported.

Toxicity

Exposure routes include food, water, inhalation, skin contact, occupational exposure in mines, mills, or sites processing uranium or contact with phosphate fertilizers.

Species	Clinical signs
Humans	Acute renal failure, anemia, coagulopathy, death, Fanconi syndrome, liver dysfunction, myocarditis, paralytic ileus, respiratory disease (pulmonary edema, pulmonary fibrosis), renal failure, rhabdomyolysis
Dogs	Vomiting
Mice	Bipartitie sternebrum, body length short, cleft palate, congenital malformations, fertility reduced, low fetal weight, food intake reduced
Rats	Neurobehavioral changes, reduced fertility, skin irritation, *Maternal exposure* resulted in *offspring*: changes in brain function, changes in ovaries, tooth malformation

Carcinogenesis, mutagenicity, embryotoxicity

Cancer has not been linked with human exposure to natural (nonenriched) uranium.
Depleted U is not carcinogenic according to National Toxicology Program (NTP), International Agency for Research on Cancer (IARC), or the EPA.
U toxicity caused developmental defects (cleft palate, bipartite sternebrum, reduced ossification) in mice.

Diagnosis

Blood, urine, hair, and body tissues analysis of U content.

Prevention

Avoid unnecessary exposure especially in the work place and through ingestion of root vegetable grown in soil high U.

Treatment

There is no specific antidote. Treat symptomatically. Use of chelators is controversial so cannot be recommended and calcium DTPA should not be used as this can increase bone deposition.
(*See Appendices 1, 2, 3*).

References

ATSDR Uranium (2020), NCBI (2020), Pavlakis et al. (1996), Domingo et al. (1989).

Vanadium (V)

V is high in soil and rock phosphates and in some potassium supplements but low in crops and vegetation.
V-steel alloys are used in manufacture of strong tools such as axles, armor plates, car gears, springs, crankshafts, cutting tools, piston rod, and also in nuclear reactors.

Clinical applications

None.

Essentiality

Human	No
Other species	No

V was considered an essential nutrient for chicks, rats, and goats based on clinical signs but these were inconclusive because of other dietary factors in the diets used.

In the body

Parameter	Location
Site of absorption	Poorly absorbed from gastrointestinal tract
Transportation	Bound to transferrin and lactotransferrin
Site(s) of excretion	Kidneys
Site(s) of accumulation	Bones, liver, kidney
Interferes with other mineral bioavailability	NR

Biological activity

V concentrations are very low in human and other vertebrate animal tissues, but tissue concentrations can greatly increase when animals are fed a high dietary V ration (sheep, rats). Higher V concentrations are found in hair samples from welders and renal failure patients, lower concentrations in patients with multiple sclerosis.
Numerous possible biological roles for Vanadium have been proposed but have not been confirmed (see review in Mertz, 1986); however, as peroxovandate it affects free radicals.

Deficiency

Species	Clinical signs
Goats	Bone abnormalities, reduced food intake, small litter size, reduced milk production, high mortality rate (kids), impaired reproduction

Toxicity

Exposure routes include ingestion in food, inhalation.

Species	Clinical signs
Humans	Airway irritation, cough, diarrhea, cramps, nausea, sore throat, vomiting, wheezing
Cattle	Ataxia, collapse, conjunctivitis, death, diarrhea, eosinophilic inflammation internal organs, hematuria, hemorrhages (abomasal, pulmonary, tracheal), listlessness, weight loss

Mice	Alveolar/bronchiolar hyperplasia, inflammation, and fibrosis, decreases in fetal growth, increases in resorptions, gross, visceral, and skeletal malformations and anomalies. diarrhea, lung cancer
Nonhuman primates	Impaired lung function ((increased resistance and decreased airflow)
Poultry (chicks)	Depressed growth rate, death
Rats	Alveolar/bronchiolar hyperplasia, inflammation, and fibrosis, diarrhea; death, dyspnea, decreases in fetal growth, increases in resorptions, and gross, visceral, and skeletal malformations and anomalies. impaired growth, low hematocrit, low hemoglobin, low mean cell volume, microcytic erythrocytosis, laryngeal degeneration, nasal lesions (hyperplasia and metaplasia), increased reticulocytes and nucleated erythrocytes, reduced survival (pups), tachypnoea, weight loss
Sheep	Anorexia, ataxia, death, diarrhea, poor growth

Carcinogenesis, mutagenicity, embryotoxicity

Vanadium is not clastogenic but it is weakly mutagenic and has marked mitogenic activity affecting chromosome distribution during mitosis. It may cause aneuploidy.
V toxicity is associated with lung cancer in mice.
Based on *in vitro* testing using the embryonic stem cell test (EST) V demonstrated weak embryotoxicity (Imai and Nakamura, 2006).

Diagnosis

Tissue analysis.

Prevention

Avoid unnecessary exposure, especially in the workplace or from contaminated pastures.

Treatment

No specific treatment, Treat symptomatically.
(*See Appendix 1 and 2*).

References

ATSDR Vanadium Toxicity (2020), Léonard and Gerber (1994), McDowell (1992p), Nielsen (1987b), Hansard et al. (1982), Hansard et al. (1978).

Ytterbium (Yb)

A rare-earth metal used in lasers, X-ray, and other radiation devices, and in stainless steel manufacture.

Clinical applications

In lasers for medical imaging.

Essentiality

Human	No
Other species	No

Parameter	Location
Site of absorption	Inhalation
Site(s) of excretion	Feces mainly, urine
Sites of accumulation	Cleared slowly from lungs after inhalation

Biological activity

None reported.

Deficiency

Not reported.

Toxicity

Species	Clinical signs	Comments
Humans	Eye irritation, pneumoconiosis, skin irritation	
Hamster	Teratogenic	Gale, 1975
Rats	Conjunctival ulceration, cardiovascular collapse, death, respiratory paralysis, ulcers to abraded skin	
Fish	Embryotoxic and developmental effects reduced survival and hatching rates, and induced tail malformations Liver catalase activity was significantly decreased	Zebrafish Goldfish

Carcinogenesis, mutagenicity, embryotoxicity

No evidence for Yb carcinogenicity or mutagenicity.
Yb is teratogenic in hamsters (Gale, 1975)and embryotoxic in fish,.

Diagnosis

History of exposure. Chemical analysis of tissues.

Prevention

Avoid unnecessary exposure especially in the work place, or by medical exposures. Avoid exposure during pregnancy.

Treatment

No specific treatment. Wash off contact exposures and treat symptomatically.

References

Damian, 2014, Gale (1975), Haley et al. (1963).

Yttrium (Y)

Y is a transition metal classified as a rare earth element. Used in lasers, fiberoptics, fluorescent lights, optical glass and metal alloys, and as a phosphor in visual displays (e.g., televisions and computer screens).

Clinical applications

The radioisotope yttrium-90 is used in drugs including Yttrium Y 90-DOTA-tyr3-octreotide and Yttrium Y 90 ibritumomab tiuxetan to treat cancers (e.g., bone, colorectal, leukemia, liver, lymphoma, ovarian, and pancreatic cancers) (Adams et al., 2004).
Yttrium-90 needles are more precise than scalpels, and are used to sever nerves in the spinal cord, and yttrium-90 also for radionuclide synovectomy rheumatoid arthritis (Fischer and Modder, 2002).
A neodymium-doped yttrium-aluminum-garnet laser has been used in robot-assisted canine prostatectomy (Gianduzzo et al., 2008).

Essentiality

Humans	No
Other species	No

Parameter	Location
Site of absorption	NR
Site(s) of excretion	NR
Sites of metabolism	NR

Biological activity

None reported.

Deficiency

None reported.

Toxicity

Water soluble compounds of yttrium are considered mildly toxic, insoluble compounds are nontoxic. Routes of exposure: eye contact, ingestion, inhalation, skin contact.

Species	Clinical signs	Comments
Humans	Lung disease: shortness of breath, cancer chest pain coughing, cyanosis, death, embolisms, possibly liver damage	*Industrial exposure* *Medical exposure*—Whole-liver 90Y radioembolization for patients with mNET results in cirrhosis and portal hypertension in >50% of treated patients
Mice	Growth impairment	
Rats	Dyspnoea, growth impaired, liver edema, pulmonary edema, pleural effusions, pulmonary hyperemia. Impaired renal function (reduced creatinine and urine production) increased liver enzymes glutamic-oxaloacetic transaminase and glutamic-pyruvate transaminase. Incidence of tumors was 33% in Y-treated rats and 14% in untreated controls.	Varies with salt

Carcinogenesis, mutagenicity, embryotoxicity

Increased cancer risk in humans and rats. The use of Y90 to treat cancer has been linked to the development of liver cirrhosis (Su et al., 2017).

Diagnosis

Known exposure. Tissue sample analysis.

Prevention

Avoid unnecessary or high dose exposure especially in the workplace and medical environments.

Treatment

No specific antidote. Wash off contact materials, avoid further exposure, treat symptomatically.

References

NIOSH Yttrium (2020), Leggett (2017), Su et al. (2017), Damian (2014), EURARE (2014), Rim et al. (2013), Gianduzzo et al., 2008, OSHA Contributors (2007), Adams et al. (2004), Fischer and Modder (2002), Schroeder and Mitchener (1971).

Zinc (Zn)

Zn is a transition metal. Used in welding, alloy production, and soldering.

Clinical applications

There are many over-the-counter medications containing Zn such as skin creams, oral preparations but the evidence for efficacy is usually lacking.

Essentiality

Humans	Yes
Species	Yes—all

In the body

Parameter	Location
Site of absorption	Small intestine. Stomach in rats. Rumen sheep.
Transportation	Bound to albumin and alpha-2 macroglobulin
Site(s) of excretion	Feces mainly. Urine.
Site(s) of accumulation	Bone, kidney, liver, muscle, pancreas, prostate, skin, uveal tract
Interferes with other mineral bioavailability	Cu

Biological activity

Zn is associated with over 200 enzymes including dehydrogenases, alkaline phosphatase, carbonic anhydrase, deoxyribonucleic acid (DNA) and ribonucleic acid (RNA) polymerase, leucine aminopeptidase and superoxide dismutase, It is important for:

- Cell replication and differentiation
- Hormone synthesis storage and secretion—insulin, corticosteroids and testosterone
- Reproductive organ development

Zinc is important for immune function, acid-base balance, digestion, growth and development, skin and hair health, genetic transcription, antioxidant activity.

Deficiency

Signs of deficiency are often due to interference with normal efficacy of utilization by the presence of high amounts of another mineral, e.g., Ca, P. It is estimated that 2 billion people worldwide have dietary Zn deficiency.
In humans Acrodermatitis enteropathica is due to inherited variants in the SLC39A4 gene encoding a zinc transporter that mediates the uptake of dietary zinc in the gut resulting in inadequate zinc absorption and signs of Zn deficiency. Similar SLC39A$-linked inherited defect resulting in Zn deficiency is seen in cattle.
Lethal acrodermatitis in Bull terriers and Miniature Bull terriers was thought to be similar but recent studies have shown that this disease is a monogenic autosomal recessive inherited disorder caused by a variant of MKLN1:c.400+3A>C leading to a splice defect in the MKLN1 gene. The precise relationship to Zn metabolism has yet to be confirmed.
Zinc-responsive dermatoses occur in dogs and in Alaskan Malamutes, Siberian Huskies, and Samoyeds has been reported as an inherited familial defect.
Other conditions that have been associated with Zn deficiency include: small intestine diseases, nutritional deficiencies, increased losses, Wilson's disease, sickle cell disease, chronic kidney disease, chronic liver disease.

Species	Clinical signs
Humans	Acne, alopecia, anemia, anorexia, basophil stippling, congenital malformations, depression (anhedonia), cheilitis (rare), impaired cognitive functions (learning and hedonic tone) cytokine suppression inflammatory cytokines (e.g., IL-1β, IL-2, IL-6, and TNF-α)) dermatitis, diarrhea, dwarfism, dystocia, eczema, eye lesions, growth retardation, hypogonadism, impaired immune function, impotence, increased infection risk (e.g., dermatitis, gastroenteritis, pneumonia, and urinary tract infections), irregular menses, irritability, labor difficulties and prolonged, lethargy, lip cracks/fissures, nausea, night blindness, oral ulcers, placental abruption, prostate problems, impaired reproduction, seborrhea, stomatitis, impaired development and function of male sex organs, delayed sexual maturity, skin rash, reduced sense of smell, reduced sense of taste, tongue coating (white), vomiting, weight loss, wound healing impairment, xerosis (dry scaly skin),
Birds	Chondrodystrophy, deformities (fetal), dermatitis (exfoliative legs and face), reduced egg production, feather abnormalities (brittle, frayed), reduced hatchability, skin scaly, swollen hocks
Buffaloes	Unthriftiness
Cats	Growth retardation, parakeratosis, seminiferous tubular degeneration, impaired testicular function
Cattle	Alopecia, bowing of hind limbs, ears—dry scaly skin, grinding of teeth, growth retardation, haircoat rough, hemorrhages (submucosal), hoof and skin cracking and fissuring, lethargy, impaired development and function of male sex organs, skin thickening and cracking around nostrils, inflammation nose and mouth, reduced milk production, mucosal overgrowth lips and dental pads, nutritional utilization reduced, edema limbs, inappetence, parakeratosis, impaired reproduction (estrus through pregnancy to parturition) unthriftiness, salivation excessive, scrotal skin inflamed and wrinkled, stiffness, increased susceptibility to infections, wound healing impaired, low serum Zn
Dogs	Conjunctivitis, generalized debility, emaciation, poor growth, skin lesions abdomen and extremities (foot pads), hyperkeratitis especially around eyes, vomiting, poor weight gain, low serum Zn
Elephants	Impaired immune response, skin lesions (hyperkeratosis)
Fish	*Catfish:* Reduced appetite, reduced growth rate, low bone calcium, low bone zinc, reduced egg production, reduced hatchability *Rainbow trout*: Cataracts, death, reduced growth, fin and skin erosions *Carp*: cataracts, poor growth, loss appetite, high mortality, skin and fin erosions
Goats	Alopecia, growth retardation, impaired development and function of male sex organs, parakeratosis, Horns soft, deformed, lack striations.
Guinea pig	Abortion, anorexia, dermatitis (scaly), premature births, gamma-globulin concentrations low, glycosaminoglycans metabolism altered, hair rough, hypersensitivity (delayed) response impaired, posture altered, skin lesions, excessive vocalisation during pregnancy, low plasma zinc,
Hamster	Growth impairment, abnormal estrus cycle, weight loss
Horses	*Foals:* alopecia, alkaline phosphatase reduced, enlarged epiphyses, reduced food intake, reduced growth rate, lameness, skin lesions (crusting, desquamation, flaking, parakeratosis, roughness, hoof, face, muzzle, nose, legs, stiffness *Adults*: effusions on legs
Mice	Adrenal hypertrophy, alopecia, emaciation, growth retardation, weight loss
Nonhuman primates	Abortions, alopecia, anorexia, apathy, appetite reduced, behavioral changes, bone formation impaired, poor bone mineralization, cognitive function impaired, debilitation, dermatitis, fetal abnormalities, unkempt hair (squirrel monkeys), plasma Zn low, play less (infants), poor growth, poor immune function, increased morbidity, parakeratosis of tongue (squirrel monkeys), skin integrity impaired, reproductive failure, skeletal development delayed, small offspring, stillbirths, tissue Zn low *Squirrel monkeys:* alopecia, growth impairment, unkempt haircoat, parakeratosis (tongue), skin lesions, cessation estrus
Pigs	Adrenal hypertrophy, appetite reduced, bone reduced strength and size, death, growth retardation, hemorrhage (young pigs—due to interference with Vitamin K), parakeratosis, stillbirths, testicular development abnormal, vomiting, wound healing impaired
Poultry	Anorexia, breathing labored, comb color, dermatitis, embryonic deformities, curvature of spine, death, growth impairment, lower feed efficacy, hyperkeratosis, low egg production, delayed egg production, posture abnormal, short thick leg bones, hock enlargement, micromelia, scaly skin, poor feathering, broken feathers, stiff gait, reproductive failure, tachypnoea, vertebral fusion, toes missing, agenesis of skeleton or limbs, weakness

Rabbits	Alopecia, dermatitis, reduced food intake, low hematocrit, hair pigment change (dark to gray), reproductive failure (nonreceptive to male, failure of ovulation), sores around mouth, wet matted hair on lower jaw and ruff, weight loss
Rats	Alopecia, anorexia, epidermal thickening, loss hair follicles, growth retardation, hyperirritability, hypoinsulinemia, impaired development and function of male sex organs. Impaired development of embryo, resorption, aspermatogenesis, fetal malformations
Sheep	Deaths (postnatal), growth retardation, Impaired development and function of male sex organs, aspermatogenesis. Impaired reproduction (estrus through pregnancy to parturition, Wool abnormalities: loose, brittle, loses crimp, alopecia, skin: thick inflamed and wrinkled, Lack of wool growth. Horns soft, deformed, lack striations. Low serum Zn

Toxicity

Toxicity occurs due to inhalation (occupational exposure), dietary supplements, denture cream, ingestion of coins, parenteral nutrition, overuse of makeup, ointments and sunscreen.
In pets Zn toxicity is common and sources are metallic objects, multivitamins, and zinc oxide creams and ointments. High Zn intake can inhibit Fe and Cu availability creating borderline deficiencies.

Species	Clinical signs
Humans	Abdominal pain, bone marrow effects, chest pain, chest tightness, coagulopathy, cough (dry throat), death, depression, diarrhea (watery), dizziness, dyspnoea, epigastric pain, fatigue, fever, hematemesis, hematuria, hemolysis, headaches, inflammation (chemical burns) to oral cavity, pharynx, trachea, oesophagus and stomach, hives and angiedema (one report), lethargy, leucocytosis, liver necrosis, muscle cramps, neurological effects, neutrophilia, thrombocytopenia, "metal fume fever" (ZnO inhalation), pancreatitis, pneumonitis (acute), pneumothorax, airway irritation, acute respiratory distress syndrome (ARDS) pulmonary fibrosis, renal injury, interstitial nephritis, acute tubular necrosis, decreased vital capacity, sweating, vomiting, increased T cells, T suppressor cells, natural killer cells in bronchoalveolar lavage fluid Zn excess can induce copper deficiency causing: anemia (sideroblastic), granulocytopenia, neutropenia, impaired immune function, adverse effect on ratio of low-density-lipoprotein to high-density-lipoprotein (LDL/HDL) cholesterol myelodysplastic syndrome, sensorimotor polyneuropathy syndrome, "swayback" (spasticity, gait abnormalities, and sensory ataxia)
Birds	Anemia, anorexia, cyanosis, death, diarrhea, feather plucking, gastrointestinal perforation, gastrointestinal ulceration, hematuria, hemoglobinuria, lethargy, peritonitis, polydipsia, polyuria, regurgitation, sepsis, weakness, weight loss
Cats	Bronchopneumonia, leukocyte infiltration into alveoli, and grayish areas with congestion. Labored breathing, upper respiratory tract obstruction.
Cattle	Ataxia, death, diarrhea, decreased food intake, decreased milk yield, weight loss
Dogs	Abdominal pain, acute renal failure, anemia (hemolytic), anorexia, azotemia, cardiac murmur (secondary to anemia), casts (granular), coagulopathy, death, dehydration, diarrhea, disseminated intravascular coagulation (DIC), duodenal ulcers, gastric ulcers, hepatic dysfunction, hypercreatininemia, hyperphosphatemia, icterus, jaundice, lameness, lethargy, melena, neurotoxicity, pallor, pancreatitis, seizures, tachycardia, tachypnoea, vocalization, vomiting, weakness
Ferrets	Anemia, lethargy, pallor, weakness (hindleg)
Fish	*Catfish:* low hematocrit, low hemoglobin, low liver copper content
Guinea pigs	*Inhalation:* Alveolitis (focal), consolidation, death, emphysema, fibrosis, decreased lung compliance, decreased functional residual capacity, impaired lung function; inflammation, macrophage infiltration
Mice	*Inhalation:* Death, increased macrophages and lymphocytes in the lungs, alveolar carcinoma
Pigs	Reduced appetite, death, impaired growth
Poultry	*Chicks:* death, depressed growth rate
Rabbits	Grayish areas with pulmonary congestion, peribronchial leukocytic infiltration, bronchial exudate composed of polymorphonuclear leukocytes
Rats	Diarrhea; death, increased LDH protein in bronchoalveolar lavage fluid, grayish areas with pulmonary congestion, peribronchial leukocytic infiltration, bronchial exudate composed of polymorphonuclear leukocytes
Reptiles	Anemia (hemolytic), anorexia, depression, lethargy, pallor
Sheep	Ataxia, death (postnatal), diarrhea, pica, poor growth, poor food intake, poor food utilization

Carcinogenesis, mutagenicity, embryotoxicity

Zn is not mutagenic or clastogenic Zinc can induce tumors after local application but is considered not to be a carcinogenic risk to man, indeed there is some evidence Zn may inhibit carcinogenesis (Kasprzak and Waalkes, 1986). Zn is required for cellular proliferation of tumors and tumor growth is retarded by Zn deficiency.
Zn is not teratogenic, however Zn deficiency may be harmful to a developing organism.

Diagnosis

Determine Zn, Cu, and ceruloplasmin levels.

Prevention

Avoid access and exposure to Zn-containing products, especially in the workplace.

Treatment

Supportive. Antiemetics, fluids, proton pump inhibitors or H2-blockers. Whole bowel irrigation. Chelation with calcium disodium edetate (CaNa2EDTA) or DTPA.
(See Appendix 1 and 2).

References

Agnew and Slesinger (2020), ATSDR Zinc Toxicity (2020), Roohani et al. (2013), NRC (2006), McDowell (1992l), Fosmire (1990), Hambidge et al. (1986), (Kasprzak and Waalkes, 1986), Léonard et al. (1986), Van Den Brock and Thoday (1986), Kane et al. (1981)

Zirconium (Zr)

Zr is a transition metal.

Uses

Zr has many uses: in nuclear power, aerospace, and chemical industries; in the manufacture of ceramics, glass, and porcelains; the synthesis of pigments, dyes, and water repellents; in tanning operations; in abrasive and polishing materials; as igniters in the manufacture of munitions and detonators; in lighter flints; in skin ointments and antiperspirants; in the manufacture of high-vacuum tubes; in iron and steel manufacture.

Clinical applications

Zr compounds have been used as treatment for dermatitis and in new ceramic orthopedic and dental implants.

Essentiality

Humans	No
Other Species	No

In the body

Parameter	Location
Site of absorption	Gastrointestinal tract. Poorly absorbed. Inhalation
Site(s) of excretion	Bile to Feces
Interferes with other mineral bioavailability	Not reported

Biological activity

None reported.

Biological effects

The expression of seven microRNAs (mRNAs) were specifically associated with Zr levels in humans (Cossellu et al., 2016) and functional target analysis showed that these miRNAs are involved in pathological mechanisms: inflammation, skeletal, and connective tissue disorders.
Some Zr salts are corrosive and cause serious chemical tissue damage.

Deficiency

Not reported.

Toxicity

Zr is regarded as a low risk for toxicity, but skin contact, inhalation, and ingestion have all resulted in toxic effects.

Species	Clinical signs
Humans	Blurred vision, circulatory collapse, contact dermatitis, disseminated intravascular coagulation, conjunctivitis, cough, dyspnoea, hemolysis, hoarseness, lachrymation, laryngeal edema, lung granulomas, mucosal inflammation, pharyngeal pain, pulmonary fibrosis, pulmonary edema, pyloric stenosis, respiratory distress, skin granulomas, stridor, thirst *Oral:* abdominal pain, achlorhydria, acute tubular necrosis, adult respiratory distress syndrome (ARDS), antral stenosis, anuria, burning pain in mouth and throat, collapse, convulsions, watery or bloody diarrhea, gastric carcinoma, gastric perforation, hematemesis, hemolysis, hematuria, hypotension, jaundice, jejunal stricture, liver damage, esophageal perforation, peritonitis, pneumonitis, protein-losing gastroenteropathy, renal failure, retching, tenesmus, vomiting,
Dogs	Reduced red blood cell count, reduced hemoglobin *Long-term exposure to zirconium tetrachloride*
Fish	Death (increased mortality rate), reduced food utilization impaired growth, increased liver enzymes activity—SGOT and SGPT
Guinea pigs	Reduced survival. Poor weight gain, chronic interstitial pneumonitis (*Zirconium tetrachloride*)
Hamsters	Poor weight gain, interstitial pneumonitis
Mice	Reduced survival time, weight loss
Rabbits	Pneumonia, peribronchial granulomas
Rats	Glycosuria, increased cholesterol, interstitial pneumonitis, reduced survival time, poor weight gain (*Zirconium tetrachloride*)

Carcinogenesis, mutagenicity, embryotoxicity

Zr ingestion has been associated with gastric carcinoma in humans. No evidence for mutagenicity or embryotoxicity.

Diagnosis

Tissue concentrations.

Prevention

Avoid access and exposure to Zr containing products especially in the workplace and home.

Treatment

No specific antidote. Treat symptomatically.
(*See Appendix 1 and 2*).

References

Harrison et al. (2020), Digitalfire Reference Library (2020), Ahmadi et al. (2017), Cossellu et al. (2016), EPA (2012), Nielsen (1987c), Spiegl et al. (1956).

Section 3

Species

Humans (like dogs, cats, pigs, and others) are a simple stomached species that rely on digestive enzymes secreted into the lumen of the tract to digest food into smaller molecules that can be absorbed across the wall of the gastrointestinal tract (mainly the small intestine) into the bloodstream or lymphatic system to be carried to the liver or other tissues for further metabolism.

Other species have evolved very different gastrointestinal anatomy (see Fig. 1) and physiology to be able to digest their food intake. Some (ruminants) relying on the microbiome in the forestomaches others (horse, rabbits, tapirs) are hindgut fermenters and rely on the microbiome in the large intestine (cecum or colon), birds have a large distensible sac (crop) to hold ingested food a glandular organ the proventriculus and a thick-walled gizzard to grind down large food particles before they enter the stomach.

In wild species available, sometimes niche, food resources have resulted in the evolution of anatomical and physiological adaptations. When in captivity it is important that the ration provided is as similar to the animals' natural diet as possible, but this is often difficult or impossible to achieve.

In species for which there is a lot of data available a full list of elements is provided with NR entered when no reports have been found in the scientific literature. For other species, a redacted list is provided.

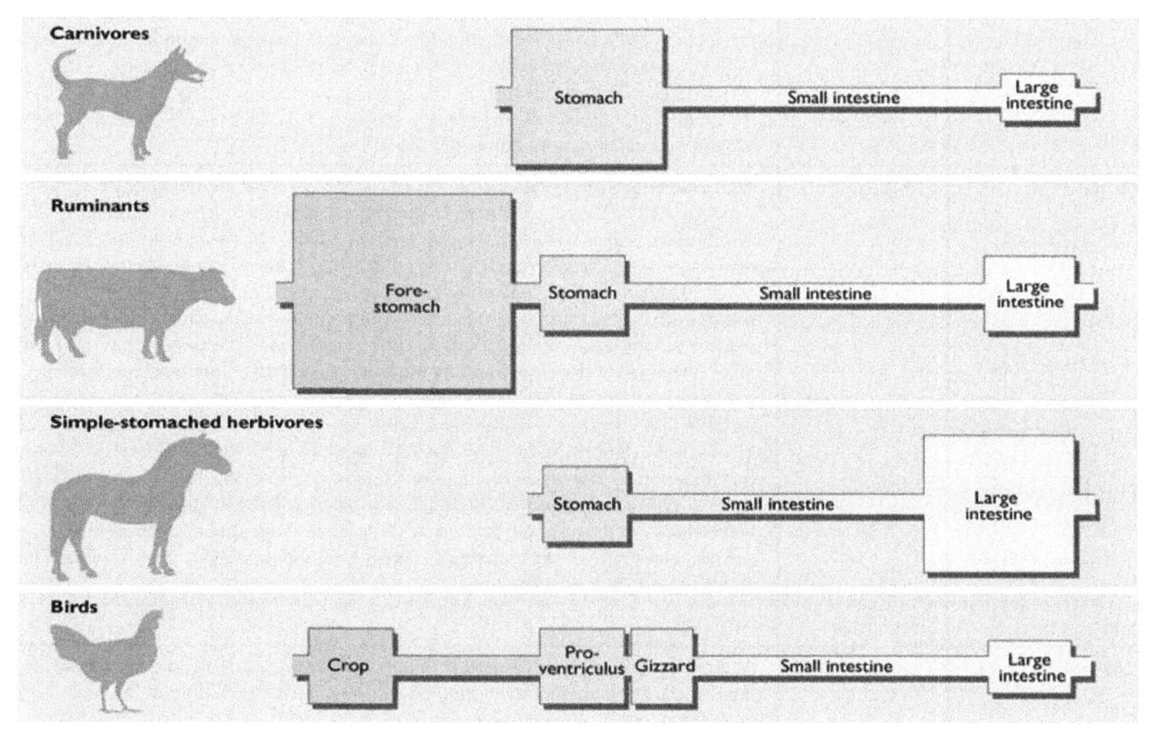

FIG. 1 Different species alimentary tracts. *(Courtesy of Physiology of Domestic Animals published by the Scandinavian Veterinary Press.)*

Clinical Signs in Humans and Animals Associated with Minerals, Trace Elements, and Rare Earth Elements. https://doi.org/10.1016/B978-0-323-89976-5.00008-6

Mammals

Simple-stomached species

Omnivores

Humans

Human dietary intakes are incredibly varied with many factors influencing what people eat including availability, beliefs, cost, learned likes/dislikes (appearance-form, color; emotions; flavor/taste, texture, source—animal, fungi, seafood), peer pressure, religious, social, tradition, taboos. A scientific basis for food selection based on existing knowledge of minimum requirements or maximum limits is rarely involved unless a qualified dietitian has compiled the ration, and the patient complies with it!

Common mineral deficiencies include Ca, Fe, Mg, K, Se, and Zn.

In addition to dietary intake and occupational/environmental exposure to toxic substances, self-administered toxic levels of mineral supplements such as Ba and Fe are often reported.

Element	Deficiency/ toxicity	Clinical signs reported
Aluminum (Al)	Toxicity	Amyotropic lateral sclerosis, Anemia, asthma, autism, bone disease (osteomalacia serum Al > 30 μg/L; due interference with P, fractures and pain), brain disease (confusion, dementia, encephalopathy (linked to dialysis and serum >80 μg/L), directional disorientation, hallucinations, myoclonus, seizures), conjunctivitis, constipation, dermatitis (contact), granulomatous enteritis, delayed gastric emptying, dyspnea, fractures; growth delayed (children), hepatorenal disease, reduced Fe absorption, lung problems (desquamative interstitial pneumonia, pneumoconiosis, pulmonary alveolar proteinosis, pulmonary fibrosis, granulomatous pneumonia, pulmonary granulomatosis after inhalation), impaired lung function (typically assessed by measuring forced expiratory volume (FEV1) and forced volume capacity (FVC), macrophagic myofasciitis, multiple sclerosis, muscle weakness (myopathy); toxic myocarditis, myocardial dysfunction myocardial hypokinesia, osteomalacia, pancreatitis, pulmonary edema (after ingestion) skeletal deformities, speech problems (dyspraxia), decreased daily sperm production, reduced sperm count and motility and increase in abnormal spermatozoa ischemic stroke due to thrombosis, left ventricular thrombosis, wheezing Al accumulation linked to Alzheimer's, breast cancer, Parkinson's disease, Crohn's disease and inflammatory bowel disease
Antimony (Sb)	Toxicity	Anorexia, arthralgia, bone marrow hypoplasia, bronchitis, cardiotoxicity (~9% of patients—ECG changes: decrease T wave height, T-wave inversion, prolonged QT, ventricular ectopics, ventricular tachycardia, torsade de pointes, ventricular fibrillation), cerebral edema, coma, conjunctivitis, convulsions, corneal burns, chronic bronchitis, cough, death, dermatitis (Antimony spots: papules, pustules around sweat and sebaceous glands with eczema and lichenification), diarrhea, dizziness, electrolyte disturbances, chronic emphysema, eye irritation, fluid loss, garlic odor, gastrointestinal signs (gastric pain, ulcers), liver damage (transient elevation of hepatocellular enzyme levels), hypotension, impaired immunity (low IgG, IgA, and IgE levels), insomnia, laryngitis, metallic taste, myalgia, myocardial depression, nausea, neuropathy (peripheral), nose bleeds, optic atrophy, palmar keratosis, pancreatitis, pharyngitis, pleural adhesions, pneumoconiosis, pneumonitis, proteinuria, pulmonary edema (rare), acute renal failure, respiratory irritation (chronic coughing), and upper airway inflammation, retinal hemorrhage, rhinitis, shock, skin spots, obstructive sleep apnea, sleeplessness, smell reduced sensitivity, stomach cramps. Throat irritation, tracheitis, uveitis, vasodilation, vomiting, weight loss, wheezing
Arsenic (As)	Deficiency	NR
Arsenic (As)	Toxicity (Acute)	Abdominal pain (colic), basophil stippling, blood volume depletion, bone marrow depression, cardiomyopathy, coma, convulsions, death, dehydration (severe), depression, dermatitis (increased capillary permeability and necrosis), diarrhea (can be bloody), fever, fluid loss (severe), gastritis, intestinal hemorrhage, hemoglobinuria, hemolysis, hepatic steatosis, hypotension, intravascular coagulation, jaundice, muscle spasms, nausea, esophagitis, oral pain (burning of the mouth), pancytopenia (severe), pulmonary edema, renal failure, renal tubular acidosis, respiratory failure, ventricular arrhythmia, vomiting

	Toxicity (Chronic)	Anemia (low Fe), behavior changes, bronchitis, cerebrovascular infarction, cirrhosis, cognitive impairment, confusion, convulsions, deep cracks palms and soles of feet, demyelination, dermatitis (exfoliative) diabetes mellitus risk, drowsiness, encephalopathy, fever, gangrene, garlic smell on breath, headaches, hearing loss (children), hepatic fibrosis, hepatomegaly, hyperkeratosis, hyperpigmentation skin, hypersalivation, hypertension, mortality increased (due hypertension),? ischemic heart disease, memory loss, mucosal lesions (inflammation throat), muscle pains, myocardial injury, nail pigmentation (linear—mee's lines), neutropenia, pericapillary hemorrhage within the white matter, nephritis, peripheral neuropathy, peripheral vascular disease, polyneuritis, recumbency, respiratory disease (restrictive and obstructive), skin lesions, throat soreness, weakness. Cancers including bone, colon, larynx, lung, prostate, skin, stomach, basal cell carcinoma, hemangiosarcoma, angiosarcoma of the liver, lymphoma, and nasopharyngeal carcinoma. Transition cell carcinomas of kidney, bladder, ureter. All urethral tumors. Adenocarcinoma of the bladder
Barium (Ba)	Toxicity	Abdominal pain, anxiety, baritosis (accumulation of Ba in the lungs after inhalation of dust) blindness, cardiac arrhythmias, cardiac arrest, death, watery diarrhea, dizziness, dyspnea, gastric pain, gastrointestinal hemorrhage, hematuria, hypertension $\pm$ hypotension, hypokalemia, hypophosphatemia, mydriasis, nausea, numbness, paralysis, rhabdomyolysis, renal damage (chronic toxicity), renal degeneration, renal insufficiency, acute renal failure, respiratory failure, seizures, sensitization, shock, shortness of breath, tremors, ventricular tachycardia, vomiting, muscle weakness
Beryllium (Be)	Toxicity	Appetite loss, chest pain, dry cough, difficulty breathing (dyspnea)/shortness of breath, fatigue, fever, granuloma formation (inflammatory nodules), joint pain, lymphadenopathy, pharyngitis, interstitial pulmonary fibrosis, night sweats, pneumonitis, rhinitis, skin rash, tracheobronchitis, weakness, weight loss
Bismuth (Bi)	Toxicity	Abdominal pain, albuminuria, aminoaciduria, Anemia, angiedema, anxiety, apathy, appetite loss, ataxia, balance loss, blurred vision, bone marrow hypoplasia, buccal hemorrhage, cognitive deficits, coma, concentration difficulty, confusion, constipation, death, delirium, delusions, dementia, depression, dermatitis, diarrhea, diplopia, dysarthria, encephalopathy [confusion, disorientation, seizures (rare)], epistaxis, erythematous rash (maculopapular), eye irritation, falls, fever, gait abnormalities, gastrointestinal symptoms, gingivitis, glomerular filtration rate (GFR) reduced, glycosuria, hematuria, halitosis, hallucinations, headache, hyperreflexia, incoordination, insomnia, irritability, joint pain, lethargy, malaise, melena (black stools), memory loss, metallic taste, muscle cramps, nausea, neuropsychiatric symptoms, memory loss, myoclonus, kidney damage (acute renal failure, glomerular, and tubular necrosis), liver damage, oliguria, paresthesia, Parkinsonism, petechie, poor oral hygiene with a blue–black gum line, proteinuria, psychosis, pyramidal signs, rectal bleeding, respiratory irritation, seizures, sleeplessness, smell sensation reduced, somnolence skin lesions (lichen planus-like rashes), taste sensation reduced, thrombocytopenia, tremor, ulcerative stomatitis, vomiting weight loss
Boron (B)	Deficiency	deprivation (<0.3 mg/day) resulted in poorer performance on tasks of motor speed and dexterity, attention, and short-term memory
Boron (B)	Toxicity	Anorexia, breathlessness, central nervous system signs (irritability, seizures), congestion, conjunctivitis, cough, death, dermatitis (exfoliative dermatitis), diarrhea, erythema, mucosal exfoliation inflammation, exfoliation gastrointestinal effects, kidney lesions (cloudy swelling and granular degeneration of renal tubular cells), lethargy, liver lesions, nausea, acute respiratory irritation, riboflavinuria, vomiting
Bromine (Br)	Deficiency	Insomnia (*on low Br intake—not confirmed to be an essential nutrient yet*)
Bromine (Br)	Toxicity	Abdominal pain, confusion, coma, cough, diarrhea, dizziness, epistaxis, feeling of oppression, hallucinations, headache, lachrymation, neurological symptoms, pneumonitis, pulmonary edema, skin burns, skin lesions, slurred speech, stupor
Cadmium (Cd)	Toxicity	Abdominal cramps, aminoaciduria, Anemia (low Fe), anosmia (loss sense of smell) breathlessness, cancer (liver, lung, prostatic), death, chest tightness, chills, cough, cyanosis, diarrhea, discoloration of teeth (yellow), dyspnea, emphysema, fatigue, fever, glucosuria, headaches, liver function impairment, muscle aches, nausea, pain (substernal), prostatic hyperplasia, pulmonary edema, renal damage: tubular epithelial cell degeneration cellular atrophy, interstitial fibrosis, and glomerular sclerosis,

		hypercalciuria, itai-itai disease (osteomalacia—linked to renal disease), neurological signs (numbness, tingling in lips, cheeks, hands, feet), proteinuria, proteins of low relative molecular mass in the urine (beta2-microglobulin and retinol-binding protein), enzymuria (*N*-acetyl-beta-glucosaminidase), increased salivation, tachycardia, vomiting, increased urinary concentration of cadmium and metallothionein
Cesium (Cs)	Toxicity	Anorexia, cardiotoxicity, death, diarrhea, ECG changes: QT interval prolonged to 650 ms. Headache, hypokalemia, hypothyroidism (Harari et al., 2015), nausea, seizures, syncope, polymorphic ventricular tachycardia (Lyon and Mayhew, 2003)
Calcium (Ca)	Deficiency	Bone demineralized, fractures (spontaneous), joint enlargement, lameness, muscle weakness, rickets, seizures (child), skeletal deformity (bowed legs), spinal curvature, tetany (child), poor tooth enamel
Calcium (Ca)	Toxicity	Abdominal pain, anxiety, arrhythmias, bone pain, calcification of soft tissues, confusion, depression, fatigue, lethargy, gastric pain, hypercalcemia, hypertension, muscle weakness, polydipsia, polyuria, vomiting
Cerium (Ce)	Toxicity	Dendriform pulmonary ossification, emphysema, endomyocardial fibrosis, itching, heat sensitivity, myocardial infarction, interstitial lung disease or pneumoconiosis, pneumonia, pulmonary fibrosis skin lesions
Chlorine (Cl)	Deficiency	Dehydration, dyspnea, diarrhea, fatigue, hypertension, metabolic alkalosis vomiting, weakness
Chlorine (Cl)	Toxicity	*Inhalation*: Chest pain, conjunctivitis, cough, death, dehydration, diarrhea, eye irritation, nasal irritation, pneumonitis (toxic), pulmonary edema, tachypnea, vomiting *Ingestion*: death, hypochloremia, esophagitis, severe gastrointestinal tract damage *Dermal contact:* Inflammation, irritation NaCl toxicity: appetite loss, cerebral hemorrhage, confusion, jitteriness; muscle twitches, nausea, seizures, coma. Death, thirst, weakness
Chromium (Cr)	Deficiency	Appetite increase, body fat increased, corneal lesions, encephalopathy, fertility reduced, glucose tolerance impaired, glycosuria, growth impaired-disorders, hypercholesterolemia, hyperglycemia, hypoglycemia, hyperinsulinemia, hypertriglyceridemia, insulin-binding reduced, insulin receptor numbers reduced increased intraocular pressure, lean muscle mass reduced, neuropathy (peripheral), nitrogen balance reduced, reduced respiratory quotient, sperm count reduced
Chromium (Cr)	Toxicity	Acidosis, allergic dermatitis (irritation, ulcers), asthma, death, diarrhea (bloody), electrolyte imbalance, liver lesions (necrosis), lung cancer (bronchiogenic carcinoma), myocardial lesions, proteinuria, renal disease (tubular necrosis), reduced respiratory function, shock, skin ulcers, increased urine β-glucuronidase, vomiting (bloody)
Cobalt (Co)	Deficiency	Anemia (megaloblastic/pernicious Anemia), goiter, neurological lesions
Cobalt (Co)	Toxicity	Anorexia, cardiac disease, high cardiac enzymes, cyanosis, death, ECG changes (low voltage), flushing, goiter, heart failure, hypotension, kidney disease, high lactic acid, lethargy, nausea, nerve damage, pericardial effusion, respiratory disease, tinnitus vasodilation, weight loss Following industrial exposure: cardiomyopathy Beer drinkers (possibly with other nutritional problems, e.g., thiamine deficiency): cardiomyopathy colloid depletion, polycythemia, thyroid hyperplasia *Children:* congestive heart failure, myxedema, thyroid hyperplasia
Copper (Cu)	Deficiency	*Rare.* Low plasma enkephalins, increased pancreatic leucine encephalin-containing peptides, reduced free leucine and methionine enkephalin-containing peptides *Infants:* Anemia, neutropenia, osteoporosis, depigmentation, neurological disturbances. *Adults:* Anorexia, cardiovascular disturbances, Anemia, bone abnormalities? ischemic heart disease? atherosclerosis, growth impairment, leukopenia, edema, pallor Menkes (hereditary due poor Cu absorption, Cu deficiency in nephrosis or inadequate protein intake), steely hair syndrome, developmental problems, brain degeneration, kinky hair, seizures, weak muscle tone (hypotonia), sagging facial features, seizures, developmental delay, and intellectual disability

Copper (Cu)	Toxicity	Blood changes, flu-like reaction, "metal fume disease" Cu accumulates in the cornea, liver, kidney, and brain, liver cirrhosis, impaired neurological processes Wilson's disease
Dysprosium (Dy)	Toxicity	NR
Erbium (Er)	Toxicity	NR
Europium (Eu)	Toxicity	Abdominal pain, eye irritation, nausea, extensive scarring on abraded skin
Fluorine (F)	Deficiency	Increased caries on low F intake—*not an essential nutrient*
Fluorine (F)	Toxicity	*Acute*: Death *Chronic*: Bone disease with exostosis, dental disease—mottling teeth and white patches on enamel (children), stained teeth (yellow/black), hypermineralization of skeleton, skeletal deformities, ligament and tendon calcification, spinal ankylosis *HF acid*: hypocalcemia, cardiac arrhythmias, severe pain, skin burns, death *F Gas*: severe irritation to eyes, nose, lungs later damage to liver and kidneys, death
Gadolinium (Gd)	Toxicity	Allergic reactions, dysgeusia, flushing, headaches, nausea, hypertension, inflammation, injection site reaction, itchiness, dizziness, nausea, nephrogenic systemic fibrosis (NSF*), parosmia, pruritus, respiratory signs, skin rash, vomiting *Contact:* Conjunctivitis, skin ulcers on abraded skin
Gallium (Ga)	Toxicity	Anemia (microcytic), azotemia, throat irritation, difficulty breathing, diarrhea, chest pain, creatinine clearance reduced, electrolyte abnormalities, low hemoglobin, hypercalciuria, hypocalcemia (if initially normocalcemic), hypokalemia, metallic taste in mouth, nausea, pulmonary edema, paralysis, pulmonary toxicity (as GaAs), renal tubular acidosis, renal toxicity, blood and urine malondialdehyde (MDA), elevated, vomiting *Acute case*: initially presented as dermatitis and appeared relatively not life-threatening, rapidly progressed to dangerous episodes of tachycardia, tremors, dyspnea, vertigo, and unexpected black-outs, irreversible cardiomyopathy
Germanium (Ge)	Toxicity	Anemia, death, dizziness, dyspnea, feinting, fatty liver, hemolysis, headache, muscle weakness, nausea, peripheral neuropathy, pulmonary toxicity, renal failure, renal tubular degeneration, syncope, vomiting
Gold (Au)	Toxicity	Alopecia, anemia (aplastic-rare) anaphylaxis (rare) bone marrow depression, dermatitis, diarrhea, encephalopathy (peripheral/cranial—rare), enterocolitis, eosinophilia (common), gastrointestinal bleeding, glomerulonephritis (common), headaches, jaundice, liver disease, metallic taste in mouth, myokymia, interstitial pneumonitis (rare), nephrotic syndrome (rare), pancreatitis (rare), pruritus, skin rashes, stomatitis, trophic nails, proteinuria, and hematological disorders, pseudocyanosis (nonblanching blue–gray skin discoloration), urticaria, vomiting
Holmium (Ho)	Toxicity	NR
Indium (In)	Toxicity	DNA damage in blood cells, death, serious lung diseases: alveolar proteinosis, emphysema, interstitial fibrosis, pulmonary fibrosis, increased 8-OHdG (leukocytes and urine)
Iodine (I)	Deficiency	Abortions, brain development impaired, congenital anomalies, cretinism (mental deficiency, deaf mutism, spastic diplegia, neurological deficits), goiter, hypothyroidism, stillbirths
Iodine (I)	Toxicity	Goiter, hypersalivation, skin ulcers ("kelp acne"), brassy taste in mouth
Iron (Fe)	Deficiency	Abortion, anemia (microcytic-hypochromic), fatigue, irritability, lethargy, listlessness, pallor, palpitations on exertion, sore tongue, angular stomatitis, cheilitis (erythema at corners of the mouth), dysphagia, koilonychia (spoon nail) prolonged cardiorespiratory recovery after exercise (women), *Children:* gastritis, gastric atrophy, gastric achlorhydria, gastrointestinal hemorrhage, enteropathy. Anorexia, reduced growth rate, increased risk of infections, reduced serum myoglobin concentrations. Cognitive deficits (attention span, intelligence, sensory perception functions, abnormal emotions and behavior)

Iron (Fe)	Toxicity	Abdominal pain, high aminotransferase, anuria, impaired cardiac and central nervous system metabolism, cardiomyopathy, coagulopathy, coma, convulsions, death (accidental intake children); diarrhea, free radicle production, hematemesis, hematochezia, hemochromatosis, hemorrhagic necrosis of gastrointestinal tract, hepatic failure, hepatic necrosis, hypoglycemia, hypotension, hypovolemia, jaundice, liver damage, acute liver failure, metabolic acidosis, multiple organ system failure, nausea, perforation of the gastrointestinal tract, peritonitis, renal failure, seizures, shock (cardiogenic), vomiting
Lanthanum (La)	Toxicity	Headache, nausea, pneumoconiosis lowers cholesterol, liver damage (high ALP, ALT, Gamma-GT, Low albumin), lowers blood pressure, appetite reduced, and risk of blood coagulation
Lead (Pb)	Toxicity	*Children*: abdominal pain, Anemia (low Fe in red blood cells), anorexia, ataxia, behavioral problems (e.g., mental retardation, selective deficits in language, cognitive function, balance, behavior, school performance), impaired consciousness, constipation, convulsions, dysarthria (slurred speech), clumsier, encephalopathy, headaches, hyperproteinemia, learning difficulties, less playful, pallor, papilledema (swelling optic disc), sluggish (lethargic), seizures, slurred speech (dysarthria), changes in kidney function, vomiting *Adults*: abdominal pain, anemia, anorexia, behavioral changes (e.g., hostility, depression, anxiety), cataracts (*increased risk*), altered consciousness, constipation, death, encephalopathy, endurance loss, fatigue, fever, hallucinations, headaches, hypertension, incoordination, insomnia, irritability, joint pain, kidney disease, lethargy, listlessness, muscle strength loss, neuropathy (peripheral—demyelination, slow conductance), optical atrophy, skill loss (recent), damage to reproductive organs, seizures, sluggishness, vomiting, wrist drop
Lithium (Li)	Deficiency	NR
Lithium (Li)	Toxicity	Abdominal pain, agitation, ataxia, cardiac ECG changes (flattened T waves, sinus node dysfunction, prolongation QT, intraventricular conduction defects, U waves), coma, confusion, consciousness changes, constipation, delirium, allergic dermatitis (rash), death, diarrhea, disorientation, drowsiness, dry mouth, dysarthria, fatigue, flatus, heart failure, hyperreflexia, hyperthermia, hyperthyroidism (rare), hypertonia, hypotension, hypothyroidism, kidney failure, lung cancer (bronchiogenic carcinoma), muscle weakness, uncontrolled muscle movements, myoclonus, nausea, nephrogenic diabetes insipidus, nephrotic syndrome, neurological signs (pyramidal, extrapyramidal, and cerebellar signs) nystagmus, polydipsia, restlessness, rigidity, seizures, skin ulcers, slurred speech, sodium-losing nephritis, stupor, tachycardia, tremors, urine concentrating ability lost, urination frequency increased, vomiting, weakness, weight change (gain or loss)
Lutetium (Lu)	Toxicity	Alopecia, nausea, pneumoconiosis, extensive scarring on abraded skin, tiredness
Magnesium (Mg)	Deficiency	Anorexia, anxiety, appetite loss, arrhythmia, cardiac arrest, Chvostek sign, dysphagia, fatigue, hyperreflexia hypocalcemia, irritability, lethargy, memory loss, muscle fasciculations, nausea, muscle weakness, hyperreflexia, hyporeflexia, incoordination, irritability, mental derangement, muscle cramps, muscle fasciculations, myographic changes, personality change, tetany, tonic–clonic seizures, tremors, Trousseau sign, vertigo, vomiting, weakness
Magnesium (Mg)	Toxicity	Abdominal cramps, arrhythmias, breathing difficulty, cardiac arrest, death, depression, diarrhea, ECG changes (prolonged PR, wide QRS, peaked T waves) flushing (face), hypermagnesemia, hypocalcemia, hyporeflexia, hypotension, ileus, lethargy, muscle weakness, nausea, respiratory depression, urinary retention, vasodilation, vomiting
Manganese (Mn)	Deficiency	*Children:* bone demineralization, impaired growth, skeletal abnormalities *Adults:* high alkaline phosphatase, ataxia, high blood calcium, low blood cholesterol, impaired glucose tolerance, hair depigmentation, high blood phosphorus, skin rash
Manganese (Mn)	Toxicity	Ataxia, central nervous system damage, confusion, fatigue, abnormal gait, hallucinations, hyperirritability, muscle spasms (dystonia), pneumonia, psychiatric abnormalities, the rigidity of trunk, skeletal abnormalities, stiffness, tremors (hands), violent acts, weakness
Mercury (Hg)	Toxicity	Abdominal cramps, abdominal pain, abnormal thoughts, abortion (wives of men chronically exposed), acrodynia (painful swelling and pink color to fingers and toes), anorexia, anuria, anxiety, ataxia (progressive), behavioral changes (excessive anger, deficits in mental concentration), brain damage (permanent), breathlessness, cerebellar ataxia, cerebral palsy (congenital), chest burning pain or tightness, choreoathetosis

		(uncontrolled jerky movements combined with slow, writhing movements), coma, concentration loss, confusion, changes in consciousness, cough (dry), death, dehydration, depression, dermatitis (rare except contact rash), diarrhea (bloody), dysarthria (slurred speech), dysmenorrhea, dysphagia, dyspnea, emotional lability, erythema, abnormal excitability, eye irritation, facial expression changes, fatigue, forgetfulness, gastrointestinal disturbances including necrosis, gingivitis, glomerulonephritis, headaches, hearing loss, hyperesthesia (extreme sensitivity), hypersalivation (ptyalism), hypertension, impaired immune function, insomnia, interstitial nephritis, irritability, kidney damage, kidney failure, lethargy, locomotor disturbances, lymphocyte aneuploidy, memory loss, metallic taste in mouth, mood swings, nausea, nerve conduction speeds slowed, neuropsychological changes, parasthesia, peritonitis, severe pharyngitis, pneumonia, pneumonitis, polyarthritis, polyneuropathy, proteinuria, pulmonary fibrosis, pulmonary edema, quick temperedness, renal failure (acute), renal tubular necrosis, respiratory distress or failure, restrictive lung disease, chronic respiratory insufficiency, seizures, shock, shortness of breath, skin irritation, sluggishness, sensory disturbances, spermatogenesis reduced, severe stomatitis, excessive shyness, tachycardia, tingling, tremors (legs, arms, tongue, lips), uremia (severe), vertigo, visual problems (e.g., loss of color vision, constriction of visual fields, tunnel vision, and blindness) vomiting, weakness
Molybdenum (Mo)	Deficiency	*Genetic defect: Type A-mutations in MOCS1 gene, Type B deficiency mutations in MOCS2 result in an inability to form molybdopterin and the molybdenum coenzyme sulfite oxidase.*: brain atrophy/lesions, early postnatal death, intellectual disability, lens dislocation, mental retardation, opisthotonos, seizures high plasma and urine sulfite, sulfate, thiosulfate, S-sulfocysteine, and taurine *Long term parenteral nutrition* *Nutritional:* coma, headache, irritability, nausea, night blindness, tachycardia, tachypnea, vomiting. High sulfite and xanthine, low sulfate and uric acid in the blood and urine, Caries incidence is higher in UK where Mo levels are low
Molybdenum (Mo)	Toxicity	*Inhalation*: cough, dyspnea, reduced lung function *Nutritional:* Arthralgia, expiratory capacity reduced (dust inhalation), cystic fibrosis (Smith et al., 2014) gout-like signs, hallucinations, hepatotoxicity (Mendy et al., 2012), hyperuricemia, hyperuricuria, joint aches, multiple sclerosis (Pitt, 1976), neurological signs, psychosis (acute), seizures, reduced testosterone (Lewis and Meeker, 2015), high tissue xanthine oxidase activity *Associations*: reduced sperm count, reduced serum testosterone concentrations; urate and cystine calculi. Mo concentration in urate and cystine calculi was 10 × that in other uroliths (Abboud, 2008)
Neodymium (Nd)	Toxicity	Abdominal cramps, increased blood clotting time, fever, chills, muscle aches, hemoglobinuria, lung embolisms, liver damage irritating to the eyes, mucous membranes, and skin, pneumoconiosis
Nickel (Ni)	Deficiency	NR
Nickel (Ni)	Toxicity	Abdominal pain, adrenal insufficiency, asthma, allergic contact dermatitis, cancer, (lung and nasal) cough, cyanosis, death, dermatitis, diarrhea, dyspnea, ECG changes (myocarditis) fever, giddiness, headache, hemianopsia, hyperglycemia, lassitude, leukocytosis, nausea, nephrotoxicity, neurasthenic signs, palpitations, pneumonitis (interstitial), prostration, pulmonary edema, shortness of breath, respiratory disease substernal pain, upper airway irritation, vertigo, vomiting, weakness
Niobium (Nb)	Toxicity	NR
Palladium (Pd)	Toxicity	Contact dermatitis, erythema (topical), eye irritation, fever, hemolysis, subcutaneous injection site reaction (discoloration, necrosis) mucositis, edema (topical), oral lichen planus, skin sensitivity reaction, stomatitis, respiratory tract inflammation
Phosphorus (P)	Deficiency	Anorexia, anxiety, bone pain, bone fragility, poor bone mineralization, breathing irregularly, poor growth, fractures, irritability, joint stiffness, numbness, poor tooth development weakness

Phosphorus (P)	Toxicity	Abdominal cramps, bronchial spasms, eyelid drooping (upper-ptosis), headaches, hypersalivation, hypocalcemia, muscular weakness, neurological damage (atrophy), pupil constriction (miosis), sensorimotor polyneuropathy, ptosis, respiratory paralysis, soft tissue calcification, sweating, twitching, vomiting, weakness, wheezing
Potassium (K)	Deficiency	Cardiac arrhythmia, impaired neuromuscular function, intestinal loss muscle tone and distension, muscle weakness, paralysis, loss of reflexes, mental confusion, soft flabby muscles, respiratory failure *K deficiency often seen with severe dehydration due diarrhea, surgery or burns*
Potassium (K)	Toxicity	Confusion, ECG changes (tall T waves absent P waves, wide QRS), hyperkalemia, lethargy, muscle weakness, palpitations, flaccid paralysis/paresthesia of hands and feet *Causes include:* oral administration (rare) renal insufficiency, trauma and tissue damage, metabolic acidosis, acute adrenal insufficiency, postoperative oliguria, blood pooling, transfusions
Praseodymium (Pr)	Toxicity	Eye irritant, lung embolisms, liver damage, pneumoconiosis, pulmonary fibrosis, skin irritant
Promethium (Pm)	Toxicity	Potentially dangerous
Rubidium (Rb)	Toxicity	Agitation, confusion, dermatitis (rash—long-term exposure), diarrhea, excitement, nausea, polyuria, vomiting, weight gain
Samarium (Sm)	Toxicity	Dyspnea, cyanosis, pneumoconiosis, progressive pulmonary fibrosis, ulcers on abraded skin *Confounded as other rare earth elements or silicon often present at same time as exposure to Sm*
Scandium (Sc)	Toxicity	Inhalation through industrial exposure can cause lung disease
Selenium (Se)	Deficiency	Cardiomyopathy (Keshan disease China), infertility (males), low glutathione oxidase, muscle pain, myalgia, osteoarthritis (Kashin–Beck disease—China, Siberia, Tibet), pancreatic degeneration, red cell fragility, weakness
Selenium (Se)	Toxicity	Abdominal pain, alleaceous breath, alopecia, anesthesia, anorexia, ataxia, bone erosions, cardiac failure, cardiomyopathy (toxic), chest pain, prolonged clotting time, coma, confusion, high serum creatinine, high serum creatine kinase (CK) activity, congestion (internal organs), cyanosis (extremities), death, delirium, dermatitis (chronic), diarrhea, disorientation, dizziness, fingernail changes (brittleness, deformity, discoloration, white spots and longitudinal streaks, loss), ECG changes (T-wave flat and inverted, high ST, prolonged QT interval), fasciculations, fatigue, flushing (face), gastroenteritis (severe), garlic smell on the breath, glycosuria, hemiplegia, hepatic degeneration, hyperreflexia, hyporeflexia, hypotension, irritability, joint pain, kidney failure, lassitude, light headedness, impaired liver function, mesenteric infarction, metabolic acidosis, metallic taste in mouth, mucosal damage (mouth, esophagus, stomach), muscle spasms, muscle tenderness, myocardial depression, myocardial infarction, myoclonus, nausea, neurological damage/dysfunction, pain in extremities, paresthesia (peripheral), ptyalism, pulmonary edema, renal insufficiency, acute respiratory distress syndrome, restlessness, high selenium in hair and nails, skin rashes, splenomegaly, tachycardia, teeth discoloration (mottling), tremors, ventricular dysrhythmias, vomiting, weakness Inhalation (Hydrogen selenide): eye pain, pneumomediastinum, restrictive and obstructive pulmonary disease, rhinorrhea, throat pain, wheezing *Selenium dioxide and selenium oxide*: forms selenius acid in presence of water in the respiratory tract: bronchospasm, diarrhea, fever, hypotension, pneumonitis (chemical), tachycardia, tachypnea, vomiting *Dermal/eye contact*: caustic burns, conjunctival edema, corneal ulcers, lacrimation, skin erythema, heat, rashes, swelling, and pain *Chronic selenosis*: pruritic scalp rash erythematous skin blisters, anesthesia, hemiplegia, paresthesia

Silicon (Si)	Deficiency	*Proposed:* atherosclerosis, hypertension, osteoarthritis (Not confirmed—Si is not an essential nutrient for humans.
Silicon (Si)	Toxicity	Increased prevalence of autoimmune diseases (rheumatoid arthritis, systemic lupus erythematosus), cardiac arrest, chronic obstructive pulmonary disease (COPD), death, kidney disease, reduced lung function, lung cancer, silicosis
Silver (Ag)	Toxicity	Diarrhea, nausea, organ discoloration (gray), skin discoloration (gray), vomiting
Sodium (Na)	Deficiency	Apathy, collapse, cramps, death, dizziness, exhaustion, headaches, dizziness, fatigue, nausea, respiratory failure, vomiting *Due to excessive sweating or prolonged vomiting or diarrhea*
Sodium (Na)	Toxicity	Chemical burns to the oral cavity, pharynx, esophagus, stomach, or small intestine, death, hypertension, vomiting
Strontium (Sr)	Toxicity	*Children:* Bone growth impairment *Radioactive isotope:* Anemia, reduced red and white blood cells and platelets, hemorrhages, leukemia, susceptibility to infections, tiredness
Sulfur (S)	Deficiency	NR
Sulfur (S)	Toxicity	Bronchiectasis, bronchitis (chronic), cough, dyspnea, emphysema, expectoration (with blood streaks), eye irritation, mucosal irritation, nasal irritation, nasal hyperplasia, nasal secretion, reduced pulmonary function, skin irritation, rhinitis, tracheobronchitis
Tellurium (Te)	Toxicity	Anorexia, dermatitis, dizziness, drowsiness, eye irritation, garlic odor on breath, headache, irritability fatigue, metallic taste, dry mouth, nausea, pulmonary edema, no sweating, vomiting, weakness
Terbium (Tb)	Toxicity	*Contact with powder:* Skin irritation, conjunctivitis *Chronic exposure inhalation dust*: Scarring of the lungs, Anemia and changes in blood cell distribution, pneumoconiosis
Thallium (Tl)	Toxicity	Abdominal pain, acne (severe), alopecia, anorexia, arrhythmia, cardiac failure, cerebellar ataxia, color blindness, coma, confusion, cranial nerve palsies, death, delirium, drowsiness (can be extreme), eye movement impaired, headaches, hemolysis red blood cells, hematemesis, hypersalivation, hypertension, inflammation of mouth, lips, gums, mental retardation, motor neuropathy, muscle aches, nausea, ophthalmoplegia (impaired function of ocular muscles) optic atrophy, organic brain syndrome, peripheral nerve damage (like walking on hot coals), psychoses, renal failure, respiratory failure, retrobulbar neuritis, seizures, sensory polyneuropathy (painful) skin lesions (dry, crusty), swallowing difficulty, tachycardia, vomiting (bloody) weakness
Thorium (Th)	Toxicity	Bone cancer, blood disease (acute myeloid leukemia), dermatitis, liver cancer (hemangiosarcoma, cholangiosarcoma), lung disease, lung cancer, pancreatic cancer
Thulium (Tm)	Toxicity	NR
Tin (Sn)	Toxicity	Abdominal pain (stomach-aches), amblyacousia, Anemia, coma, confusion, conjunctivitis, convulsions, dermatitis, diarrhea, dizziness, cholangitis, hallucinations, headache, hepatotoxicity; hypersomnia, hypodynamia, inappetence, nausea, neurotoxicity (CNS edema), psychomotor disturbances, psychotic behavior, renal damage, restlessness, somnipathy, tinnitus, tremor, twitches, vomiting
Titanium (Ti)	Toxicity	Bone loss around dental implants, hypersensitivity (chronic fatigue syndrome, muscle pain, skin rashes), yellow nail syndrome
Tungsten (W)	Toxicity	Cancer (lung), dermatitis, memory loss, increased mortality (due cancer), pulmonary fibrosis, sensory deficits *Acute:* Rare anuria, nausea, seizures, clouded consciousness, coma, encephalopathy, moderate renal failure, acute tubular necrosis, hypocalcemia *Inhalation or contact—may be due cobalt exposure rather than Tungsten itself* *Soldiers drinking alcohol from discharged gun barrels*
Uranium (U)	Toxicity	Anemia, acute renal failure, coagulopathy, death, liver dysfunction, Fanconi syndrome, myocarditis, paralytic ileus, respiratory disease (pulmonary edema, pulmonary fibrosis), renal failure, rhabdomyelysis
Vanadium (V)	Toxicity	Abdominal cramps, airway irritation, cough, diarrhea, nausea, vomiting, wheezing
Ytterbium (Yb)	Toxicity	Eye irritation, pneumoconiosis, skin irritation

Yttrium (Y)	Toxicity	*Lung disease*: chest pain, shortness of breath, coughing, cyanosis, embolisms cancer
Zinc (Zn)	Deficiency	Acne, alopecia, anemia, anorexia, congenital malformations, depression (anhedonia), cheilitis (rare), impaired cognitive functions (learning and hedonic tone) cytokine suppression inflammatory cytokines (e.g., IL-1β, IL-2, IL-6, and TNF-α), dermatitis, diarrhea, dwarfism, dystocia, eczema, eye lesions, growth retardation, hypogonadism, impaired immune function, impotence, increased infection risk (e.g., dermatitis, gastroenteritis, pneumonia, and urinary tract infections), irregular menses, irritability, labor difficulties and prolonged, lethargy, lip cracks/fissures, nausea, night blindness, oral ulcers, placental abruption, prostate problems, impaired reproduction, seborrhea, stomatitis, impaired development and function of male sex organs, delayed sexual maturity, skin rash, reduced sense of smell, reduced sense of taste, tongue coating (white), vomiting, weight loss, wound healing impairment, xerosis (dry scaly skin),
Zinc (Zn)	Toxicity	Abdominal pain, bone marrow effects, chest pain, chest tightness, coagulopathy, cough (dry throat), death, depression, diarrhea (watery), dizziness, dyspnea, epigastric pain, fatigue, fever, hematemesis, hematuria, hemolysis, headaches, inflammation (chemical burns) to oral cavity, pharynx, trachea, esophagus and stomach, hives and angioedema (one report), lethargy, leucocytosis, liver necrosis, muscle cramps, neurological effects, neutrophilia, thrombocytopenia, "metal fume fever" (ZnO inhalation), pancreatitis, pneumonitis (acute), pneumothorax, airway irritation, acute respiratory distress syndrome (ARDS) pulmonary fibrosis, renal injury, interstitial nephritis, acute tubular necrosis, sweating, decreased vital capacity, vomiting, increased T cells, T suppressor cells, natural killer cells in bronchoalveolar lavage fluid Zn excess can induce copper deficiency causing: Anemia (sideroblastic), granulocytopenia, neutropenia, impaired immune function, adverse effect on ratio of low-density-lipoprotein to high-density-lipoprotein (LDL/HDL) cholesterol myelodysplastic syndrome, sensorimotor polyneuropathy syndrome, "swayback" (spasticity, gait abnormalities, and sensory ataxia)
Zirconium (Zr)	Toxicity	Blurred vision, circulatory collapse, contact dermatitis, disseminated intravascular coagulation, conjunctivitis, cough, dyspnea, hemolysis, hoarseness, lachrymation, laryngeal edema, lung granulomas, mucosal inflammation, pharyngeal pain, pulmonary fibrosis, pulmonary edema, pyloric stenosis, respiratory distress, skin granulomas, stridor, thirst Oral: abdominal pain, achlorhydria, acute tubular necrosis, adult respiratory distress syndrome (ARDS), antral stenosis, anuria, burning pain in mouth and throat, collapse, convulsions, watery or bloody diarrhea, gastric carcinoma, gastric perforation, hematemesis, hemolysis, aematuria, hypotension, jejunal stricture, liver damage with jaundice, esophageal perforation, peritonitis, pneumonitis, protein-losing gastroenteropathy, renal failure, retching, tenesmus, vomiting
Not Applicable (Element is not essential in this species)	Deficiency	Al, Sb, Ba, Be, Bi, Br, Cd, Cs, Ce, Dy, Er, Eu, Gd, Ga, Ge, Au, Ho, In, La, Pb, Li, Lu, Hg, Nd, Nb, Pd, Pr, Pm, Rh, Rb, Sm, Sc, Ag, Sr, Te, Tb, Tl, Th, Tm, Sn, Ti, W, U, V, Yb, Y, Zr

Bears

Bears (Urside) have simple stomachs. Availability of food in the wild affects reproductive performance (body size, litter size, the inter-litter interval, and population densities). Bears often hibernate for months during the winter and need to build up reserves of fat and lean muscle mass to survive even though their metabolic rate is greatly reduced.
Brown bears (*Ursus arctos*) are omnivores: in some geographical regions predominantly herbivores, in others predominantly carnivores. In one study (López-Alfaro et al., 2015) food sources for Brown bears were varied including ants, berries, green vegetation, meat, pine nuts, roots, trout, and false-truffles.
Dietary meat intake directly affects body size and population density and, in captive Polar Bears, there is seasonal variability in meat intake—declining in autumn. Polar bears are primarily carnivores living mainly on ringed seals, other seals (bearded and harp), whales (white and narwhal), walrus, carrion, reindeer, sea birds, and vegetation (Lintzenich et al., 2006).
Malayan Sun Bears consume a wide variety of foods including 10 plant species. Animal origin foods included mammals, insects, spiders, and others.
Giant Panda Bears (*Ailuropoda melanoleuca*) do not hibernate and are omnivores although bamboo constitutes 99% of their dietary intake.
Sun bears are mainly frugivorous and sloth bears insectivorous but both species can adapt to utilize available food resources.
As with all obligate carnivores for bears (such as Polar Bears), muscle meat alone does not meet the minimum requirement for minerals including Ca and P and additional food sources or supplementation are necessary to avoid the development of metabolic bone disease (Chesney and Hedberg, 2011).
Very little is known about the specific mineral and trace element requirements of Bears and guidelines are based on studies in other species such as dogs and cats. Zoos frequently offer captive Bears rations that vary greatly from the guidelines that have been established. (Lintzenich et al., 2006).

Clinical signs in bears

Limited data available. Many studies have been performed to analyze the tissue content of minerals and trace elements, but there are few reports of mineral, trace elements or rare earth elements-associated disease in bears.

Element	Deficiency/ toxicity	Clinical signs reported
Calcium (Ca)	Deficiency	*Suspected in captive Polar Bear cubs that develop classical signs of metabolic bone disease—maybe due low Vitamin D:* poor bone mineralization, fragile bones, fractures, rickets, wide metaphyses (Chesney and Hedberg, 2011)
Not Applicable (Element is not essential in this species)	Deficiency	Al, Sb, Ba, Be, Bi, Br, Cd, Cs, Ce, Dy, Er, Eu, Gd, Ga, Ge, Au, Ho, In, La, Pb, Li, Lu, Hg, Nd, Nb, Pd, Pr, Pm, Rh, Rb, Sm, Sc, Ag, Sr, Te, Tb, Tl, Th, Tm, Sn, Ti, W, U, V, Yb, Y, Zr

References

Sethy and Chauhan (2018), López-Alfaro et al. (2015), Chesney and Hedberg (2011), Stalenberg (2010), Lintzenich et al. (2006).

Coatis

Coatis belong to the family Procyonide (comprises 14 species in 8 genera of small carnivores, including raccoons, and kinkajou). They live in South and Central America and in southern United States. They are omnivores and eat mainly ground litter, invertebrates [millipedes, spiders (tarantulas), gastropods)] and fruit, but also small vertebrates (crocodile eggs, birds, birds' eggs, lizards, rodents).
Clinical signs. Limited data available.
Iron overload has been reported in captive Coatis (Nasua spp.) (Clauss et al., 2006).

Element	Clinical signs	Comments
Iron toxicosis	Iron overload with hemosiderosis/ hemochromatosis	Due to high Fe in captive diets containing vertebrate products (dog/cat food, prey—high in heme) wild coatis natural diet is mainly wild fruits and invertebrates and rarely vertebrate prey

References

Clauss et al. (2006).

Dogs

Although classified anatomically as carnivores, and whilst wild dogs and wolves remain carnivores domesticated dogs are now truly omnivores and have evolved mechanisms to be able to utilize starches present in plant materials. They can obtain all their minimum nutritional requirements from rations based on animal carcase or plant material. So, dogs can be fed vegetarian rations.

The trend to feed just raw meat to dogs despite the lack of scientific evidence to support such behavior is resulting in many mineral-related clinical problems due to the inadequate amounts of calcium, copper, iron, phosphorus that muscle meat contains.

Element	Deficiency/ toxicity	Clinical signs reported
Aluminium (Al)	Toxicity	Altered mentation, ataxia, decreased activity, convulsions, low hemoglobin, acute kidney injury, lethargy, decrease in MCV, microcytosis, muscle twitches, neurological signs (depressed menace response, normal cranial reflexes, obtundation, decreased pupillary light response, paraparesis, reduced patellar reflexes, poor pupillary light, reduced withdrawal reflexes), diffuse cerebral and peripheral neuropathy or junctionopathy, recumbency, reduced testes weight, reduced sperm quality, tetraparesis, tremor, weakness
Antimony (Sb)	Toxicity	NR
Arsenic (As)	Deficiency	NR
Arsenic (As)	Toxicity (Acute)	Abdominal pain, collapse, death, diarrhea, hypersalivation, hypothermia, lethargy, pulse weak, staggering, tachycardia, vomiting, weakness *Dermal contact*: blisters, cracking, bleeding, infections, swelling
	Toxicity (Chronic)	Renal tubular lesions (mild degeneration and vacuolization)
Barium (Ba)	Toxicity	Acidemia, arrhythmias (ventricular), collapse, severe hypokalemia, loss of muscle tone, muscle twitches, flaccid paralysis, impaired limb reflexes, respiratory failure, tachypnea
Beryllium (Be)	Toxicity	Granulomatous lung disease *Inhalation Beryllium oxide* (Haley et al., 1989)
Bismuth (Bi)	Toxicity	NR
Boron (B)	Deficiency	NR
Boron (B)	Toxicity	Testicular degeneration, arrested spermatogenesis, vomiting *1170μg/L B for 2 years*
Bromine (Br)	Toxicity	NR
Cadmium (Cd)	Toxicity	Osteomalacia
Cesium (Cs)	Toxicity	NR
Calcium (Ca)	Deficiency	*Common in dogs fed an all-meat diet:* bone cortices thin, bone undermineralized, fractures, feed efficacy reduced, the growth rate slowed, nutritional secondary hyperparathyroidism, osteomalacia, rickets, tooth loss, reproductive failure, seizures, tetany, vertebral compression
Calcium (Ca)	Toxicity	Angular deformities, bone increased-mineralization, growth impaired, osteochondrosis, skeletal abnormalities, stunted growth, tooth loss
Cerium (Ce)	Toxicity	Prothrombin levels and blood coagulation time were significantly increased (Graca et al., 1964).
Chlorine (Cl)	Deficiency	Ataxia, hypochloremia, hypokalemia, metabolic acidosis, weight gain ceased, weakness
Chlorine (Cl)	Toxicity	NR
Chromium (Cr)	Deficiency	NR
Chromium (Cr)	Toxicity	NR
Cobalt (Co)	Deficiency	NR
Cobalt (Co)	Toxicity	Cardiomyopathy

Copper (Cu)	Deficiency	Achromotrichia, anemia, hyperextension of the phalanges, poor keratinization *Unpublished study* on sudden death in Grayhounds: Death, major blood vessel rupture
Copper (Cu)	Toxicity	Death, hepatotoxicosis, vomiting; interference with Fe and Zn Hepatic accumulation of Cu is seen in an autosomal recessive disease in Bedlington, Skye, and West Highland White Terriers
Dysprosium (Dy)	Toxicity	NR
Erbium (Er)	Toxicity	NR
Europium (Eu)	Toxicity	NR
Fluorine (F)	Deficiency	NR
Fluorine (F)	Toxicity	Reduced growth rate
Gadolinium (Gd)	Toxicity	NR
Gallium (Ga)	Toxicity	NR
Germanium (Ge)	Toxicity	NR
Gold (Au)	Toxicity	NR
Holmium (Ho)	Toxicity	NR
Indium (In)	Toxicity	NR
Iodine (I)	Deficiency	Alopecia, apathy, coat dull and dry, cretinism (mental deficiency, deaf-mutism, spastic diplegia, neurological deficits), drowsiness, dullness, goiter, hairlessness, myxoedema of skin, skeletal deformities, low thyroid hormone levels (T3 and T4), timidity, weight gain
Iodine (I)	Toxicity	Aggressiveness, hyperthyroidism, panting, restlessness, tachycardia, weight loss *Reported most often after feeding raw foods*
Iron (Fe)	Deficiency	Anemia (microcytic, hypochromic), diarrhea, poor growth, lethargy, low hematocrit, low hemoglobin, hematochezia, melena, reduced muscle myoglobin concentrations, pallor, low transferrin, weakness
Iron (Fe)	Toxicity	Coagulopathy, death, depression, diarrhea, gastrointestinal damage, gastric obstruction, hematemesis, hematochezia, hypotension, lethargy, liver failure, metabolic acidosis, renal failure (acute), shock (hypovolemia), skin discoloration and edema (at injection site), tachycardia
Lanthanum (La)	Toxicity	NR
Lead (Pb)	Toxicity	Abdominal pain, anxiety, ataxia, blindness, encephalopathy, lethargy, seizures, tremors, vomiting
Lithium (Li)	Deficiency	NR
Lithium (Li)	Toxicity	Body mass loss, dehydration, diarrhea, polydipsia polyuria, seizures, weakness
Lutetium (Lu)	Toxicity	Increased clotting and prothrombin times, increased blood pressure and heart rate, death
Magnesium (Mg)	Deficiency	Anorexia, aortic mineralization, ataxia, hyperextension of carpal joints, hypocalcemia, hypomagnesemia, lameness, hindleg paralysis, muscle necrosis, poor weight gain, rhabdomyolysis, seizures, vomiting, weight loss *? mitral valve prolapse in Cavalier King Charles Spaniels* (Pederson and Mow, 1998)
Magnesium (Mg)	Toxicity	NR
Manganese (Mn)	Deficiency	Angular deformities of the limbs, fatty liver, reduced growth, reproduction impaired
Manganese (Mn)	Toxicity	NR
Mercury (Hg)	Toxicity	Abdominal pain, anuria, anxiety, ataxia, blindness, depression, bloody diarrhea, death, extensor rigidity, glycosuria, head tilt, hypocalcemia, hypokalemia, hypoproteinemia, incoordination, lymphocytopenia, metabolic acidosis, motor disabilities, nervousness, nystagmus (horizontal) opisthotonos, proteinuria, loss of pupillary reflexes, kidney damage (inability to urinate, abdominal swelling), acute renal tubular necrosis, tetraparesis, uremia (severe), visual disturbances, vomiting, loss of withdrawal reflexes
Molybdenum (Mo)	Deficiency	NR

Molybdenum (Mo)	Toxicity	NR
Neodymium (Nd)	Toxicity	NR
Nickel (Ni)	Deficiency	NR
Nickel (Ni)	Toxicity	Anemia (mild), growth depressed, hypersalivation, vomiting *2500μg/g food* (Ambrose et al., 1976)
Niobium (Nb)	Toxicity	*IV injection*: glomerular filtration rate (GFR) decreased, the maximal rate of tubular secretion of *p*-aminohippurate decreased, maximal rate of tubular reabsorption of glucose decreased, injuries in both proximal and distal tubules, the granular pigment in tubules (cortex and medulla), casts, epithelial cells flattened, cellular necrosis. *Potassium niobite given IV single dose*
Palladium (Pd)	Toxicity	NR
Phosphorus (P)	Deficiency	Angular deformities in long bones, anorexia, emaciation, poor growth, depressed myocardial performance, pica, seizures, rhabdomyolysis, rickets, swelling of joints, tremors
Phosphorus (P)	Toxicity	NR
Potassium (K)	Deficiency	Bradycardia, cardiac lesions, cardiac output reduced, dehydration, growth impairment, hypotension, muscle potassium low, paralysis (hindlegs), renal blood flow reduced, renal lesions, restlessness, stroke volume reduced, ventroflexion of the neck, weakness
Potassium (K)	Toxicity	NR
Praseodymium (Pr)	Toxicity	Deaths, drop in blood pressure and heart rate, increase in respiratory rate, increased prothrombin time. *Intravenous injection* (Graca et al., 1964) Variable results between different salts
Promethium (Pm)	Toxicity	NR
Rubidium (Rb)	Toxicity	NR
Samarium (Sm)	Toxicity	*Acute toxicity study:* death, circulatory failure (reduced heart rate, reduced blood pressure)
Scandium (Sc)	Toxicity	NR
Selenium (Se)	Deficiency	Anorexia, coma, depression, dyspnea, edema (ventral), muscle pallor, muscle degeneration, skeletal muscle discoloration (white streaks), intestinal muscle discoloration (brownish-yellow) due to lipofuscinosis, myopathy, renal white mineral deposits at the corticomedullary junction, subendocardial necrosis in cardiac ventricular muscles
Selenium (Se)	Toxicity	Anemia (microcytic, hypochromic), ascites, reduced food intake, reduced growth rate, low hemoglobin, liver damage (atrophy, cirrhosis, necrosis)
Silicon (Si)	Deficiency	NR
Silicon (Si)	Toxicity	Silicate uroliths. Usually from ingesting soil.
Silver (Ag)	Toxicity	NR
Sodium (Na)	Deficiency	Alopecia, dehydration, exhaustion, increase hematocrit, increased hemoglobin concentration, fatigue, mucous membranes dry and tacky, polydipsia, polyuria, restlessness, retarded growth, dry skin, tachycardia
Sodium (Na)	Toxicity	Abdominal pain, reduced food intake, gastritis, hematemesis, hypersalivation (ptyalism), pharyngeal edema, vomiting
Strontium (Sr)	Toxicity	NR
Sulfur (S)	Deficiency	NR
Sulfur (S)	Toxicity	*Experimental*: Sodium bisulfite LD50 244 mg/kg body weight
Tellurium (Te)	Toxicity	NR
Terbium (Tb)	Toxicity	NR
Thallium (Tl)	Toxicity	NR

Thorium (Th)	Toxicity	NR
Thulium (Tm)	Toxicity	NR
Tin (Sn)	Toxicity	NR
Titanium (Ti)	Toxicity	NR
Tungsten (W)	Toxicity	NR
Uranium (U)	Toxicity	Vomiting
Vanadium (V)	Toxicity	NR
Ytterbium (Yb)	Toxicity	NR
Yttrium (Y)	Toxicity	NR
Zinc (Zn)	Deficiency	Conjunctivitis, generalized debility, emaciation, poor growth, skin lesions abdomen and extremities(foot pads), hyperkeratitis especially around eyes, vomiting, poor weight gain, low serum Zn
Zinc (Zn)	Toxicity	Abdominal pain, acute renal failure, anemia (hemolytic), anorexia, azotemia, cardiac murmur (secondary to anemia), casts (granular), coagulopathy, death, dehydration, diarrhea, disseminated intravascular coagulation (DIC), duodenal ulcers, gastric ulcers, hepatic dysfunction, hypercreatininemia, hyperphosphatemia, icterus, jaundice, lameness, lethargy, melena, neurotoxicity, pallor, pancreatitis, seizures, tachycardia, tachypnea, vocalization, vomiting, weakness
Zirconium (Zr)	Toxicity	Reduced red blood cell count, reduced hemoglobin. *Long term exposure to zirconium tetrachloride*
Not Applicable (Element is not essential in this species)	Deficiency	Al, Sb, Ba, Be, Bi, Br, Cd, Cs, Ce, Dy, Er, Eu, Gd, Ga, Ge, Au, Ho, In, La, Pb, Li, Lu, Hg, Nd, Nb, Pd, Pr, Pm, Rh, Rb, Sm, Sc, Ag, Sr, Te, Tb, Tl, Th, Tm, Sn, Ti, W, U, V, Yb, Y, Zr

Foxes

Red foxes are omnivores and consume a wide range of foods including beetles, berries, birds, chickens, crayfish, crickets, caterpillars, fruits, grasses, grasshoppers, mice, rabbits, squirrels.
Limited information is available.

Element	Deficiency/ toxicity	Clinical signs reported
Calcium (Ca)	Deficiency	Angular deformity of long bones, bones poorly calcified, bones soft, bones vascularized, facial bones enlarged, gums swollen, hypocalcemia, joint swelling, lameness, muzzle enlarged, spasms, rickets, stiffness (rear legs), tooth loss, walk on pasterns
Iron (Fe)	Deficiency	Anemia, depigmentation of underfur
Mercury (Hg)	Toxicity	Deaths
Phosphorus (P)	Deficiency	Angular deformity of legs, bones undermineralized, joint enlargement, lameness, undershot jaw
Not Applicable (Element is not essential in this species)	Deficiency	Al, Sb, Ba, Be, Bi, Br, Cd, Cs, Ce, Dy, Er, Eu, Gd, Ga, Ge, Au, Ho, In, La, Pb, Li, Lu, Hg, Nd, Nb, Pd, Pr, Pm, Rh, Rb, Sm, Sc, Ag, Sr, Te, Tb, Tl, Th, Tm, Sn, Ti, W, U, V, Yb, Y, Zr

References

NRC (1982).

Gerbils

There are 14 species of gerbils and jirds of the genus Meriones distributed in the wild in North Africa, the Middle East, and central and eastern Asia They have adapted to living in arid conditions and they can concentrate urine and have low water requirements.
Limited data available.

Element	Deficiency/ toxicity	Clinical signs reported
Aluminium (Al)	Toxicity	Disrupted prostate development
Boron (B)	Toxicity	Testicular degeneration
Magnesium (Mg)	Deficiency	Alopecia, death, seizures, weight loss
Sodium (Na)	Deficiency	Alopecia (NaCl deficiency)
Sodium (Na)	Toxicity	Reduced appetite
Not Applicable (Element is not essential in this species)	Deficiency	Al, Sb, Ba, Be, Bi, Br, Cd, Cs, Ce, Dy, Er, Eu, Gd, Ga, Ge, Au, Ho, In, La, Pb, Li, Lu, Hg, Nd, Nb, Pd, Pr, Pm, Rh, Rb, Sm, Sc, Ag, Sr, Te, Tb, Tl, Th, Tm, Sn, Ti, W, U, V, Yb, Y, Zr

Guinea pigs

Guinea pigs are simple-stomached herbivores. They originate from the South American continent and in their wild natural habitat, they consume large quantities of green vegetation and fruits. Their stomach is lined with glandular epithelium. Gram-positive bacteria in the small intestine contribute to nutrition by synthesizing essential nutrients that are reingested through coprophagy. The guinea pig has a large cecum. The Ca, P, K, and Mg requirements of the guinea pig reflect interactions among these elements. They have a higher requirement for Ca and P than rats—the reason unknown.
Essential minerals—requirement per kg food for growth (NRC, 1995).

Essential mineral	Daily intake (g/mg)	Comments
Calcium	8.0g	Requirements for calcium, phosphorus, magnesium and potassium seem to reflect interactions among them
Chloride	0.5g	From the estimate for rats fed purified diet
Copper	6.0mg	
Iodine	150µg	Based on rat
Iron	50.0mg	Estimated
Magnesium	1.0g	
Manganese	40mg	
Molybdenum	150µg	Based on rat
Phosphorus	4.0g	
Potassium	5.0g	
Selenium	150µg	Based on rat
Sodium	0.5g	
Zinc	20.0mg	

Element	Deficiency/toxicity	Clinical signs reported
Aluminium (Al)	Toxicity	NR
Antimony (Sb)	Toxicity	NR
Arsenic (As)	Deficiency	NR
Arsenic (As)	Toxicity (Acute)	NR
	Toxicity (Chronic)	NR
Barium (Ba)	Toxicity	NR
Beryllium (Be)	Toxicity	*Inhalation:* Granulomatous lung disease, fibrosis
Bismuth (Bi)	Toxicity	NR
Boron (B)	Deficiency	NR
Boron (B)	Toxicity	NR
Bromine (Br)	Toxicity	NR
Cadmium (Cd)	Toxicity	NR
Cesium (Cs)	Toxicity	NR
Calcium (Ca)	Deficiency	Dental enamel hypoplasia, rickets, weight loss (*and P deficiency*)
Calcium (Ca)	Toxicity	Soft tissue calcification (*and P toxicity*)
Cerium (Ce)	Toxicity	NR

Element	Deficiency/ toxicity	Clinical signs reported
Chlorine (Cl)	Deficiency	NR
Chlorine (Cl)	Toxicity	NR
Chromium (Cr)	Deficiency	Glucose intolerance
Chromium (Cr)	Toxicity	NR
Cobalt (Co)	Deficiency	NR
Cobalt (Co)	Toxicity	NR
Copper (Cu)	Deficiency	*Young*: growth retardation, cardiovascular defects, and severe abnormalities of the central nervous system including agenesis of cerebellar folia, cerebral edema, and delayed myelination, high liver copper concentrations
Copper (Cu)	Toxicity	NR
Dysprosium (Dy)	Toxicity	Epilation, nodule formation
Erbium (Er)	Toxicity	Nodules (*at intradermal injection site*)
Europium (Eu)	Toxicity	*Subcutaneous injection site:* Nodules
Fluorine (F)	Deficiency	NR
Fluorine (F)	Toxicity	NR
Gadolinium (Gd)	Toxicity	NR
Gallium (Ga)	Toxicity	NR
Germanium (Ge)	Toxicity	NR
Gold (Au)	Toxicity	NR
Holmium (Ho)	Toxicity	NR
Indium (In)	Toxicity	NR
Iodine (I)	Deficiency	NR
Iodine (I)	Toxicity	NR
Iron (Fe)	Deficiency	NR
Iron (Fe)	Toxicity	NR
Lanthanum (La)	Toxicity	NR
Lead (Pb)	Toxicity	NR
Lithium (Li)	Deficiency	NR
Lithium (Li)	Toxicity	NR
Lutetium (Lu)	Toxicity	*Intradermal injection site:* mineralization, subcutaneous nodules
Magnesium (Mg)	Deficiency	Decreased activity, anemia, ataxia, poor growth, hair loss, stiffness (hind limbs), teeth discoloration and erosion, soft tissue calcification, tetany, elevated serum phosphorus
Magnesium (Mg)	Toxicity	NR
Manganese (Mn)	Deficiency	Angular deformity (bowing) forelegs, ataxia (neonatal), balance loss, body righting reflexes delayed development, bone growth retarded, forelimb shortening, glucose tolerance impaired, head retraction, head tremors, hyperreactivity to stimuli, litter size reduced, neonatal mortality, otolith abnormalities, pancreatic aplasia or hypoplasia, premature birth, skull deformities, vestibular syndrome
Manganese (Mn)	Toxicity	NR
Mercury (Hg)	Toxicity	Death, neurotoxicity, ototoxicity
Molybdenum (Mo)	Deficiency	NR

Element	Deficiency/ toxicity	Clinical signs reported
Molybdenum (Mo)	Toxicity	Death, growth retardation, contact sensitization (topical or intradermal exposure), weight loss
Neodymium (Nd)	Toxicity	NR
Nickel (Ni)	Deficiency	NR
Nickel (Ni)	Toxicity	Adenomatous alveolar lesions (*inhalation*) *Subcutaneous injection:* heme oxygenase activity increased
Niobium (Nb)	Toxicity	NR
Palladium (Pd)	Toxicity	NR
Phosphorus (P)	Deficiency	Dental enamel hypoplasia, rickets, weight loss (*and Ca deficiency*)
Phosphorus (P)	Toxicity	Soft tissue calcification, (*and Ca toxicity*) nephrocalcinosis
Potassium (K)	Deficiency	Death, membrane potentials in striated muscle cells from young guinea pigs fed a potassium-deficient diet were higher, increase in Na+, K+-ATPase activity in heart muscle cells
Potassium (K)	Toxicity	NR
Praseodymium (Pr)	Toxicity	NR
Promethium (Pm)	Toxicity	NR
Rhodium (Rh)	Toxicity	NR
Rubidium (Rb)	Toxicity	NR
Samarium (Sm)	Toxicity	NR
Scandium (Sc)	Toxicity	NR
Selenium (Se)	Deficiency	NR
Selenium (Se)	Toxicity	NR
Silicon (Si)	Deficiency	NR
Silicon (Si)	Toxicity	NR
Silver (Ag)	Toxicity	NR
Sodium (Na)	Deficiency	NR
Sodium (Na)	Toxicity	NR
Strontium (Sr)	Toxicity	NR
Sulfur (S)	Deficiency	NR
Sulfur (S)	Toxicity	NR
Tellurium (Te)	Toxicity	NR
Terbium (Tb)	Toxicity	NR
Thallium (Tl)	Toxicity	NR
Thorium (Th)	Toxicity	NR
Thulium (Tm)	Toxicity	NR
Tin (Sn)	Toxicity	NR
Titanium (Ti)	Toxicity	NR
Tungsten (W)	Toxicity	NR
Uranium (U)	Toxicity	NR
Vanadium (V)	Toxicity	NR
Ytterbium (Yb)	Toxicity	NR

Element	Deficiency/ toxicity	Clinical signs reported
Yttrium (Y)	Toxicity	NR
Zinc (Zn)	Deficiency	Abortion, anorexia, dermatitis (scaly), gamma-globulin levels low, glycosaminoglycans metabolism altered, hair rough, impaired ability to elicit a hypersensitivity reaction (delayed), posture abnormal, premature births, excessive, skin lesions, vocalization during pregnancy, low plasma zinc concentration
Zinc (Zn)	*Toxicity*	*Inhalation:* Alveolitis (focal), consolidation, death, emphysema, fibrosis, decreased lung compliance, decreased functional residual capacity, impaired lung function; inflammation, macrophage infiltration
Zirconium (Zr)	Toxicity	Reduced survival. Poor weight gain, chronic interstitial pneumonitis *Zirconium tetrachloride*
Not Applicable (Element is not essential in any species)	Deficiency	Al, Sb, Ba, Be, Bi, Br, Cd, Cs, Ce, Dy, Er, Eu, Gd, Ga, Ge, Au, Ho, In, La, Pb, Li, Lu, Hg, Nd, Nb, Pd, Pr, Pm, Rh, Rb, Sm, Sc, Ag, Sr, Te, Tb, Tl, Th, Tm, Sn, Ti, W, U, V, Yb, Y, Zr

References

NRC (1995), Anderson (1987b), Breazile and Brown (1976), Navia and Hunt (1976), Falk et al. (1974).

Hamsters

Hamsters belong to the family Muride, subfamily Cricetine, which consists of 7 genera and 18 species. Their home environment consists of clay deserts, shrub-covered plains, forested steppes, and/or cultivated fields.

They have an unusual stomach with two compartments: a keratinized, nonglandular forestomach (cardiac) which contains a higher number of microorganisms, separated from a glandular region (pyloric) by sphincter-like muscular folds functionally similar to the rumen of herbivores, they also have a large cecum. They eat every 2h.

Limited data available.

Element	Deficiency/ toxicity	Clinical signs reported
Arsenic (As)	Toxicity (Acute)	NR
Arsenic (As)	Toxicity (Chronic)	NR
Barium (Ba)	Toxicity	NR
Beryllium (Be)	Toxicity	*Inhalation Be salts:* Death, bronchiolar-alveolar tumors, adenomas Granulomatous lung lesions
Cadmium (Cd)	Toxicity	*Inhalation:* Alveolar hyperplasia, hemorrhagic ovarian necrosis, interstitial fibrosis *Injection during pregnancy:* Fetal deformities (craniofacial, eyes, limbs)
Calcium (Ca)	Deficiency	NR
Calcium (Ca)	Toxicity	NR
Chlorine (Cl)	Deficiency	NR
Chlorine (Cl)	Toxicity	NR
Chromium (Cr)	Deficiency	NR
Chromium (Cr)	Toxicity	*Fetus:* cleft palate, increased resorption rate, growth retardation, hydrocephalus, skeletal ossification impaired *Maternal:* death, renal tubular necrosis, reduced weight gain
Cobalt (Co)	Deficiency	NR
Cobalt (Co)	Toxicity	NR
Copper (Cu)	Deficiency	NR
Copper (Cu)	Toxicity	NR
Iodine (I)	Deficiency	Goiter, impaired growth
Iodine (I)	Toxicity	NR
Iron (Fe)	Deficiency	Pregnancy: low-maternal weight gain, high frequency of prenatal mortality
Iron (Fe)	Toxicity	NR
Lead (Pb)	Toxicity	NR
Magnesium (Mg)	Deficiency	NR
Magnesium (Mg)	Toxicity	NR
Manganese (Mn)	Deficiency	NR
Manganese (Mn)	Toxicity	NR
Mercury (Hg)	Toxicity	NR
Molybdenum (Mo)	Deficiency	Impaired growth
Molybdenum (Mo)	Toxicity	NR

Nickel (Ni)	Toxicity	*Inhalation:* alveolar septal fibrosis, atelectasia, bronchial hyperplasia, death, emphysema, granulomatous pneumonia, interstitial pneumonia, squamous metaplasia, decrease in tracheal ciliary activity, tracheal mucosal degeneration, *Intratracheal injection* Lung tumors (adenoma, adenocarcinoma) *Subcutaneous injection:* heme oxygenase activity increased *Intramuscular injection:* sarcomas
Phosphorus (P)	Deficiency	NR
Phosphorus (P)	Toxicity	NR
Potassium (K)	Deficiency	NR
Potassium (K)	Toxicity	NR
Selenium (Se)	Deficiency	Impaired growth
Selenium (Se)	Toxicity	NR
Sodium (Na)	Deficiency	NR
Sodium (Na)	Toxicity	NR
Sulfur (S)	Deficiency	NR
Sulfur (S)	Toxicity	NR
Thorium (Th)	Toxicity	Carcinogenic
Vanadium (V)	Toxicity	NR
Ytterbium (Yb)	Toxicity	Teratogenic
Zinc (Zn)	Deficiency	Growth impairment, abnormal estrus cycle, weight loss
Zinc (Zn)	Toxicity	NR
Zirconium (Zr)	Toxicity	Poor weight gain, interstitial pneumonitis
Not Applicable (Element is not essential in any species)	Deficiency	Al, Sb, Ba, Be, Bi, Br, Cd, Cs, Ce, Dy, Er, Eu, Gd, Ga, Ge, Au, Ho, In, La, Pb, Li, Lu, Hg, Nd, Nb, Pd, Pr, Pm, Rh, Rb, Sm, Sc, Ag, Sr, Te, Tb, Tl, Th, Tm, Sn, Ti, W, U, V, Yb, Y, Zr

Mice

Mice are simple-stomached rodent omnivores.

Element	Deficiency/ toxicity	Clinical signs reported
Aluminium (Al)	Toxicity	Less active, behavioral changes (altered escape behavior, delayed puberty and adulthood), impaired fetal bone development, fetal micronucleated erythrocytes, increased fetal abnormalities, intestinal inflammation, learning and memory disorders, reduced estradiol production, motor disturbances (grip strength), reduced neural stem cells, impaired cell proliferation, and neuroblast differentiation, reflexes slowed, increased startle responses *Signs of P deficiency as Al binds to P.*
Antimony (Sb)	Toxicity	NR
Arsenic (As)	Deficiency	NR
Arsenic (As)	Toxicity	Organomegaly (kidney, liver, spleen), hyperglycemia, elevated enzymes [serum alkaline phosphatase (ALP), alanine aminotransferase (ALT), lactate dehydrogenase (LDH)] uric acid
Barium (Ba)	Toxicity	Death, kidney damage, decreases in body weight, decreased survival, renal tubule dilatation, hyaline cast formation, interstitial fibrosis, glomerulosclerosis
Beryllium (Be)	Toxicity	Bone tumors
Bismuth (Bi)	Toxicity	NR
Boron (B)	Deficiency	NR
Boron (B)	Toxicity	*Inhalation diborane gas:* pulmonary congestion, bleeding, and edema, *Oral:* hepatic necrosis gastric hyperplasia and dysplasia, renal lesions, testicular degeneration *Embryo:* skeletal malformations (e.g., rib agenesis, fused rib, cervical rib, reduced rib length) Reductions in fetal body weights
Bromine (Br)	Deficiency	Impaired growth (*on low Br diet—not confirmed as an essential nutrient yet*)
Bromine (Br)	Toxicity	NR
Cadmium (Cd)	Toxicity	Alveolar hyperplasia, cleft lip and palate, club foot, death, fetal developmental abnormalities (craniofacial, exencephaly, eyes, limbs), hemorrhagic ovarian necrosis, hydrocephalus, lung tumors, micrognathia, microphthalmia, osteomalacia (femur), pulmonary interstitial fibrosis, tail dysplasia, testicular necrosis, weight loss
Cesium (Cs)	Toxicity	Deaths, nasal discharge
Calcium (Ca)	Deficiency	Poor weight gain, poor bone mineralization
Calcium (Ca)	Toxicity	NR
Cerium (Ce)	Toxicity	Altered hepatic levels of some cytochrome (CYP) P450, congestion of abdominal organs, death, gastroenteritis including mucosal necrosis, hepatocellular necrosis, pulmonary and tracheal hemorrhage, hepatic congestion, renal damage, splenic hypertrophy, lower white blood cell, and platelet counts, reticulocytes reduced. CD3 and CD8 lymphocytes decreased, IgM decreased. Liver fatty degeneration, alanine aminotransferase high, increased liver metallothionein and glutathione
Chlorine (Cl)	Deficiency	NR
Chlorine (Cl)	Toxicity	NR
Chromium (Cr)	Deficiency	Aortic plaques, appetite increase, glucose intolerance, growth disorders, hypercholesterolemia, longevity reduced
Chromium (Cr)	Toxicity	Death (post inhalation), fetal abnormalities (anencephaly, exencephaly) *after intraperitoneal injection*, granuloma formation (*inhalation*), growth impaired, kidney damage, liver damage, nasal papillomas, nasal septum perforation, pulmonary adenomas, sarcomas *at the intramuscular implant site* *Intraperitoneal injection*: reduced litter size, reduced sperm count, abnormal sperm

Cobalt (Co)	Deficiency	NR
Cobalt (Co)	Toxicity	Hypothermia
Copper (Cu)	Deficiency	Anemia, impaired humoral immune response (antibody production)
Copper (Cu)	Toxicity	NR
Dysprosium (Dy)	Toxicity	*Dy injected intradermally:* Tissue fibrosis, calcification, ataxia, labored breathing, lacrimation stretching of limbs while walking, writhing
Erbium (Er)	Toxicity	*Subcutaneous injection site*: Local calcification, fibrosis, multinucleated giant cell accumulation *Acute toxicity study:* ataxia, labored breathing, stretching of limbs while walking, lachrymation, writhing
Europium (Eu)	Toxicity	Arched back, ataxia, death, epiphora, labored breathing, neurotoxicity stretching of the hind limbs while walking, writhing *Neurological signs:* reduced motor activity, disorientation, loss of motor coordination, rapid breathing, tachycardia, and infrequent tremor
Fluorine (F)	Deficiency	NR
Fluorine (F)	Toxicity	NR
Gadolinium (Gd)	Toxicity	Pneumonia was increased in most of the exposed groups but was not related to the duration of exposure. Calcification of the lung
Gallium (Ga)	Toxicity	Anemia (microcytic), bone marrow hyperplasia, growth suppression, hemosiderosis (liver, spleen) reduced life span, plasma cell hyperplasia (lymph nodes), sperm count and motility low, testicular atrophy weight loss. *Blood results*: high platelets, neutrophils, alanine aminotransferase, sorbitol dehydrogenase
Germanium (Ge)	Toxicity	Reduced lifespan, fatty degeneration liver
Gold (Au)	Toxicity	NR
Holmium (Ho)	Toxicity	*Acute toxicity:* arched back, ataxia (loss of muscular coordination), labored breathing, lachrymation stretching of limbs while walking, writhing
Indium (In)	Toxicity	Alveolar proteinosis and alveolar macrophage infiltration, thickened pleural wall
Iodine (I)	Deficiency	Goiter
Iodine (I)	Toxicity	NR
Iron (Fe)	Deficiency	Anemia, reduced birth weights, reduced litter sizes
Iron (Fe)	Toxicity	NR
Lanthanum (La)	Toxicity	Hypoglycemia, hypotension, degeneration of the spleen and hepatic damage, low uric acid, high creatinine, low Ca, low P, low BUN
Lead (Pb)	Toxicity	NR
Lithium (Li)	Deficiency	NR
Lithium (Li)	Toxicity	Reduced growth, kidney damage, liver damage
Lutetium (Lu)	Toxicity	Arched back, ataxia (loss of muscular coordination), labored breathing, sedation stretching of limbs while walking, and excessive tearing, writhing. *IV:* Impaired blood clotting
Magnesium (Mg)	Deficiency	Convulsions, death
Magnesium (Mg)	Toxicity	NR
Manganese (Mn)	Deficiency	Ataxia (congenital), fat deposition in liver, otolith deformity, impaired melanin synthesis (in pallid mice)
Manganese (Mn)	Toxicity	
Mercury (Hg)	Toxicity	Behavioral changes in neonates, hind leg weakness and degenerative changes in the corpus striatum, cerebral cortex, thalamus, and hypothalamus. Renal tumors
Molybdenum (Mo)	Deficiency	NR

Molybdenum (Mo)	Toxicity	*Inhalation:* hyaline degeneration of nasal epithelium, squamous metaplasia of nasal planum, increased cancer (alveolar/bronchiolar adenomas or carcinomas)
Neodymium (Nd)	Toxicity	Acute toxicity included ataxia (loss of muscular coordination), labored breathing, sedation, and stretching of limbs while walking, writhing
Nickel (Ni)	Deficiency	NR
Nickel (Ni)	Toxicity	Reduced feed intake *Subcutaneous injection:* heme oxygenase activity increased *Intramuscular or subcutaneous injection*: local fibrosarcomas, sarcomas, immunosuppression *Inhalation:* inflammation, fibrosis, hyperplasia alveolar macrophages, necrotizing pneumonia, degeneration of the bronchiolar epithelium, atrophy of the olfactory epithelium, hyperplasia of the bronchial and mediastinal lymph nodes, thymus atrophy *Intratracheal:* pulmonary hemorrhage
Niobium (Nb)	Toxicity	Fatty liver, growth suppressed, reduced lifespan
Palladium (Pd)	Toxicity	Amyloidosis, carcinogenic, death, longer life span (males), reduced testes weight
Phosphorus (P)	Deficiency	NR
Phosphorus (P)	Toxicity	NR
Potassium (K)	Deficiency	Coat dull, emaciation, tail dry and scaly, weight loss
Potassium (K)	Toxicity	NR
Praseodymium (Pr)	Toxicity	*Acute toxicity* arched back, ataxia, labored breathing, sedation, stretching of limbs while walking, walking on toes, writhing
Promethium (Pm)	Toxicity	NR
Rhodium (Rh)	Toxicity	Carcinogenic
Rubidium (Rb)	Toxicity	NR
Samarium (Sm)	Toxicity	*Acute toxicity IP injection*: abdominal cramps, diarrhea, labored breathing, lethargy, muscular spasms, growth and fertilization rates decreased, LDH enzyme activity decreased in testes
Scandium (Sc)	Toxicity	NR
Selenium (Se)	Deficiency	NR
Selenium (Se)	Toxicity	NR
Silicon (Si)	Deficiency	NR
Silicon (Si)	Toxicity	NR
Silver (Ag)	Toxicity	NR
Sodium (Na)	Deficiency	NR
Sodium (Na)	Toxicity	NR
Strontium (Sr)	Toxicity	Growth depression, rickets, death (at very high doses) *Radioactive isotope:* Anemia, cancers (bone, leukemia, lung, nose) reduced red and white blood cell counts and platelets, hemorrhages, susceptibility to infections, thinning of the cornea, thinning of the skin
Sulfur (S)	Deficiency	NR
Sulfur (S)	Toxicity	NR
Tellurium (Te)	Toxicity	NR
Terbium (Tb)	Toxicity	*Tb Administered IV*: Lung gamma-glutamyl transpeptidase (gamma-GTP) activity was increased; alkaline phosphatase (ALP) was increased; lipid peroxidation increased, pulmonary superoxide dismutase (SOD), catalase (CAT), and glutathione peroxidase (GSH-Px) activities were all decreased
Thallium (Tl)	Toxicity	Death (*used in rodenticide*)

Thorium (Th)	Toxicity	Oxidative stress (in liver, spleen, bone) increased lipid peroxidation, decreased activity antioxidant enzymes (superoxide dismutase, catalase), death
Thulium (Tm)	Toxicity	*After injection:* degeneration of the liver and spleen
Tin (Sn)	Toxicity	Brain spongiosis, growth depression, food efficiency reduced, pancreatic atrophy, renal calcification, testicular degeneration, reduced alkaline phosphatase activity in bone and liver, low hemoglobin
Titanium (Ti)	Toxicity	NR
Tungsten (W)	Toxicity	Reduced lifespan
Uranium (U)	Toxicity	Congenital malformations (cleft palate, poor bone ossification, bipartite sternebre), reduced fertility, reduced fetal weight, reduced fetal length, reduced food intake, poor weight gain,
Vanadium (V)	Toxicity	Alveolar/bronchial hyperplasia, inflammation and fibrosis, cancer (lung), fetal skeletal and visceral malformations and anomalies laryngeal degeneration, nasal hyperplasia
Ytterbium (Yb)	Toxicity	NR
Yttrium (Y)	Toxicity	Growth impairment
Zinc (Zn)	Deficiency	Adrenal hypertrophy, alopecia, emaciation, growth retardation, weight loss
Zinc (Zn)	Toxicity	Death, increased macrophages and lymphocytes in the lungs, alveolar carcinoma
Zirconium (Zr)	Toxicity	Reduced survival time, weight loss
Not Applicable (Element is not essential in this species)	Deficiency	Al, Sb, Ba, Be, Bi, Br, Cd, Cs, Ce, Dy, Er, Eu, Gd, Ga, Ge, Au, Ho, In, La, Pb, Li, Lu, Hg, Nd, Nb, Pd, Pr, Pm, Rh, Rb, Sm, Sc, Ag, Sr, Te, Tb, Tl, Th, Tm, Sn, Ti, W, U, V, Yb, Y, Zr

Barai et al. (2017), NRC (1995).

Meerkat

Meerkats (*Suricata suricatta*) are primarily insectivores eat invertebrates, but also consume egg, small vertebrate prey, and vegetables.
Limited information available.

Element	Deficiency/ toxicity	Clinical signs
Calcium	Deficiency	Fractures, metabolic bone disease, nutritional secondary hyperparathyroidism, undermineralized bone

Reference

Aza mongoose, Meerkat, and Fossa (Herpestide/Eupleride), Novikova et al. (2020).

Mongoose

Mongooses diet:

a. Dwarf mongooses (*Helogale parvula*) eggs, fruit, invertebrate, small vertebrates
b. Banded mongooses' small vertebrates (birds, rats, snakes) and eggs (bird, reptile) invertebrates, fruit

Limited information available.

Element	Deficiency/toxicity	Clinical signs
Chromium (Cr)	Toxicity	Impaired reproductive function

Reference

Andleeb et al. (2019) and AZA Mongoose, Meerkat and Fossa Guide (2020).

Non-human primates

Primates have evolved diverse digestive tract anatomical systems in response to environmental food availability. Primates that consume animal material (faunivores)—which include invertebrates (e.g., insects), reptiles, and other life forms as well as mammals—have a simpler and shorter gastrointestinal tract than those that consume plant material. Most primates are also frugivores, but none exist on a solely fruit-based diet.

Primate folivores have specially adapted foregut or hind gut to accommodate bacteria which degrade the plant materials by fermentation. Our knowledge base is so incomplete that formulating a complete and balanced ration is very difficult. In the wild, some primates indulge in soil or rock-eating (geophagy) or bones to obtain essential minerals and trace elements. Green leaves and gums can be high in Ca content whereas muscle, invertebrates, nuts, and seeds can be high in P content. In wild Mountain Gorillas juveniles self-select food ingredients to ensure they take in more minerals than females or silverbacks (Rothman et al., 2008).

Callitrichids are omnivores that eat insects and plants (flowers, fruits). There are differences between species:

Species	Natural food sources
Marmosets	Plant exudates (gum arabic) (45%), fruits (16%) insects (39%),
Tamarins	Fruits (35%) insects (45%), exudates (10%), nectar (7%), seeds (3%)

Exudates are fermented by the cecal microbiome in marmosets.

Element	Deficiency/ toxicity	Clinical signs reported
Aluminium (Al)	Toxicity	NR
Antimony (Sb)	Toxicity	NR
Arsenic (As)	Deficiency	NR
Arsenic (As)	Toxicity	NR
Barium (Ba)	Toxicity	NR
Beryllium (Be)	Toxicity	*Inhalation: Be ores. Squirrel monkeys:* Death, inflammatory cells in lungs (macrocytes, lymphocytes, plasma cells) *Inhalation: Be sulfate Rhesus monkeys*: Lung tumors
Bismuth (Bi)	Toxicity	NR
Boron (B)	Deficiency	NR
Boron (B)	Toxicity	NR
Bromine (Br)	Toxicity	NR
Cadmium (Cd)	Toxicity	Renal tubular dysfunction *1.2 mg/kg bw per day for 9 years*
Cesium (Cs)	Toxicity	NR
Calcium (Ca)	Deficiency	Motor neurone damage (*cynomolgus monkeys*), nutritional secondary hyperparathyroidism (*Lemurs*: angular deformity long bones, bone undermineralized, high blood P, low blood Ca, high ALP, reduced mobility), soft tissue mineralization *Tarsiers*: poor reproductive performance
Calcium (Ca)	Toxicity	NR
Cerium (Ce)	Toxicity	NR
Chlorine (Cl)	Deficiency	NR
Chlorine (Cl)	Toxicity	NR
Chromium (Cr)	Deficiency	Corneal lesions, Glucose intolerance
Chromium (Cr)	Toxicity	NR
Cobalt (Co)	Deficiency	NR

Cobalt (Co)	Toxicity	NR
Copper (Cu)	Deficiency	Cardiac weakness, blood vessel weakness, increased plasma cholesterol
Copper (Cu)	Toxicity	NR
Dysprosium (Dy)	Toxicity	NR
Erbium (Er)	Toxicity	NR
Europium (Eu)	Toxicity	NR
Fluorine (F)	Deficiency	NR
Fluorine (F)	Toxicity	NR
Gadolinium (Gd)	Toxicity	NR
Gallium (Ga)	Toxicity	NR
Germanium (Ge)	Toxicity	NR
Gold (Au)	Toxicity	NR
Holmium (Ho)	Toxicity	NR
Indium (In)	Toxicity	NR
Iodine (I)	Deficiency	*Marmosets:* Birth weight lower, slower growth rate, sparse haircoat
Iodine (I)	Toxicity	NR
Iron (Fe)	Deficiency	Anemia (Rhesus)
Iron (Fe)	Toxicity	Hemosiderosis (*lemurs*)
Lanthanum (La)	Toxicity	NR
Lead (Pb)	Toxicity	NR
Lithium (Li)	Deficiency	NR
Lithium (Li)	Toxicity	NR
Lutetium (Lu)	Toxicity	NR
Magnesium (Mg)	Deficiency	Low Ca, cardiovascular (ECG changes due low K or low Ca), gastrointestinal signs, irritability (*Rhesus*), neuromuscular signs
Magnesium (Mg)	Toxicity	NR
Manganese (Mn)	Deficiency	Poor righting reflexes
Manganese (Mn)	Toxicity	NR
Mercury (Hg)	Toxicity	Abortions, low conception rate, reduced visual sensitivity, restricted visual field, impairment of spatial vision, intention tremors, somasethic impairment, incoordination
Molybdenum (Mo)	Deficiency	NR
Molybdenum (Mo)	Toxicity	NR
Neodymium (Nd)	Toxicity	NR
Nickel (Ni)	Deficiency	NR
Nickel (Ni)	Toxicity	Reduced feed intake
Niobium (Nb)	Toxicity	NR
Palladium (Pd)	Toxicity	NR
Phosphorus (P)	Deficiency	NR
Phosphorus (P)	Toxicity	NR
Potassium (K)	Deficiency	NR
Potassium (K)	Toxicity	NR
Praseodymium (Pr)	Toxicity	NR

Promethium (Pm)	Toxicity	NR
Rhodium (Rh)	Toxicity	NR
Rubidium (Rb)	Toxicity	NR
Samarium (Sm)	Toxicity	NR
Scandium (Sc)	Toxicity	NR
Selenium (Se)	Deficiency	*Squirrel monkeys:* alopecia, cardiomyopathy (*only if low protein as well*) hepatic degeneration, listlessness, myopathy, weight loss
Selenium (Se)	Toxicity	*Cynomolgus monkeys:* anorexia, cataract (prenatal exposure), cheilitis, death, dermatitis, hypothermia, menstrual disturbances, vomiting, weight loss, xerosis
Silicon (Si)	Deficiency	NR
Silicon (Si)	Toxicity	NR
Silver (Ag)	Toxicity	NR
Sodium (Na)	Deficiency	NR
Sodium (Na)	Toxicity	NR
Strontium (Sr)	Toxicity	NR
Sulfur (S)	Deficiency	NR
Sulfur (S)	Toxicity	NR
Tellurium (Te)	Toxicity	NR
Terbium (Tb)	Toxicity	NR
Thallium (Tl)	Toxicity	NR
Thorium (Th)	Toxicity	NR
Thulium (Tm)	Toxicity	NR
Tin (Sn)	Toxicity	NR
Titanium (Ti)	Toxicity	NR
Tungsten (W)	Toxicity	NR
Uranium (U)	Toxicity	NR
Vanadium (V)	Toxicity	NR
Ytterbium (Yb)	Toxicity	NR
Yttrium (Y)	Toxicity	NR
Zinc (Zn)	Deficiency	Abortions, alopecia, anorexia, apathy, appetite reduced, behavioral changes, bone formation impaired, poor bone mineralization, cognitive function impaired, debilitation, dermatitis, fetal abnormalities, unkempt hair (squirrel monkeys), plasma Zn low, play less (infants), poor growth, poor immune function, increased morbidity, parakeratosis of tongue (squirrel monkeys), skin integrity impaired, reproductive failure, skeletal development delayed, small offspring, stillbirths, tissue Zn low *Squirrel monkeys:* alopecia, growth impairment, unkempt haircoat, parakeratosis (tongue), skin lesions, cessation estrus
Zinc (Zn)	Toxicity	*Galvanized cages:* Depigmentation of hair (steely gray) (achromotrichia), alopecia, anorexia, weakness (*due low Cu*)
Zirconium (Zr)	Toxicity	NR
Not Applicable (Element is not essential in these species)	Deficiency	Al, Sb, Ba, Be, Bi, Br, Cd, Cs, Ce, Dy, Er, Eu, Gd, Ga, Ge, Au, Ho, In, La, Pb, Li, Lu, Hg, Nd, Nb, Pd, Pr, Pm, Rh, Rb, Sm, Sc, Ag, Sr, Te, Tb, Tl, Th, Tm, Sn, Ti, W, U, V, Yb, Y, Zr

References

Cancelliere et al. (2014), Rothman et al. (2008), Susanne and Ann-Kathrin (2005), NRC (2003), Tomson and Lotshaw (1978), Roberts and Kohn (1993).

Pigs

Pigs are simple-stomached ominvores.

Element	Deficiency/ toxicity	Clinical signs reported
Aluminium (Al)	Toxicity	Growth retardation, muscle atrophy—due hypophosphatemia, osteomalacia
Antimony (Sb)	Toxicity	Pulmonary lesions: Focal fibrosis, adenomatous hyperplasia, multinucleated giant cells, cholesterol clefts, pneumonocyte hyperplasia, and pigmented macrophages
Arsenic (As)	Deficiency	NR
Arsenic (As)	Toxicity	NR
Barium (Ba)	Toxicity	NR
Beryllium (Be)	Toxicity	NR
Bismuth (Bi)	Toxicity	NR
Boron (B)	Deficiency	NR
Boron (B)	Toxicity	NR
Bromine (Br)	Toxicity	NR
Cadmium (Cd)	Toxicity	NR
Cesium (Cs)	Toxicity	NR
Calcium (Ca)	Deficiency	NR
Calcium (Ca)	Toxicity	Hemorrhage (young pigs—due to interference with Vitamin K), parakeratosis
Cerium (Ce)	Toxicity	NR
Chlorine (Cl)	Deficiency	NR
Chlorine (Cl)	Toxicity	NR
Chromium (Cr)	Deficiency	Body fat increased, hypercholesterolemia, hyperinsulinemia, reduced lean muscle mass
Chromium (Cr)	Toxicity	NR
Cobalt (Co)	Deficiency	Reduced body weight gain, reduced litter size, poor survivability, mild anemia (uncommon)
Cobalt (Co)	Toxicity	NR
Copper (Cu)	Deficiency	(*Uncommon*) Anemia (microcytic hypochromic), angular deformity of forelimbs, ataxia, blood vessel rupture, cardiomegaly, endochondral ossification impaired, hemoglobin low, hyperflexion of hocks, mean corpuscular volume low, osteoblast activity reduced
Copper (Cu)	Toxicity	High Cu can destroy natural tocopherols (Vitamin E)
Dysprosium (Dy)	Toxicity	NR
Erbium (Er)	Toxicity	NR
Europium (Eu)	Toxicity	NR
Fluorine (F)	Deficiency	NR
Fluorine (F)	Toxicity	Anorexia, bones soften and overgrow, death, dental erosion and excess wear, kidney damage, lactation reduced, thirst reduced due oral pain, poor weight gain
Gadolinium (Gd)	Toxicity	NR
Gallium (Ga)	Toxicity	NR
Germanium (Ge)	Toxicity	NR
Gold (Au)	Toxicity	NR
Holmium (Ho)	Toxicity	NR
Indium (In)	Toxicity	NR

Iodine (I)	Deficiency	Deaths (neonatal), dwarfism, hairlessness, lethargy, reproductive failure, shortened limb bones, skin dry rough and thickened, stillbirths, thyroid enlargement, weakness
Iodine (I)	Toxicity	Decreased growth, reduced feed intake, reduced hemoglobin levels, reduced liver Fe content
Iron (Fe)	Deficiency	*Common.* Anemia (hypochromic-microcytic), ascites, breathing labored, cardiac dilation, death, ears limp, edema (subcutaneous), eyelids droop, fatty liver, haircoat coarse dull and rough, hemoglobin low, hepatomegaly, inappetence, increased infection risk (especially enteritis due to *E. coli* and respiratory infections), impaired immune function, listlessness, lung collapse, mortality rate increased, mucus membranes pale, muscle myoglobin low, pulmonary edema, tail hangs limp, thoracic fluid, ventroflexion of the neck, weight gain rate reduced
Iron (Fe)	Toxicity	Convulsions, death, diarrhea, dyspnea, shivering, incoordination, hyperpnea, jaundice, pallor, tetanic posterior paralysis
Lanthanum (La)	Toxicity	NR
Lead (Pb)	Toxicity	NR
Lithium (Li)	Deficiency	NR
Lithium (Li)	Toxicity	NR
Lutetium (Lu)	Toxicity	NR
Magnesium (Mg)	Deficiency	Ataxia, collapse, death, poor growth, hyperirritability, hypocalcemia, muscle twitches, stepping syndrome, weakness
Magnesium (Mg)	Toxicity	Depressed weight gain/growth
Manganese (Mn)	Deficiency	Angular deformity limbs, fat deposition in liver, joint enlargement (hocks), lameness, shortened limbs
Manganese (Mn)	Toxicity	NR
Mercury (Hg)	Toxicity	Anorexia, ataxia, blindness, coma, crystalluria (white), bloody diarrhea, death, gastrointestinal tract necrosis, paralysis (partial), kidney damage, uremia, wandering, weight
Molybdenum (Mo)	Deficiency	NR
Molybdenum (Mo)	Toxicity	NR
Neodymium (Nd)	Toxicity	NR
Nickel (Ni)	Deficiency	*Minipigs*: Impaired growth rate, skin scales and crusts, delayed estrus, poor mineralization of skeleton
Nickel (Ni)	Toxicity	Diarrhea, growth impairment, poor haircoat, melena
Niobium (Nb)	Toxicity	NR
Palladium (Pd)	Toxicity	NR
Phosphorus (P)	Deficiency	NR
Phosphorus (P)	Toxicity	NR
Potassium (K)	Deficiency	*Experimental:* reduced appetite, ataxia, bradycardia, emaciation, reduced growth rate, rough haircoat, lethargy, increases PQRS intervals
Potassium (K)	Toxicity	Bradycardia
Praseodymium (Pr)	Toxicity	NR
Promethium (Pr)	Toxicity	NR
Rhodium (Rh)	Toxicity	NR
Rubidium (Rb)	Toxicity	NR
Samarium (Sm)	Toxicity	NR
Scandium (Sc)	Toxicity	NR
Selenium (Se)	Deficiency	Death, hepatosis, myopathy, mulberry heart disease,

Selenium (Se)	Toxicity	Alopecia, anorexia, cachexia, central nervous system depression, conception rate low, coma, death, depression, emaciation, emesis, increased glutathione peroxidase levels, hoof lesions (cracks, deformities), lethargy, postnatal mortality high, focal symmetrical poliomyelomalacia, respiratory distress, rough skin, reproductive failure, reduced conception rate, paralysis (posterior), High SGOT, high SGPT, poor weight gain, weight loss
Silicon (Si)	Deficiency	NR
Silicon (Si)	Toxicity	NR
Silver (Ag)	Toxicity	NR
Sodium (Na)	Deficiency	Reduced appetite, reduced efficacy of food utilization, poor growth rate, licking metal cages, performance reduced, low body weight in neonates
Sodium (Na)	Toxicity	NR
Strontium (Sr)	Toxicity	NR
Sulfur (S)	Deficiency	NR
Sulfur (S)	Toxicity	NR
Tellurium (Te)	Toxicity	NR
Terbium (Tb)	Toxicity	NR
Thallium (Tl)	Toxicity	NR
Thorium (Th)	Toxicity	NR
Thulium (Tm)	Toxicity	NR
Tin (Sn)	Toxicity	NR
Titanium (Ti)	Toxicity	NR
Tungsten (W)	Toxicity	NR
Uranium (U)	Toxicity	NR
Vanadium (V)	Toxicity	NR
Ytterbium (Yb)	Toxicity	NR
Yttrium (Y)	Toxicity	NR
Zinc (Zn)	Deficiency	Adrenal hypertrophy, appetite reduced, bone poorly mineralized—reduced size and strength, death, hemorrhage (young pigs—due to interference with Vitamin K), parakeratosis, stillbirths, testicular development abnormal, vomiting, wound healing impaired
Zinc (Zn)	Toxicity	Reduced appetite, impaired growth, death
Zirconium (Zr)	Toxicity	NR
Not Applicable (Element is not essential in this species)	Deficiency	Al, Sb, Ba, Be, Bi, Br, Cd, Cs, Ce, Dy, Er, Eu, Gd, Ga, Ge, Au, Ho, In, La, Pb, Li, Lu, Hg, Nd, Nb, Pd, Pr, Pm, Rh, Rb, Sm, Sc, Ag, Sr, Te, Tb, Tl, Th, Tm, Sn, Ti, W, U, V, Yb, Y, Zr

Rats

Element	Deficiency/ toxicity	Clinical signs reported
Aluminium (Al)	Toxicity	High ALP, ALT, AST, reduced androgen receptor protein expression, low bilirubin, bone disease, suppressed erythropoiesis, congenital heart defects in offspring: septal defects, conotruncal defects and right ventricular outflow, high blood cortisol, high creatinine, high blood epinephrine, hepatotoxicity, impaired glucose tolerance, hyperglycemia, hypoinsulinemia, hypoparathyroidism, hypoproteinemia, hypothyroidism (T3 and T4), hyperlipidemia, hypercholesterolemia and hypertriglyceridemia, low luteinizing hormone, muscle necrosis at the injection site, neuronal death in the hippocampus, pancreatic islet cell necrosis, small intestine epithelial degeneration, goblet cell proliferation and lymphocyte infiltration in the mucosa of the small intestine, low pregnancy rate, impairment of spermatogenesis and increase in sperm malformation rate. Low testosterone, uremia, uterine resorption, reduced viable offspring
Antimony (Sb)	Toxicity	Blood urea nitrogen (BUN) high, creatinine high, electrolytes (Na and K) high, renal damage (structural), reproductive failure, bodyweight low *Lung*: tumors, interstitial inflammation, fibrosis, granulomatous changes, alveolar-wall cell hypertrophy and hyperplasia, and cuboidal and columnar cell metaplasia. Degenerating macrophages, alveolitis, multinucleated giant cells, high aspartate transaminase concentrations
Arsenic (As)	Deficiency	High mortality rate
Arsenic (As)	Toxicity	Impaired growth
Barium (Ba)	Toxicity	Death, gastroenteritis, kidney damage, decreases in body weight, decreased survival, decreased number of sperm, decreased percentage of motile sperm and decreased osmotic resistance of sperm. *Laboratory findings:* eosinophilia, erythropenia, leukocytosis, neutrophilia
Beryllium (Be)	Toxicity	*Inhalation Be ores:* Lung tumors (mainly bronchioalveolar, adenocarcinomas and adenomas) *Intratracheal Be metal and various salts:* Lung neoplasia (mainly adenocarcinomas and adenomas) *Intravenous:* Death, hypoglycemia *Intranasal*: Necrotizing, hemorrhagic pulmonitis, intraalveolar fibrosis, chronic inflammation *Oral in food:* Rickets, inhibits alkaline phosphatase *Be carbonate: Severe rickets not ameliorated by Vitamin D*
Bismuth (Bi)	Toxicity	*Inhalation:* Foreign body inflammatory response in lungs
Boron (B)	Deficiency	Growth impairment, low hemoglobin, reduced immune response (antibody production), reduced red cell production (red cell count)
Boron (B)	Toxicity	1170 µg/L *B for 2 years*: hair changes (coarse), delayed maturation prepubescent hair, incisors lack pigmentation, scaly tails, desquamation pads on feet, abnormal nails (long), aspermia, impaired ovarian development, testicular atrophy, hemorrhage from eyes, low hemoglobin, low hematocrit. Fetuses: decreases in body weight and increases in the occurrence of skeletal malformations
Bromine (Br)	Toxicity	NR
Cadmium (Cd)	Toxicity	Albuminuria, anemia, behavioral changes (reduced activity, decreased acquisition of avoidance behavior), carcinogenicity (Cd orally: leukemia, prostate, testis, and hematopoeitic system; Cd inhalation: lung; Cd injection: at injection site (rhabdomyosarcomas; fibrosarcomas) and interstitial-cell tumors of the testes, pancreatic Islet cell tumors), fetal developmental abnormalities if high Cd during pregnancy(included exencephaly, hydrocephaly, cleft lip and palate, microphthalmia,

		micrognathia, club foot, and dysplastic tail) death, reduced calcium and phosphorus absorption from the intestine, lung tumors, inhibition of renal conversion of 25-hydroxycholecalceferol to 1,25-dihydroxycholecalceferol (only if on a low ca diet), reduced femoral bone density, desquamation and necrosis of gastric and intestinal mucosa, hemorrhagic ovarian necrosis, ovarian necrosis, periportal liver-cell necrosis, proteinuria, pulmonary edema, renal tubular necrosis, testicular necrosis. *Sarcomas and carcinomas formed after injection or inhalation of Cd salts* *Tumor formation could be inhibited if there was Zn deficiency*
Cesium (Cs)	Toxicity	Abdominal adhesions, deaths, gastrointestinal hemorrhage, nasal discharges, (*after IP injection*),
Calcium (Ca)	Deficiency	Bone undermineralized, food intake reduced, growth impaired, hemorrhage (internal), joint swelling, lactation failure, paralysis (hind legs), rickets, reproduction impaired
Calcium (Ca)	Toxicity	NR
Cerium (Ce)	Toxicity	Ataxia, death, dyspnea, lethargy, prostration, neutrophilia, lymphoid hyperplasia. *Post-IV injection:* increased ALT, AST, and SDH fatty liver, fatty degeneration, and necrosis, damage to mitochondria, and invaginations of the nuclear membrane in hepatocytes, decrease cell viability, increased lactate dehydrogenase, total protein, and alkaline phosphatase. Leukocytosis, percentage neutrophils, and concentrations of proinflammatory cytokines elevated (Srinivas et al., 2011). Liver damage, decreased liver weight, dose-dependent hydropic degeneration, hepatocyte enlargement, sinusoidal dilation, and nuclear enlargement. Low albumin, the sodium-potassium ratio, and triglyceride levels were decreased Nalabotu et al. (2011)
Chlorine (Cl)	Deficiency	Reduced appetite, reduced growth, renal lesions
Chlorine (Cl)	Toxicity	NR
Chromium (Cr)	Deficiency	Aortic plaques, appetite increase, corneal lesions, fertility reduced, glucose intolerance, glycosuria, growth disorders, hypercholesterolemia, hyperinsulinemia, lean muscle mass reduced, longevity reduced, sperm count reduced
Chromium (Cr)	Toxicity	Bronchial squamous metaplasia, growth impaired, kidney damage (proximal convoluted tubules, increases in urinary β-glucuronidase, lysozyme, glucose and protein), liver damage, lung tumors (implantation) Sarcomas (rhabdomyosarcomas, spindle-cell sarcomas and fibrosarcomas at subcutaneous or intramuscular injection site), nasal septum necrosis and perforation
Cobalt (Co)	Deficiency	Reduced thyroid hormone synthesis
Cobalt (Co)	Toxicity	(*Experimental*) Blood volume increased, bone marrow hyperplasia, death LD50 (200–500 mg/kg body weight), myocardial hypoxic contracture reduced, polycythemia, reticulocytosis
Copper (Cu)	Deficiency	Anemia, achromotrichia, behavioral changes, cardiac function impaired, cholesterol levels high, fetal death, fetal resorption, high phospholipids, high serum triglyceride
Copper (Cu)	Toxicity	NR
Dysprosium (Dy)	Toxicity	Loss urine concentrating ability, renal vascular resistance increased
Erbium (Er)	Toxicity	NR
Europium (Eu)	Toxicity	Reduced creatinine excretion, (damage to glomerular basement membrane), decreased glomerular filtration rate (GFR). *N*-acetyl-β-D-glucosaminidase (NAG) high (adverse effect on proximal convoluted tubule)
Fluorine (F)	Deficiency	NR
Fluorine (F)	Toxicity	NR
Gadolinium (Gd)	Toxicity	Abdominal cramps, alanine aminotransferase (ALT) albumin low, high, aspartate aminotransferase (AST) high, breathing labored, cholesterol high, diarrhea, hypoglycemia, lactate dehydrogenase (LDH) high, lethargy, organ weight (liver and spleen) increased, platelet numbers low, prothrombn time (PTT) prolonged, soft tissue mineralization (lung, kidney, mononuclear phagocytes), necrosis (liver, spleen), triglycerides high, white blood cell count high *Acute toxicity study Spencer* et al. (*1997*)

Gallium (Ga)	Toxicity	Growth suppression, hypocalcemia, life span reduced, nephrotoxicosis *In lungs:* inflammation, necrosis, neutrophil infiltration, and fibrosis (inhalation)
Germanium (Ge)	Toxicity	Anemia, reduced erythropoiesis, reduced lifespan, fatty degeneration liver, renal dysfunction (tubular disease, reduced creatinine clearance, high BUN, and P), weight loss
Gold (Au)	Toxicity	NR
Holmium (Ho)	Toxicity	inflammatory and fibrotic responses in the skin
Indium (In)	Toxicity	Rough coat, weight loss *Inhalation:* bronchioloalveolar hyperplasia, alveolar wall fibrosis and thickened pleural wall, alveolar proteinosis and infiltrations of alveolar macrophages and inflammatory cells **Carcinogenic effect** seen bronchioloalveolar adenomas and carcinomas
Iodine (I)	Toxicity	NR
Iron (Fe)	Deficiency	Achromotrichia, anemia, cecal enlargement, cardiomegaly, increased infection risk, splenomegaly. *Laboratory changes*: high serum triglycerides, low serum folic acid, low muscle myoglobin, low tissue cytochrome *c* concentrations
Iron (Fe)	Toxicity	NR
Lanthanum (La)	Toxicity	Behavioral changes (impaired spatial learning and memory), morphological and histological changes in the hippocampus, neural degeneration, lung lesions (granulation, giant cells), stomach hyperkeratosis, gastric eosinocyte infiltration, reduced sense of smell, reduced body weight, Increased enzymes alkaline phosphate (ALP) and alanine aminotransferase (ALT), low blood urea (BUN) lowered creatinine
Lead (Pb)	*Deficiency*	Anemia (hypochromic, microcytic), growth depression, reduced hepatic Fe stores *Insufficient evidence to support a claim for essentiality*
Lead (Pb)	Toxicity	NR
Lithium (Li)	Deficiency	NR
Lithium (Li)	Toxicity	Reduced growth, kidney damage, liver damage
Lutetium (Lu)	Toxicity	NR
Magnesium (Mg)	Deficiency	Convulsions, death, hyperirritability, soft tissue mineralization (kidneys), failure to suckle young, vasodilation
Magnesium (Mg)	Toxicity	NR
Manganese (Mn)	Deficiency	Angular deformity (bowing) forelegs, ataxia (neonatal), balance loss, body-righting reflexes delayed development, bone growth retarded, convulsions (epileptiform), epiphyseal dysplasia, forelimb shortening, glucose tolerance impaired, head retraction, head tremors, hyperreactivity to stimuli, insulin synthesis impaired, libido lack, male sterility, neonatal mortality, otolith abnormalities, ovulation defective, poor survivability, reproductive abnormalities, skull deformity (doming), skeletal abnormalities (short radius, ulna, tibia, and fibula), sperm lack, sterility (male), testicular degeneration, vaginal orifice delayed development, vestibular syndrome
Manganese (Mn)	Toxicity	Hyperglycemia, hypoinsulinemia
Mercury (Hg)	Toxicity	Degeneration of the cerebellum and dorsal route ganglia, and clinical signs of neurotoxicity
Molybdenum (Mo)	Deficiency	NR
Molybdenum (Mo)	Toxicity	*Oral ingestion:* Alkaline phosphatase in liver decreased activity, alkaline phosphatase in small intestine and kidney increased activity, increases in serum alanine aminotransferase (ALT) and aspartate aminotransferase (AST) activities, albuminuria, Anemia, anorexia, high blood urea, bone strength reduction, caries, cartilaginous dysplasia, creatininuria, death, dental enamel lesions, impaired endochondral ossification, diarrhea, diuresis (polyuria), impaired endochondral ossification, fetal death rate increased, fetal resorption rate increased, fetal size decreased, growth retardation, growth plates widened, hemorrhages and tears in tendons and ligaments, increases in urinary kallikrein (distal tubule enzyme), kidney damage (degeneration, increased renal lipid, hyperplasia renal proximal tubules), limb shortening, liver glucose-6-phosphatase

		activity decreased, neurocyte degeneration (cerebral cortex and hippocampus), impaired estrus cycles, abnormal oocysts, male sterility, mandibular/maxillary exostoses, milk production reduced, estrus prolonged, spermatogenesis impaired, seminiferous tubule degeneration, survivability decreased, testicular damage, weight loss *Inhalation:* alveolar epithelial metaplasia, histiocytic cellular infiltration, hyaline degeneration of nasal epithelium, pulmonary inflammation, squamous metaplasia of nasal planum *Note: low confidence in the hepatic and renal changes reported in these studies, treat findings with caution. There are studies showing conflicting results for the effect of Mo toxicity on reproductive function. See ATSDR (2020c)*
Neodymium (Nd)	Toxicity	NR
Nickel (Ni)	Deficiency	Impaired growth rate
Nickel (Ni)	Toxicity	Reduced feed intake, *Inhalation*: lung adenomas (inhalation), squamous metaplasia, inflammation, necrotizing pneumonia, degeneration of the bronchiolar epithelium, atrophy of the olfactory epithelium, hyperplasia of the bronchial and mediastinal lymph nodes, hyperplasia alveolar macrophages, fibrosis, thymus atrophy *Intratracheal injection:* alveolitis, lung tumors (mainly squamous cell carcinomas and adenocarcinomas) *Intrapleural injection*: spindle-cell sarcomas, mesothelioma, rhabdomyosarcomas *Subcutaneous injection (dental materials):* sarcomas hem oxygenase activity increased (many organs), lipid peroxidation increased (liver kidney lung) *Intramuscular injection:* injection site fibrosarcomas, rhabdomyosarcomas, sarcomas, suppression of natural killer cell activity *Intraperitoneal injections*: carcinomas, mesothelioma, rhabdomyosarcomas, sarcomas, myocardial, degeneration liver necrosis, renal failure (proximal tubular degeneration), seminiferous tubular degeneration *Intravenous injection:* sarcomas *Subperiosteal or intramedullary injection:* bone tumors *Intrarenal injection:* erythrocytosis, renal-cell carcinomas *Intratesticular injection*: testicular sarcomas *Intraocular injection:* Various tumors, e.g., melanoma, gliomas, retinoblastomas *Dermal application:* epidermal acanthosis, epidermal atrophy, disordered epidermal cells, hyperkeratinization, liver necrosis
Niobium (Nb)	Toxicity	*IV injection*: increased thirst, increased urine volume, decreased urinary specific gravity and osmolality. Urinary sodium excretion increased initially then decreased. Lost urine-concentrating ability, failed to respond to injection of antidiuretic hormone, delayed peak urine excretion rate after a water load, proximal and distal tubules showed tubular dilation, hyaline casts, deposits of granular pigment. Occasional necrosis, regeneration, and early fibrosis. Increased Cu, Mn and Zn in organs
Palladium (Pd)	Toxicity	Anemia, anorexia, ataxia, cardiovascular effects, conjunctivitis, convulsions (clonic–tonic), death, emaciation, exudates, gastrointestinal hemorrhage, gastric ulcers, glomerulosclerosis, hematomas, reduced hepatic enzyme activity, keratoconjunctivitis, ketonuria, liver damage, peritonitis (intraperitoneal administration), proteinuria, pulmonary hemorrhage, pulmonary inflammation, renal damage, sluggishness, testicular damage, tiptoe gait, urination reduced, water intake reduced, weight loss
Phosphorus (P)	Deficiency	NR
Phosphorus (P)	Toxicity	Calcinosis, nephrocalcinosis
Potassium (K)	Deficiency	Coat rough, death, edema, food intake reduced, poor growth rate. Lesions in many organ tissues
Potassium (K)	Toxicity	NR
Praseodymium (Pr)	Toxicity	*IV injection:* Hepatotoxicity (fatty degeneration altered hepatic microsomal lipid content, decreased RNA polymerase activity, reduced gluconeogenesis, reduced drug metabolizing enzymes)

Promethium (Pm)	Toxicity	NR
Rhodium (Rh)	Toxicity	NR
Rubidium (Rb)	Toxicity	NR
Samarium (Sm)	Toxicity	Decreased growth, liver damage, liver weight/body weight ratio increased, reduced antioxidant enzyme superoxide dismutase (in liver and brain), increase in liver malondialdehyde (MDA)
Scandium (Sc)	Toxicity	Oral LD50 is at 4 g/kg body weight (*Experimental*)
Selenium (Se)	Deficiency	NR
Selenium (Se)	Toxicity	High bilirubin, cataracts, cirrhosis, reduced food intake, increased glutathione peroxidase levels, reduced growth, low hemoglobin, hepatic atrophy, impaired reproduction
Silicon (Si)	Deficiency	Growth impaired
Silicon (Si)	Toxicity	*Inhalation:* pneumonitis, pulmonary fibrosis
Silver (Ag)	Toxicity	Signs of Vit E/Se deficiency—see Se
Sodium (Na)	Deficiency	Reduced appetite, corneal lesions, undermineralized (soft) bones, growth impaired, infertility (males), delayed sexual maturity (females), death
Sodium (Na)	Toxicity	NR
Strontium (Sr)	Toxicity	Growth depression, rickets, death (at very high doses) *Radioactive isotope:* Anemia, cancers (bone, leukemia, lung, nose) reduced red and white blood cell counts and platelets, hemorrhages, susceptibility to infections, thinning of the cornea, thinning of the skin
Sulfur (S)	Deficiency	NR
Sulfur (S)	Toxicity	*Experimental:* Sodium bisufite LD50 650–740 mg/kg body weight
Tellurium (Te)	Toxicity	Alopecia, garlic odor on breath, erythema and edema of feet, muscle necrosis (cardiac and skeletal muscle) paralysis (hind legs—temporary) *Teratogenic effects*: exophthalmia, hydrocephalus, ocular hemorrhage, small kidneys, edema, umbilical hernia, retained testicles
Terbium (Tb)	Toxicity	Growth suppression death cardiovascular collapse, respiratory paralysis, irritant to intact skin, ulceration of abraded skin, corneal damage, conjunctivitis
Thallium (Tl)	Toxicity	Death (*used as rodenticide*)
Thorium (Th)	Toxicity	NA
Thulium (Tm)	Toxicity	Cardiovascular collapse, onjunctival ulcers, death, growth suppression, respiratory paralysis, skin ulcers on abraded skin
Tin (Sn)	Toxicity	NR
Titanium (Ti)	Toxicity	NR
Tungsten (W)	Toxicity	Reduced lifespan, nephrotoxicity, weight loss
Uranium (U)	Toxicity	Neurobehavioral changes, reduced fertility, skin irritation, *Maternal exposure* offspring: changes in brain function, changes in ovaries, tooth malformation.
Vanadium (V)	Toxicity	Alveolar/bronchiolar hyperplasia, inflammation, and fibrosis, impaired neuro behavior performance (open field and active avoidance tests), death, diarrhea, dyspnea, decreases in fetal growth, fetal resorptions, impaired growth, low hematocrit, low hemoglobin, hypertension, low mean cell volume, laryngeal degeneration, microcytic erythrocytosis, nasal lesions (hyperplasia, metaplasia), increased reticulocytes and nucleated erythrocytes, skeletal and visceral malformations and anomalies, reduced survivability (pups), tachypnea, weight loss
Ytterbium (Yb)	Toxicity	Conjunctival ulceration, cardiovascular collapse, death, respiratory paralysis, ulcers to abraded skin

Yttrium (Y)	Toxicity	Dyspnea, growth impaired, liver edema, pulmonary edema, pleural effusions, pulmonary hyperemia. Impaired renal function (reduced creatinine and urine production) increased liver enzymes glutamic-oxaloacetic transaminase and glutamic-pyruvate transaminase Incidence of tumors was 33% in Y-treated rats and 14% in untreated controls
Zinc (Zn)	Deficiency	Alopecia, anorexia, epidermal thickening, loss hair follicles, growth retardation, hyperirritability, hypoinsulinemia, impaired development and function of male sex organs. Impaired development of embryo, resorption, aspermatogenesis, fetal malformations
Zinc (Zn)	Toxicity	Diarrhea, death, increased LDH protein in bronchoalveolar lavage fluid, grayish areas with pulmonary congestion, peri-bronchial leukocytic infiltration, bronchial exudate composed of polymorphonuclear leukocytes
Zirconium (Zr)	Toxicity	Cholesterol high, glycosuria, interstitial pneumonitis, reduced survival time, poor weight gain,
Not Applicable (Element is not essential in this species)	Deficiency	Al, Sb, Ba, Be, Bi, Br, Cd, Cs, Ce, Dy, Er, Eu, Gd, Ga, Ge, Au, Ho, In, La, Pb, Li, Lu, Hg, Nd, Nb, Pd, Pr, Pm, Rh, Rb, Sm, Sc, Ag, Sr, Te, Tb, Tl, Th, Tm, Sn, Ti, W, U, V, Yb, Y, Zr

References

Rashedy et al. (2013), NRC (1995), McDowell (1992a).

Squirrels

Squirrels (Sciuride) are omnivorous rodents consisting of over 200 species. They consume nuts, leaves, roots, seeds, other plants, and small animals, insects, caterpillars.
Corn, nuts and seeds are all low in Ca content and high in P so captive squirrels fed just these ingredients are prone to develop metabolic bone disease.

Species	Main dietary intake
Flying squirrels	Mainly hypogeous mycorrhizal fungi (truffles), birds eggs, buds, carrion, flowers, insects, nestlings, nuts, tree sap
Fox squirrels	Mostly vegetarian: tree seeds (hickory, oak, pine, walnut), tree buds, bulbs, roots, fungi and various crops (corn, fruit, oats, soybeans, wheat). Rarely, insects, bird eggs
Gray squirrels	Predominantly vegetarian eating acorns, berries, buds, flowers, some fungi, seeds, walnuts and other nuts. Rarely, they can eat shed antlers, birds, eggs, bones, insects, frogs, and tree sap
Ground squirrels	Buds, flowers forbs and grasses
Red squirrels	Mainly coniferous tree seeds, berries, buds, flowers, fruit, fungi, nuts, and shoots, Rarely, bird eggs and tree sap
Tree squirrels	Seeds (including nuts), also bark, bones, buds, fruit, fungi, leaves. Also, invertebrates (beetles, caterpillars, and various larve), soil

Mineral associated diseases in squirrels

Limited data available.

Element	Clinical signs
Calcium deficiency	Aggressiveness, angular deformity of bones (limbs, spine, tail), anorexia, bones poorly calcified (weak), growth impairment, reluctance to move (lethargy), labored breathing, metabolic bone disease (abnormal bone growth, facial deformities (malocclusion) fractures, swollen joints) muscle spasms, paralysis, seizures, excessive sleeping, tremors, twitching (muscles under skin, toes, legs, tail weight loss
Phosphorus excess	Metabolic bone disease

Nichols et al. (2018), Garriga et al. (2004).

Obligate carnivores

Cats

Domesticated cats have adapted to be able to utilize starch from plant material but, unlike dogs which are now omnivores, cats are still obligate carnivores and cannot obtain all the essential nutrients that they need [including arachidonic acid, vitamin A (as cats cannot convert β-carotene into retinol), vitamin D and Taurine] from plants alone, so they cannot be fed vegetarian or vegan diets.

Element	Deficiency/ toxicity	Clinical signs reported
Aluminum (Al)	Toxicity	Neurotoxic
Antimony (Sb)	Toxicity	NR
Arsenic (As)	Deficiency	NR
Arsenic (As)	Toxicity	NR
Barium (Ba)	Toxicity	NR
Beryllium (Be)	Toxicity	NR
Bismuth (Bi)	Toxicity	NR
Boron (B)	Deficiency	NR
Boron (B)	Toxicity	NR
		NR
Bromine (Br)	Toxicity	Paradoxical sleep. *Br as 1-methylheptyl-gamma-bromoacetoacetate given IV*
Cadmium (Cd)	Toxicity	NR
Cesium (Cs)	Toxicity	NR
Calcium (Ca)	Deficiency	*Common in cats fed an all meat diet:* bone cortices thin, bone pain, bones undermineralized, fractures (spontaneous), lameness, nutritional secondary hyperparathyroidism, rickets, spinal curvature. (*Includes domesticated and captive wild cats*)
Calcium (Ca)	Toxicity	Increased bone mineral content, serum calcium concentrations increased (total and ionized), reduced food intake, growth retardation
Cerium (Ce)	Toxicity	NR
Chlorine (Cl)	Deficiency	Hypochloridemia, hypokalemia due excess renal losses, sodium increased in renal fluids
Chlorine (Cl)	Toxicity	NR
Chromium (Cr)	Deficiency	NR
Chromium (Cr)	Toxicity	*Inhalation:* Bronchitis, pneumonia
Cobalt (Co)	Deficiency	NR
Cobalt (Co)	Toxicity	NR
Copper (Cu)	Deficiency	Achromotrichia, abortion, anemia, ataxia, congenital deformities (kinked tails, carpi), cannibalism, poor keratinization, poor reproductive performance (increased time to conception), poor weight gain
Copper (Cu)	Toxicity	Interference with Fe and Zn bioavailability
Dysprosium (Dy)	Toxicity	NR
Erbium (Er)	Toxicity	NR
Europium (Eu)	Toxicity	NR
Fluorine (F)	Deficiency	NR
Fluorine (F)	Toxicity	NR

Gadolinium (Gd)	Toxicity	NR
Gallium (Ga)	Toxicity	NR
Germanium (Ge)	Toxicity	NR
Gold (Au)	Toxicity	NR
Holmium (Ho)	Toxicity	NR
Indium (In)	Toxicity	NR
Iodine (I)	Deficiency	Alopecia, abnormal Ca metabolism, death, fetal resorption, goiter
Iodine (I)	Toxicity	Lachrymation, nasal discharge, salivation, skin scales
Iron (Fe)	Deficiency	Anemia, diarrhea, lethargy, low hemoglobin, low hematocrit, hematuria, melena, pallor, weakness, weight loss
Iron (Fe)	Toxicity	Diarrhea, vomiting
Lanthanum (La)	Toxicity	NR
Lead (Pb)	Toxicity	Anorexia, excitation, seizures
Lithium (Li)	Deficiency	NR
Lithium (Li)	Toxicity	NR
Lutetium (Lu)	Toxicity	Death, cardiovascular collapse with respiratory paralysis. *IV: Acute toxicity study*
Magnesium (Mg)	Deficiency	Aortic calcification, anorexia, arrhythmias, ataxia, behavioral changes, convulsions, depression, poor growth, hyperextension carpi, hyperirritability, hyperreflexia, hypomagnesemia, muscle twitches, muscle weakness, tachycardia, tetany, torsades de pointes (a tachycardia, or fast heart rhythm that originates in one of the ventricles of the heart), soft tissue calcification
Magnesium (Mg)	Toxicity	Struvite urolithiasis
Manganese (Mn)	Deficiency	NR
Manganese (Mn)	Toxicity	NR
Mercury (Hg)	Toxicity	Anorexia, ataxia (cerebellar), loss of balance, blindness, death, fetal anomalies, motor incoordination, loss of nerve cells with replacement by reactive and fibrillary gliosis, hypermetria, neuronal necrosis, nystagmus (vertical), impaired reflexes, e.g., righting diminished sensitivity to pain, proprioceptive deficits, rigidity (hindlegs) difficulty walking, tonic–clonic seizures, tremors, uncontrolled vocalization (howling), vomiting
Molybdenum (Mo)	Deficiency	NR
Molybdenum (Mo)	Toxicity	*Intravenous injection:* hypertension, impaired neurological system (CNS)
Neodymium (Nd)	Toxicity	NR
Nickel (Ni)	Deficiency	NR
Nickel (Ni)	Toxicity	NR
Niobium (Nb)	Toxicity	NR
Palladium (Pd)	Toxicity	*IV administration:* Bronchospasm, increased serum histamine
Phosphorus (P)	Deficiency	*Rare due to carnivorous diet.* Anemia (hemolytic), ataxia, metabolic acidosis, dental enamel hypoplasia, rickets, weight loss (*and Ca deficiency*)
Phosphorus (P)	Toxicity	Dehydration, depression, hyperphosphatemia, metabolic acidosis
Potassium (K)	Deficiency	Anorexia, ataxia, cardiac failure, poor growth, hypercreatininemia, immune function depressed, muscle weakness, depressed reflexes, ventroflexion of neck—*Usually due to K losses in urine or feeding a vegetarian diet*
Potassium (K)	Toxicity	NR
Praseodymium (Pr)	Toxicity	NR
Promethium (Pr)	Toxicity	NR
Rhodium (Rh)	Toxicity	NR

Rubidium (Rb)	Toxicity	NR
Samarium (Sm)	Toxicity	NR
Scandium (Sc)	Toxicity	NR
Selenium (Se)	Deficiency	*Kittens:* Reduced GSH-px concentrations in blood
Selenium (Se)	Toxicity	NR
Silicon (Si)	Deficiency	NR
Silicon (Si)	Toxicity	NR
Silver (Ag)	Toxicity	NR
Sodium (Na)	Deficiency	Anorexia, impaired growth, polydipsia, polyuria, high hematocrit, low urine specific gravity, high plasma aldosterone
Sodium (Na)	Toxicity	NR
Strontium (Sr)	Toxicity	NR
Sulfur (S)	Deficiency	NR
Sulfur (S)	Toxicity	NR
Tellurium (Te)	Toxicity	NR
Terbium (Tb)	Toxicity	NR
Thallium (Tl)	Toxicity	NR
Thorium (Th)	Toxicity	NR
Thulium (Tm)	Toxicity	NR
Tin (Sn)	Toxicity	NR
Titanium (Ti)	Toxicity	NR
Tungsten (W)	Toxicity	NR
Uranium (U)	Toxicity	NR
Vanadium (V)	Toxicity	NR
Ytterbium (Yb)	Toxicity	NR
Yttrium (Y)	Toxicity	NR
Zinc (Zn)	Deficiency	Growth retardation, parakeratosis, seminiferous tubular degeneration, impaired testicular function (testicular atrophy)
Zinc (Zn)	Toxicity	Bronchopneumonia, leukocyte infiltration into alveoli, and grayish areas with congestion, labored breathing, upper respiratory tract, obstruction.
Zirconium (Zr)	Toxicity	NR
Not Applicable (Element is not essential in this species)	Deficiency	Al, Sb, Ba, Be, Bi, Br, Cd, Cs, Ce, Dy, Er, Eu, Gd, Ga, Ge, Au, Ho, In, La, Pb, Li, Lu, Hg, Nd, Nb, Pd, Pr, Pm, Rh, Rb, Sm, Sc, Ag, Sr, Te, Tb, Tl, Th, Tm, Sn, Ti, W, U, V, Yb, Y, Zr

References

NRC (2005).

Big cats

Captive wild cats share the nutrition-related diseases that are commonly seen in domesticated species—notably metabolic bone disease due to nutritional secondary hyperparathyroidism caused by feeding an exclusive muscle-meat based ration, and other mineral and trace element deficiencies (see table above).
Copper deficiency has been reported in captive Cheetahs, with similar reports in Lions and Snow Leopards (Kaiser et al., 2014).

Element	Deficiency/toxicity	Clinical signs reported
Copper	Deficiency	Neurological signs (ataxia, hind limb paralysis and paresis)
		In captive Cheetahs, Lions, and Snow Leopards—due to degenerative spinal cord lesions)

References

Kaiser et al. (2014), German (2010), NRC (2006), Daniel et al. (1986), Kane et al. (1981).

Ferrets

Ferrets and other mustelids are simple stomached and have a very short intestinal tract with no cecum or ileocolic valve so food transit time is very short. They are obligate carnivores. Pet ferrets are fed complete commercially available foods. Limited data is available.

Element	Deficiency/ toxicity	Clinical signs reported
Calcium (Ca)	Deficiency	Abduction of forelegs, bones pliable and soft, spontaneous fractures, hyperplastic parathyroid glands, nutritional secondary hyperparathyroidism (metabolic bone disease), osteoporosis, skeletal deformities in kits, tooth loss.
Mercury (Hg)	Toxicity	Anorexia, paroxysmal convulsions, death, incoordination, tremors, weight loss
Sodium (Na)	Toxicity	Brain edema, cerebellar coning, death, depression, meningitis (nonsuppurative), periodic choriform spasms
Zinc (Zn)	Toxicity	Anemia, lethargy, pallor, weakness (hindleg)
Not Applicable (Element is not essential in this species)	Deficiency	Al, Sb, Ba, Be, Bi, Br, Cd, Cs, Ce, Dy, Er, Eu, Gd, Ga, Ge, Au, Ho, In, La, Pb, Li, Lu, Hg, Nd, Nb, Pd, Pr, Pm, Rh, Rb, Sm, Sc, Ag, Sr, Te, Tb, Tl, Th, Tm, Sn, Ti, W, U, V, Yb, Y, Zr

References

Johnson-Delaney (2014a).

Mink

Mink are essentially carnivores but eat a wide range of foods.
Limited information available.

Element	Deficiency/ toxicity	Clinical signs reported
Calcium (Ca)	Deficiency	Abnormal bone growth
Calcium (Ca)	Toxicity	Angular deformity of legs, bones undermineralized, joint enlargement, lameness, undershot jaw
Iodine (I)	Toxicity	Reduced: number of females whelping, number of offspring birth weight and viability
Iron (Fe)	Deficiency	Anemia, death, emaciation, rough fur, lack of fur pigmentation (called "cotton fur" or "cotton pelt"), retarded growth
Lead (Pb)	Toxicity	Anorexia, convulsions, dehydration, ocular discharge (mucopurulent), muscular incoordination, stiffness, trembling
Manganese (Mn)	Deficiency	A condition called "screwneck" in pastel mink, otolith abnormalities, abnormal postural reflexes
Mercury (Hg)	Toxicity	Anorexia, ataxia, blindness, chewing, circling, convulsions, death, dysphonia, head tremors, incoordination, paralysis, falling to recumbency, salivation, tremors, vocalization (high pitched), vomiting, weight loss
Phosphorus (P)	Deficiency	Abnormal bone growth
Sodium (Na)	Deficiency	Nursing sickness
Sodium (Na)	Toxicity	? Impaired reproductive performance, reduced water intake
Not Applicable (Element is not essential in this species)	Deficiency	Al, Sb, Ba, Be, Bi, Br, Cd, Cs, Ce, Dy, Er, Eu, Gd, Ga, Ge, Au, Ho, In, La, Pb, Li, Lu, Hg, Nd, Nb, Pd, Pr, Pm, Rh, Rb, Sm, Sc, Ag, Sr, Te, Tb, Tl, Th, Tm, Sn, Ti, W, U, V, Yb, Y, Zr

References

Aulerich et al. (1974).

Various nutritional adaptations

Bats

Most bats are insectivores, and only vampire bats need blood as a food resource. Because of the risk of rabies, these species should not be kept in captivity.
Insectivore bats: foods commonly include blowfly larve, crickets, fruit flies, and mealworms. These are all low in calcium content and when farmed should be fed a high calcium diet for at least 3 days so their intestine is high in calcium when fed to the bats.
Frugivorous bats: are usually fed mixed fruits (apples, bananas, grape, melon, oranges, papaya, pear) and vegetables (cooked carrot, sweet potato) high iron content needs to be avoided to prevent iron overload and toxicity.
Some bats also consume pollen and nectar (e.g., Tube-Lipped Nectar Bat).
A small number of bats consume vertebrates (small birds, fish, frogs, lizards) and vampire bats are specially adapted to consume blood from living mammals.
It is important that captive bats are accurately identified and fed appropriately for their species.

Iron toxicity

Iron storage disease (hemochromatosis) is a frequent cause of liver disease and mortality in captive Egyptian fruit bats (*Rousettus egyptiacus*) (Farina et al., 2005).
In one case report of fatalities a monodicalcium phosphate supplement contained 11,860 mg/kg Fe dry matter (DM), the diet provided a very high dose of 400 mg/kg DM and, as the diet also contained a high amount of vitamin C which increases Fe absorption it was not surprising that Fe toxicity resulted.

Element	Deficiency/toxicity	Clinical signs
Iron (Fe)	Toxicity	Death, hemochromatosis (*captive fruit bats*), liver disease

References

Roswag et al. (2011).

Herbivores

Ruminants

Ruminants have a unique multichambered "stomach" with four compartments (reticulum, rumen, omasum, and abomasum) which support the fermentation of food materials before they progress to the small and large intestines for absorption. Camelids also possess a multicompartmented stomach, but they have three compartments.

A key reference for this group of animals is NRC (2014) Nutrient Requirements of Small Ruminants (Sheep, Goats, Cervids, and New World Camelids) species included are sheep (*Ovis aries*), goats (*Capra hircus*), cervids (white-tailed deer [*Odocoileus virginianus*], red deer [*Cervus elaphus*], wapiti/American elk [*Cervus elaphus*], and caribou/reindeer [Rangifer tarandus]), as well as New World camelids (alpacas [*Lama glama*] and llamas [Vicugna pacos]).

Minerals are supplied by food intake, water, and geophagy and factors include:

- Plant species
- Soil mineral content
- Soil pH
- Season
- Climatic conditions
- Stage of growth
- Water content
- Atmosphere

Acid soils (pH <6.0) may reduce plant absorption of Ca, Mg, P, K, Se, S, and increase absorption of B, Cu, Fe, Mn, and Zn. Alkaline soil can result in plants containing excess amounts of Mo and Se. Some plants accumulate high NaCl content. Mineral content of plants may be increased by:

- Air pollution from industries, vehicle emissions, or volcanic eruptions: Cd, F, Pb, S
- Pastures treated with sewage may contain toxic levels of Cu, Fe and Mn
- High rainfall in coastal regions increases I, Na

Buffaloes

Buffaloes are herbivore ruminants with alimentary tracts similar to cattle, they chew the cud, and ingested plant material is digested by bacteria in the rumen. Historically, domesticated buffaloes have been fed according to our knowledge of cattle nutrition, however, there are many differences in physiology between the species, for example, the rumen microbiome is considerably different (Franzolin and Dehority, 1999) and our knowledge about optimum mineral and trace element requirements is still lacking.

Nutritional deficiency is common in some parts of the World where soils are deficient of essential minerals or trace elements and vegetation is also deficient. Sometimes deficiency is caused by interference from high amounts of other minerals, e.g., Zn.

Whilst the nutritional requirements of Buffaloes are essentially the same as for other large ruminants like cattle, they appear to be more sensitive to some deficiencies such as P deficiency which affects Buffalo in India on pastures low in P whereas local cattle are not affected (Habib et al., 2004).

Element	Deficiency/ toxicity	Clinical signs reported
Calcium (Ca)	Deficiency	Bone fragility, fractures, reduced fertility, delayed maturity, stunted growth, reduced milk production, osteoporosis (adults), paralysis, rickets (calves), unthriftiness
Cobalt (Co)	Deficiency	Unthriftiness (calves)
Copper (Cu)	Deficiency	Anemia, anestrus, ataxia, poor body condition score, leukoderma, increased malondialdehyde (MDA) increased nitric oxide (NO), low serum progesterone level, unthriftiness (calves), vitiligo, **decreased:** catalase (CAT), ascorbic acid (ASCA), superoxide dismutase (SOD), reduced glutathione (GSH-R), total antioxidant capacity (TAC); zinc (Zn), iron (Fe), copper (Cu), ceruloplasmin
Iron (Fe)	Deficiency	Unthriftiness (calves)
Lead (Pb)	Toxicity	Blindness, head pressing, violent movement, salivation
Magnesium (Mg)	Deficiency	Death, depression, head shaking, opisthotonos, muscle fasciculations, internal organ congestion and hemorrhages, over-reaction to stimuli,
Molybdenum (Mo)	Toxicity	Cu deficiency, neutropenia
Phosphorus (P)	Deficiency	Anorexia, death, emaciation, growth retardation, reduced fertility, fractures, hemoglobinuria, hypophosphatemia, joint stiffness, limb abduction, reduced meat production, reduced milk production, neck extension, reluctance to move, paralysis (hindleg), pica, prostration, stiffness weakness
Selenium (Se)	Deficiency	Unthriftiness (calves)
Selenium (Se)	Toxicity	Emaciation, death, hoof and horn deformities, detached hooves, gangrenous of extremities, skin cracks, skin sloughing, recumbency
Zinc (Zn)	Deficiency	Unthriftiness (calves)

References

Gapat et al. (2016), Khanal and Knight (2010), Ahmed et al. (2009), Mude and SyaamaSundar (2009), Sharma et al. (2008), Habib et al. (2004), Randhawa et al. (2002), Franzolin and Dehority (1999), Ghosh et al. (1993), Gupta et al. (1982).

Camels

Camels are highly adapted to their arid environment and poor mineral containing food so, not surprisingly, they have some peculiarities in their trace element metabolism so they can increase absorption (Cu, Zn) and have higher storage capacity (Cu) when minerals are scarce, and higher tolerance to excess intake of minerals (Ca, P) and electrolytes (Na) and an ability to maintain normal enzyme activity (ceruloplasmin, superoxide-dismutase) in deficient periods.

Element	Deficiency/ toxicity	Clinical signs reported
Aluminium (Al)	Toxicity	NR
Antimony (Sb)	Toxicity	NR
Arsenic (As)	Deficiency	NR
Arsenic (As)	Toxicity (Acute)	NR
Arsenic (As)	Toxicity (Chronic)	NR
Barium (Ba)	Toxicity	NR
Beryllium (Be)	Toxicity	NR
Bismuth (Bi)	Toxicity	NR
Boron (B)	Deficiency	NR
Boron (B)	Toxicity	NR
Bromine (Br)	Toxicity	NR
Cadmium (Cd)	Toxicity	NR
Cesium (Cs)	Toxicity	NR
Calcium (Ca)	Deficiency	*Calves (1–12 months):* angular deformity of limbs, anorexia, arched back, death (due to starvation), epiphyseal plate widened, facial deformity, fibrous osteodystrophy, fractures (spontaneous) joint pain, stiff gait, difficulty to move, joint swelling (forelimbs), lameness, osteoporosis, stiff gait, weakness (*and P deficiency*) *Adults:* reduced appetite, bone fragility, emaciation, lameness, osteomalacia, pica (*and P deficiency*)
Calcium (Ca)	Toxicity	Hypercalcemia, soft tissue calcification, urolithiasis (*and P excess*)
Cerium (Ce)	Toxicity	NR
Chlorine (Cl)	Deficiency	NR
Chlorine (Cl)	Toxicity	NR
Chromium (Cr)	Deficiency	NR
Chromium (Cr)	Toxicity	NR
Cobalt (Co)	Deficiency	NR
Cobalt (Co)	Toxicity	NR
Copper (Cu)	Deficiency	*Young calves (4–6 months):* ataxia, angular deformity, bone demineralization, poor growth, limb incoordination (forelegs) *Adult camels:* anemia, loss of pigment in hair, rough haircoat, infertility (temporary), reduced milk production, weakness
Copper (Cu)	Toxicity	Hemoglobinuria and jaundice, elevated levels of liver enzymes such as sorbitol dehydrogenase, gamma glutamyle transferase, and glutamic oxaloacetic transaminase are indicative of liver tissue damage and precede death
Dysprosium (Dy)	Toxicity	NR
Erbium (Er)	Toxicity	NR
Europium (Eu)	Toxicity	NR

Fluorine (F)	Deficiency	NR
Fluorine (F)	Toxicity	NR
Gadolinium (Gd)	Toxicity	NR
Gallium (Ga)	Toxicity	NR
Germanium (Ge)	Toxicity	NR
Gold (Au)	Toxicity	NR
Holmium (Ho)	Toxicity	NR
Indium (In)	Toxicity	NR
Iodine (I)	Deficiency	*Pregnant:* congenital goiter, stillbirths, weakness in newborn *Adults:* abortion, anestrus, loss of libido (males), stillborn, weak calves
Iodine (I)	Toxicity	NR
Iron (Fe)	Deficiency	Fever, hemorrhagic disease, increased susceptibility to infections (severe mange, heavy tick infestation, trypanosomiasis), ruminal lactic acidosis, low serum iron (40 μg/100 mL),
Iron (Fe)	Toxicity	NR
Lanthanum (La)	Toxicity	NR
Lead (Pb)	Toxicity	NR
Lithium (Li)	Deficiency	NR
Lithium (Li)	Toxicity	NR
Lutetium (Lu)	Toxicity	NR
Magnesium (Mg)	Deficiency	NR
Magnesium (Mg)	Toxicity	NR
Manganese (Mn)	Deficiency	NR
Manganese (Mn)	Toxicity	NR
Mercury (Hg)	Toxicity	NR
Molybdenum (Mo)	Deficiency	NR
Molybdenum (Mo)	Toxicity	NR
Neodymium (Nd)	Toxicity	NR
Nickel (Ni)	Deficiency	NR
Nickel (Ni)	Toxicity	NR
Niobium (Nb)	Toxicity	NR
Palladium (Pd)	Toxicity	NR
Phosphorus (P)	Deficiency	*Calves (1–12 months):* angular deformity of limbs, anorexia, arched back, death (due to starvation), epiphyseal plate widened, facial deformity, fibrous osteodystrophy, fractures (spontaneous) joint pain, stiff gait, difficult to move, joint swelling (forelimbs), lameness, osteoporosis, stiff gait, weakness (*and Ca deficiency*) *Adults:* reduced appetite, bone fragility, emaciation, lameness, osteomalacia, pica (*and Ca deficiency*)
Phosphorus (P)	Toxicity	Bones undermineralized, urolithiasis
Potassium (K)	Deficiency	NR
Potassium (K)	Toxicity	NR
Praseodymium (Pr)	Toxicity	NR
Promethium (Pm)	Toxicity	NR
Rhodium (Rh)	Toxicity	NR
Rubidium (Rb)	Toxicity	NR

Samarium (Sm)	Toxicity	NR
Scandium (Sc)	Toxicity	NR
Selenium (Se)	Deficiency	Anorexia, breathing difficulty, degenerative myocarditis, fever, reduced fertility, stiff gait, low hemoglobin, heart disturbance, lethargy, muscular deficiency, pain when moving, poor racing performance, recumbency, respiratory disturbances, tachycardia, tachypnea, white muscle disease
Selenium (Se)	Toxicity	Alopecia, anorexia, diarrhea, dyspnea, fissured pads, hair color change, hypersalivation, lymphadenopathy, hepatic degeneration, hepatomegaly, impaired vision, muscle pallor, internal organ congestion and necrosis, pancreatic atrophy, weakness, weight loss
Silicon (Si)	Deficiency	NR
Silicon (Si)	Toxicity	NR
Silver (Ag)	Toxicity	NR
Sodium (Na)	Deficiency	NR
Sodium (Na)	Toxicity	NR
Strontium (Sr)	Toxicity	NR
Sulfur (S)	Deficiency	NR
Sulfur (S)	Toxicity	NR
Tellurium (Te)	Toxicity	NR
Terbium (Tb)	Toxicity	NR
Thallium (Tl)	Toxicity	NR
Thorium (Th)	Toxicity	NR
Thulium (Tm)	Toxicity	NR
Tin (Sn)	Toxicity	NR
Titanium (Ti)	Toxicity	NR
Tungsten (W)	Toxicity	NR
Uranium (U)	Toxicity	NR
Vanadium (V)	Toxicity	Abomasal hemorrhage, collapse, conjunctivitis, death, hematuria, inflamed (eosinophilic) internal organs, listlessness, pulmonary hemorrhage, tracheal hemorrhage
Ytterbium (Yb)	Toxicity	NR
Yttrium (Y)	Toxicity	NR
Zinc (Zn)	Deficiency	NR
Zinc (Zn)	Toxicity	NR
Zirconium (Zr)	Toxicity	NR
Not Applicable (Element is not essential in this species)	Deficiency	Al, Sb, Ba, Be, Bi, Br, Cd, Cs, Ce, Dy, Er, Eu, Gd, Ga, Ge, Au, Ho, In, La, Pb, Li, Lu, Hg, Nd, Nb, Pd, Pr, Pm, Rh, Rb, Sm, Sc, Ag, Sr, Te, Tb, Tl, Th, Tm, Sn, Ti, W, U, V, Yb, Y, Zr

References

Qureshi et al. (2020), Kuria et al. (2013), Al-Juboori and Korian (2011), Faye and Seboussi (2009), Abdulwahhab (2003), Corbera et al. (2003).

Cattle

Many pastures may be deficient and fail to meet the basic nutritional requirements of cattle, for example, in one study (Knowles and Grace, 2014) analysis of 1106 pastures in New Zealand the Cu, Na, and P requirements of cattle were met by only 25%, 78%, and 87% of pastures, respectively. Furthermore, 24% of pastures were deficient in Se.

Element	Deficiency/ toxicity	Clinical signs reported
Aluminium (Al)	Toxicity	NR
Antimony (Sb)	Toxicity	NR
Arsenic (As)	Deficiency	NR
Arsenic (As)	Toxicity	Ataxia, hair coat changes, hypocalcemia, hypoglycemia, inflammation of upper respiratory tract membranes, inflammation of eyes, weight loss
Barium (Ba)	Toxicity	NR
Beryllium (Be)	Toxicity	NR
Bismuth (Bi)	Toxicity	NR
Boron (B)	Deficiency	NR
Boron (B)	Toxicity	Food intake reduced, growth impairment, edema (limbs), low hematocrit (PCV), low hemoglobin, low plasma phosphate
Bromine (Br)	Toxicity	NR
Cadmium (Cd)	Toxicity	NR
Cesium (Cs)	Toxicity	NR
Calcium (Ca)	Deficiency	*Dairy:* fractures (spontaneous), retarded growth, osteoporosis, osteomalacia
Calcium (Ca)	Toxicity	NR
Cerium (Ce)	Toxicity	NR
Chlorine (Cl)	Deficiency	Anorexia, cardiovascular depression, constipation, emaciation, reduced food intake, lethargy, milk yield reduced, pica, polydipsia (mild), polyuria (mild), eye defects (severe) and reduced respiration rates, blood and mucus in feces, reduced water intake, weight loss, alkalosis (severe), hypochloremia, secondary hypokalemia, hyponatremia, and uremia.
Chlorine (Cl)	Toxicity	NR
Chromium (Cr)	Deficiency	Humoral immune response decreased, hypercholesterolemia, morbidity increased
Chromium (Cr)	Toxicity	Inflammation and ulceration of abomasum, stomach, and rumen
Cobalt (Co)	Deficiency	Anemia, emaciation, listlessness, pallor, muscle wastage (marasmus), reduced appetite, poor growth, rough haircoat, loss of hair color, unthriftiness, thickening of the skin, body weight loss, increased susceptibility to infections/parasites, death, reduced milk production, weak calves, poor survivability, low blood glucose, low red cell count, low hemoglobin, impaired neutrophil activity, fatty liver, hemosiderin in spleen, bone marrow hypoplasia *Calves*: poor appetite, poor growth, poor condition, muscle weakness, demyelination of peripheral nerves
Cobalt (Co)	Toxicity	Reduced appetite, reduced body weight gain, anemia, death, polycythemia, polyuria, salivation, shortness of breath, increased hemoglobin, increased packed cell volume
Copper (Cu)	Deficiency	Anemia, ataxia (neonates), bone swellings above joints, bones weak and fragile, diarrhea (severe—especially if high Mo), fractures (spontaneous), depressed growth, hair pigmentation loss, heart failure, infertility (delayed/depressed estrus, reduced conception rate, retained placenta), lameness, milk production reduced, myocardial fibrosis, congenital skeletal abnormalities (rickets)
Copper (Cu)	Toxicity	NR
Dysprosium (Dy)	Toxicity	NR

Erbium (Er)	Toxicity	NR
Europium (Eu)	Toxicity	NR
Fluorine (F)	Deficiency	NR
Fluorine (F)	Toxicity	Acetonemia (adults), anemia, chewing (constant), collapse, death (neonatal), dental mottling with pits and streaked with pigmentation (yellow, green, brown, black) feed intake decreased, fertility reduced, gastroenteritis, hyperesethesia, impaired growth, joint thickening and ankylosis, lameness, milk production decreased, muscle tremors, periosteal hyperostosis, pupillary dilation, ruminal statis, stiffness, renal tubular fibrosis and degeneration, renal mineralization, tetany, weakness, weight loss
Gadolinium (Gd)	Toxicity	NR
Gallium (Ga)	Toxicity	NR
Germanium (Ge)	Toxicity	NR
Gold (Au)	Toxicity	NR
Holmium (Ho)	Toxicity	NR
Indium (In)	Toxicity	NR
Iodine (I)	Deficiency	Abortions, alopecia, blindness, death, food intake reduced, gestation prolonged, goiter, hairlessness, abnormal estrus cycling, arrested fetal development, fetal resorption, hypothyroidism, milk yield reduced and poor quality, fetal membrane retention, male infertility (lack of libido and poor semen quality), stillbirths, weakness, weak offspring
Iodine (I)	Toxicity	NR
Iron (Fe)	Deficiency	*Rare*: Calves, anemia, appetite reduced, breathing labored, low hemoglobin, low liver nonheme Fe content, low serum Fe, low weight gain, listlessness, atrophy of tongue papilla, pallor, low performance, increased susceptibility to infections
Iron (Fe)	Toxicity	Abdominal pain, anorexia, death, depression, dyspnea, hepatic degeneration, jaundice (buccal mucosa, lips), renal congestion
Lanthanum (La)	Toxicity	NR
Lead (Pb)	Toxicity	Common. *Acute:* blindness, convulsions, death (can be sudden), depression, hyperirritability, hypersalivation, muscle twitches, teeth grinding, depressed reflexes *Chronic:* Anemia, growth retardation, low hemoglobin
Lithium (Li)	Deficiency	NR
Lithium (Li)	Toxicity	Inflammation and ulceration of stomach, abomasum and rumen
Lutetium (Lu)	Toxicity	NR
Magnesium (Mg)	Deficiency	Ataxia, convulsions, collapse, death, excitability, feed intake reduced, called grass-tetany, hypersalivation, hypocalcemia (calves), hypomagnesemia, ketosis, lethargy, muscular tremors, uncontrolled limb movements stiffness
Magnesium (Mg)	Toxicity	Degeneration of rumen papille, diarrhea, nephroliths (calves—calcium appetite), weakness
Manganese (Mn)	Deficiency	*Calves:* malformed crooked limbs, joint enlargement, short forelimbs, difficulty standing, adults: abortion, birth weight low, conception rate low, estrus delayed or depressed, stillbirths
Manganese (Mn)	Toxicity	NR
Mercury (Hg)	Toxicity	Ataxia, convulsions, dyspnea, muscle tremors, neuromuscular incoordination, oral ulcers, paresis, proteinuria, renal failure
Molybdenum (Mo)	Deficiency	NR
Molybdenum (Mo)	Toxicity	Anorexia, anemia, poor condition, death, depigmentation hair, diarrhea (severe—especially if low Cu), reduced growth, reduced fertility (conception rate), lameness, loss of libido, mandibular exostoses, osteoporosis, delayed puberty, skeletal deformities, spermatogenesis impaired, staring coat, high serum phosphatase activity, testicular damage, weight loss depigmentation due low Cu Excess Mo causes fatal Cu deficiency. Molybdate and sulfide in the rumen form thiomolybdates (Gould and Kendall, 2011)

Neodymium (Nd)	Toxicity	NR
Nickel (Ni)	Deficiency	NR
Nickel (Ni)	Toxicity	Reduced feed intake
Niobium (Nb)	Toxicity	NR
Palladium (Pd)	Toxicity	NR
Phosphorus (P)	Deficiency	Poor growth, hypophosphatemia, inappetence, poor milk production, poor fertility, osteomalacia, pica, rickets (calves), unthriftiness
Phosphorus (P)	Toxicity	Hypocalcemia, hypophosphatemia, bone resorption (osteomalacia), urinary calculi, diarrhea, abdominal discomfort (mild, inhibits vitamin D synthesis, reduced reproductive performance
Potassium (K)	Deficiency	*Experimental*: reduced food intake, anorexia, death, impaired growth, reduced water intake, reduced efficiency of food utilization, pica, licking objects, reduced milk yield, hypokalemia, muscle weakness, nervous disorders, stiffness, reduced pliability of hide, emaciation, intracellular acidosis, degeneration vital organs, weakness, high hematocrit
Potassium (K)	Toxicity	*Experimental:* death, diarrhea with mucus, excess salivation, muscle tremors, excitability, hyperkalemia, high PCV food intake and milk yield reduced, polydipsia, polyuria, hyperkaliuria *Calves*: death
Praseodymium (Pr)	Toxicity	NR
Promethium (Pm)	Toxicity	NR
Rhodium (Rh)	Toxicity	NR
Rubidium (Rb)	Toxicity	NR
Samarium (Sm)	Toxicity	NR
Scandium (Sc)	Toxicity	NR
Selenium (Se)	Deficiency	Abortions, cardiac changes (frequency, rhythm), sudden death, dyspnea, fetal heart lesions, metritis, muscular dystrophy, myopathy, ovarian cysts, recumbency, impaired reproduction, retained placenta, stiffness, stillbirths, weakness, white muscle disease (calves)
Selenium (Se)	Toxicity	Abortion, anorexia, ataxia, body condition loss, cardiac atrophy, coma, conception rate reduced, constipation, coronary band inflammation, death, decubital ulcers, depression, diarrhea, dyspnea, emaciation, excitability, garlic smell on the breath, gastrointestinal irritation, glutathione peroxidase activity increased in granulosa cells, hypersalivation, incoordination, lameness (joint erosion); horizontal grooves or cracks in the hooves; hoof horn malformation, occasional sloughing of hooves; incoordination, inappetence, infertility, mastitis, polioencephalomalacia (on histology), pulmonary edema, paralysis, phagocytosis (white cells) reduced, reluctance to move, long periods of recumbency; reproductive failure, tachycardia, severe unthriftiness, vision defects *India:* gangrene of distal extremities (below the tarsus or carpus, tips of ears, muzzle, the tip of tail, or tongue) occurs occasionally.
Silicon (Si)	Deficiency	NR
Silicon (Si)	Toxicity	NR
Silver (Ag)	Toxicity	NR
Sodium (Na)	Deficiency	Loss of appetite, dehydration, rough haircoat, reduced milk production, pica, shivering, incoordination, cardiac arrhythmia, death, unthriftiness, weakness, weight loss
Sodium (Na)	Toxicity	Anhydremia, anorexia, collapse, udder edema, reduced milk yield, increased water intake, reduced water intake (if very saline water), weight loss
Strontium (Sr)	Toxicity	NR
Sulfur (S)	Deficiency	NR
Sulfur (S)	Toxicity	Anorexia, blindness, coma, enteritis (severe) halitosis (hydrogen sulfide) muscle twitches, reduced rumen activity, nervousness, polioencephalomalacia-like Syndrome, respiratory distress. Depressed milk production peritonitis, petechial hemorrhages (organs, e.g., renal), recumbency

Tellurium (Te)	Toxicity	NR
Terbium (Tb)	Toxicity	NR
Thallium (Tl)	Toxicity	NR
Thorium (Th)	Toxicity	NR
Thulium (Tm)	Toxicity	NR
Tin (Sn)	Toxicity	NR
Titanium (Ti)	Toxicity	NR
Tungsten (W)	Toxicity	NR
Uranium (U)	Toxicity	NR
Vanadium (V)	Toxicity	Ataxia, diarrhea, weight loss, death
Ytterbium (Yb)	Toxicity	NR
Yttrium (Y)	Toxicity	NR
Zinc (Zn)	Deficiency	Alopecia, bowing of hind limbs, ears—dry scaly skin, grinding of teeth, growth retardation, haircoat rough, hemorrhages (submucosal), hoof and skin cracking and fissuring, lethargy, impaired development and function of male sex organs, skin thickening and cracking around nostrils, inflammation nose and mouth, reduced milk production, mucosal overgrowth lips and dental pads, nutritional utilization reduced, edema limbs, inappetence, parakeratosis, impaired reproduction (estrus through pregnancy to parturition) unthriftiness, salivation excessive, scrotal skin inflamed and wrinkled, stiffness, increased susceptibility to infections, wound healing impaired, low serum Zn
Zinc (Zn)	Toxicity	Ataxia, diarrhea, decreased food intake, decreased milk yield, weight loss, death
Zirconium (Zr)	Toxicity	NR
Not Applicable (Element is not essential in this species)	Deficiency	Al, Sb, Ba, Be, Bi, Br, Cd, Cs, Ce, Dy, Er, Eu, Gd, Ga, Ge, Au, Ho, In, La, Pb, Li, Lu, Hg, Nd, Nb, Pd, Pr, Pm, Rh, Rb, Sm, Sc, Ag, Sr, Te, Tb, Tl, Th, Tm, Sn, Ti, W, U, V, Yb, Y, Zr

References

NRC (2001).

Deer

There is very little information published about mineral requirements and mineral-related disease in Deer. The NRC has incorporated some Deer data in their publication on Small Ruminants (NRC, 2014). Antler growth significantly increases mineral requirements—Ca, Cl, F, P.

Farmed Deer are routinely provided with mineral supplements to ensure an adequate supply of Ca and P and other essential minerals—especially during spring and summer. However, clinical signs and deaths associated with Cu deficiency have been reported in a farmed Red deer herd that had access to Cu-containing salt licks, whereas direct administration of Cu supplements to individuals did correct the deficiency.

High Mo or S can interfere with Cu availability.

There is limited data available.

Element	Deficiency/ toxicity	Clinical signs reported
Calcium (Ca)	Deficiency	Delayed antler growth and velvet shedding
Cobalt (Co)	Deficiency	Death, emaciation
Copper (Cu)	Deficiency	Ataxia (enzootic), low conception rate, deaths, diarrhea, emaciation, poor calf growth rate, dull light-colored hair coats, hemosiderosis (liver). osteochondrosis, unthriftiness, low weights (adult hinds)
Iodine (I)	Deficiency	Goiter
Molybdenum (Mo)	Toxicity	Impaired growth, low Cu status (blood and liver)
Phosphorus (P)	Deficiency	Delayed antler growth and velvet shedding
Selenium (Se)	Deficiency	Conception rate low, white muscle disease, nutritional myopathy, unthriftiness
Selenium (Se)	Toxicity	Anorexia, death, lameness, myocardial fibroplasia, myocardial mineralization, myocardial necrosis, weight loss
Not Applicable (Element is not essential in this species)	Deficiency	Al, Sb, Ba, Be, Bi, Br, Cd, Cs, Ce, Dy, Er, Eu, Gd, Ga, Ge, Au, Ho, In, La, Pb, Li, Lu, Hg, Nd, Nb, Pd, Pr, Pm, Rh, Rb, Sm, Sc, Ag, Sr, Te, Tb, Tl, Th, Tm, Sn, Ti, W, U, V, Yb, Y, Zr

References

Handeland et al. (2008), Vikøren et al. (2005), Grace et al. (2005), Handeland and Bernhoft (2004), Bernhoft et al. (2002), Wilson and Grace (2001), Handeland and Flåøyen (2000), Fyffe (1996), Thompson et al. (1994), Mackintosh et al. (1989), Wilson (1989), Knox et al. (1987), Clark and Hepburn (1986), Wilson et al. (1979), Barlow and Butler (1964).

Giraffes

Giraffes are ruminants that are highly adapted to eating the woody branches of trees especially trees of *Commiphora* and *Terminalia*.
Limited information is available.

Element	Deficiency/toxicity	Clinical signs
Phosphorus (P)	Toxicity	Urolithiasis

References

Wolfe et al. (2000).

Goats

Goats are small ruminants with a wide range of natural foods. A lot of research has been conducted and demonstrated some unique nutrient requirements. They have essential dietary requirements for: Co, Cr, Cu, F, I, Fe, Mn, Mo, Se, Zn, and now several elements which are not regarded as essential in other species: As, Br, Cd, Li, Pb, Ni, Si, Sn, V (Lamand, 1981; Haenlein, 1987; Kessler, 1991).

Element	Deficiency/ toxicity	Clinical signs reported
Aluminium (Al)	Deficiency	Abortion, impaired growth, incoordination, reduced life expectancy, weakness (hindlegs)
Aluminium (Al)	Toxicity	NR
Antimony (Sb)	Toxicity	NR
Arsenic (As)	Deficiency	Abortion, death (sudden death in lactating goats) dermatitis, reduced food intake, poor growth, poor milk production, mitochondrial damage (in cardiac muscle, skeletal muscle and liver), poor survivability of kids, skeletal lesions
Arsenic (As)	Toxicity	NR
Barium (Ba)	Toxicity	NR
Beryllium (Be)	Toxicity	NR
Bismuth (Bi)	Toxicity	NR
Boron (B)	Deficiency	NR
Boron (B)	Toxicity	NR
Bromine (Br)	Deficiency	Abortions, conception rate reduced, fatty liver, fat in skeletal muscle, reduced food intake, reduced fertility, high gamma glutamyl transferase, impaired growth, poor weight gain, low hematocrit, low hemoglobin, reduced life expectancy, high triglycerides
Bromine (Br)	Toxicity	NR
Cadmium (Cd)	Deficiency	Abortion, conception impaired, death, lethargy, mitochondrial damage (especially in liver and kidneys), mobility reduced, muscle weakness, myasthenia, ventroflexion of the neck
Cadmium (Cd)	Toxicity	NR
Cesium (Cs)	Toxicity	NR
Calcium (Ca)	Deficiency	Osteomalacia (adults), rickets (kids)
Calcium (Ca)	Toxicity	NR
Cerium (Ce)	Toxicity	NR
Chlorine (Cl)	Deficiency	NR
Chlorine (Cl)	Toxicity	NR
Chromium (Cr)	Deficiency	NR
Chromium (Cr)	Toxicity	NR
Cobalt (Co)	Deficiency	Death, emaciation
Cobalt (Co)	Toxicity	Anemia, grow impairment, fatty degeneration of liver, unthriftiness, weight loss
Copper (Cu)	Deficiency	Anemia (hypochromic microcytic), ataxia (neonatal), bone fractures, deformed articular cartilage, cardiovascular disorders, diarrhea, impaired endochondral ossification, abnormal bone growth, impaired immunity, increased susceptibility to infections, infertility, osteoporosis, depigmentation of skin and hair
Copper (Cu)	Toxicity	Death, hemoglobinuria, hemolysis, jaundice, liver damage, high liver enzymes (sorbitol dehydrogenase, gamma glutamyle transferase, and glutamic oxaloacetic transaminase)
Dysprosium (Dy)	Toxicity	NR

Erbium (Er)	Toxicity	NR
Europium (Eu)	Toxicity	NR
Fluorine (F)	Deficiency	Reduced food intake, impaired growth, joint deformity, lifespan reduced, increased mortality (kids), skeletal deformity, thymus hypoplasia
Fluorine (F)	Toxicity	Increased food intake, increased intrauterine and postnatal death rate (mortality), growth retarded, joint deformities (older), skeletal deformities (older). Thymus hypoplasia (fetus)
Gadolinium (Gd)	Toxicity	NR
Gallium (Ga)	Toxicity	NR
Germanium (Ge)	Toxicity	NR
Gold (Au)	Toxicity	NR
Holmium (Ho)	Toxicity	NR
Indium (In)	Toxicity	NR
Iodine (I)	Deficiency	Impaired brain development, goiter, hairless young, infertility, (poor conception rate), low milk production, skin disorders, sterility, stillbirths, susceptibility to cold stress
Iodine (I)	Toxicity	NR
Iron (Fe)	Deficiency	NR
Iron (Fe)	Toxicity	May cause Cu deficiency
Lanthanum (La)	Toxicity	NR
Lead (Pb)	Toxicity	NR
Lithium (Li)	Deficiency	Abortion, low birth weight, impaired growth, decreased milk production, shorter lifespan, reduced fertility, reduced weight gain, low serum dehydrogenases, aldolases and liver monoamine oxidase
Lithium (Li)	Toxicity	Ataxia, death, depression, diarrhea
Lutetium (Lu)	Toxicity	NR
Magnesium (Mg)	Deficiency	Convulsions, tetany
Magnesium (Mg)	Toxicity	Death, diarrhea, reduced feed intake, lethargy, disturbed locomotion, poor performance
Manganese (Mn)	Deficiency	Abortion, angular deformities, ataxia, birth weight low, conception rate low, joint swelling (tarsus), estrus delayed or depressed, stillbirths
Manganese (Mn)	Toxicity	NR
Mercury (Hg)	Toxicity	NR
Molybdenum (Mo)	Deficiency	Abortion, low conception rates, poor growth, increased mortality (kids)
Molybdenum (Mo)	Toxicity	Elevated liver, serum, and kidney molybdenum, reduced semen quality
Neodymium (Nd)	Toxicity	NR
Nickel (Ni)	Deficiency	Abortions, reduced conception rates, growth rate impaired, low hemoglobin, milk production reduced, reduced survivability, skin lesions (scales and crusts—parakeratosis), skeletal lesions, lower testicular weight
Nickel (Ni)	Toxicity	NR
Niobium (Nb)	Toxicity	NR
Palladium (Pd)	Toxicity	NR
Phosphorus (P)	Deficiency	Bone undermineralized, conception rate low, reduced food intake, reduced milk production, estrus suppression, pica, skeletal deformities, reduced weight gain
Phosphorus (P)	Toxicity	Bones undermineralized, urolithiasis
Potassium (K)	Deficiency	Arched back, emaciation, reduced feed intake, reduced growth, muscle paralysis, rhabdomyolysis, stiffness
Potassium (K)	Toxicity	Hypocalcemia (milk fever), hypomagnesemia (grass tetany)

Praseodymium (Pr)	Toxicity	NR
Promethium (Pm)	Toxicity	NR
Rhodium (Rh)	Toxicity	NR
Rubidium (Rb)	Deficiency	Abortion, lower birth weights, lower weaning weights, high mortality *NOT corroborated*
Rubidium (Rb)	Toxicity	NR
Samarium (Sm)	Toxicity	NR
Scandium (Sc)	Toxicity	NR
Selenium (Se)	Deficiency	Cardiac changes (frequency, rhythm), death, muscular dystrophy, recumbency, stiffness, weakness, white muscle disease
Selenium (Se)	Toxicity	NR
Silicon (Si)	Deficiency	NR
Silicon (Si)	Toxicity	NR
Silver (Ag)	Toxicity	NR
Sodium (Na)	Deficiency	Loss of appetite, emaciation, reduced food intake, reduced efficacy of food utilization, poor growth shaggy dull haircoat, pica, poor weight gain, wobbly gait
Sodium (Na)	Toxicity	NR
Strontium (Sr)	Toxicity	NR
Sulfur (S)	Deficiency	Death, emaciation, reduced feed intake, reduced dietary digestibility and utilization, hypersalivation, lachrimation, poor production, high blood urea
Sulfur (S)	Toxicity	Death, diarrhea, dyspnea, halitosis (hydrogen sulfide smell on breath), muscle tremors
Tellurium (Te)	Toxicity	NR
Terbium (Tb)	Toxicity	NR
Thallium (Tl)	Toxicity	NR
Thorium (Th)	Toxicity	NR
Thulium (Tm)	Toxicity	NR
Tin (Sn)	Toxicity	NR
Titanium (Ti)	Toxicity	NR
Tungsten (W)	Toxicity	NR
Uranium (U)	Toxicity	NR
Vanadium (V)	Deficiency	Abortion, bone abnormalities, reduced food intake, small litter size, reduced milk production, high mortality rate, reduced survival (kids), impaired reproduction
Vanadium (V)	Toxicity	NR
Ytterbium (Yb)	Toxicity	NR
Yttrium (Y)	Toxicity	NR
Zinc (Zn)	Deficiency	Alopecia, growth retardation, impaired development and function of male sex organs, parakeratosis, horns soft, deformed, lack striations.
Zinc (Zn)	Toxicity	NR
Zirconium (Zr)	Toxicity	NR
Not Applicable (Element is not essential in this species)	Deficiency	Al, Sb, Ba, Be, Bi, Cs, Ce, Dy, Er, Eu, Gd, Ga, Ge, Au, Ho, In, La, Li, Lu, Hg, Nd, Nb, Pd, Pr, Pm, Rh, Rb, Sm, Sc, Ag, Sr, Te, Tb, Tl, Th, Tm, Sn, Ti, W, U, Yb, Y, Zr

References

NRC (2014), Anke et al. (2005a, b, c), Kessler (1991), Haenlein (1987), Lamand (1981).

Sheep

Many pastures may be deficient and fail to meet the basic nutritional requirements of sheep, for example in a New Zealand study (Knowles and Grace, 2014) that analyzed 1106 pastures the Co requirement of sheep was met by only 54% of pastures and 24% of pastures were deficient in Se.

Element	Deficiency/ toxicity	Clinical signs reported
Aluminium (Al)	Toxicity	Signs of P deficiency as Al binds to P. See P
Antimony (Sb)	Toxicity	NR
Arsenic (As)	Deficiency	NR
Arsenic (As)	Toxicity	NR
Barium (Ba)	Toxicity	NR
Beryllium (Be)	Toxicity	NR
Bismuth (Bi)	Toxicity	NR
Boron (B)	Deficiency	NR
Boron (B)	Toxicity	NR
Bromine (Br)	Toxicity	NR
Cadmium (Cd)	Toxicity	NR
Cesium (Cs)	Toxicity	NR
Calcium (Ca)	Deficiency	Milk fever, osteomalacia (adults), rickets (lambs)
Calcium (Ca)	Toxicity	NR
Cerium (Ce)	Toxicity	NR
Chlorine (Cl)	Deficiency	NR
Chlorine (Cl)	Toxicity	NR
Chromium (Cr)	Deficiency	NR
Chromium (Cr)	Toxicity	NR
Cobalt (Co)	Deficiency	Anemia (normocytic, normochromic), low body condition score, death, depression, emaciation, lachrymation, lethargy, listlessness, muscle wastage (marasmus), reduced appetite, poor growth, head pressing, increased susceptibility to parasites, neonatal deaths, pallor, reduced milk production, poor survivability, low red cell count, low hemoglobin, fatty liver, poor fertility, hemosiderin in the spleen, bone marrow hypoplasia, hypocobalminemia (lambs), impaired immune response, low liver B12 (lambs), tear staining, unthriftiness, poor response to vaccination, aimless wandering, weak lambs, body weight loss
Cobalt (Co)	Toxicity	Anemia, reduced appetite, body weight gain reduced, death, growth impairment, fatty degeneration of liver, unthriftiness, weight loss
Copper (Cu)	Deficiency	Abnormal wool (lack crimp, steely or stringy (straight) hair, abortion, achromotrichia, anemia
Copper (Cu)	Toxicity	Death, hemoglobinuria, hemolysis, jaundice, liver damage, high liver enzymes (sorbitol dehydrogenase, gamma glutamyl transferase, and glutamic oxaloacetic transaminase)
Dysprosium (Dy)	Toxicity	NR
Erbium (Er)	Toxicity	NR
Europium (Eu)	Toxicity	NR
Fluorine (F)	Deficiency	NR
Fluorine (F)	Toxicity	Death (neonatal). Fertility reduced, impaired growth, joint thickening and ankylosis, lameness, mandibular hyperostosis, long bone periosteal hyperostosis, stiffness
Gadolinium (Gd)	Toxicity	NR
Gallium (Ga)	Toxicity	NR

Germanium (Ge)	Toxicity	NR
Gold (Au)	Toxicity	NR
Holmium (Ho)	Toxicity	NR
Indium (In)	Toxicity	NR
Iodine (I)	Deficiency	Goiter, poor brain development, reduced wool production, abnormal estrus cycling, arrested fetal development, resorption, abortion, stillbirths, prolonged gestation, weak offspring, fetal membrane retention, male infertility (poor semen quality)
Iodine (I)	Toxicity	anorexia, depression, hypothermia, poor weight gain
Iron (Fe)	Deficiency	*Lambs*: Anemia
Iron (Fe)	Toxicity	High alanine aminotransferase (ALT, high serum alkaline phosphatase (ALP) anorexia, high blood urea nitrogen (BUN), circulatory failure, low serum copper, high creatinine, death diarrhea, depression, hemosiderosis, hepatic degeneration, high serum iron, myocardial degeneration, high serum phosphorus, pulmonary edema, renal degeneration, respiratory failure, splenic degeneration, high total iron binding capacity (TIBC), weight loss May cause Cu deficiency
Lanthanum (La)	Toxicity	NR
Lead (Pb)	Toxicity	Abortions, death, hearing loss, hydronephrosis, osteoporosis, sterility, stiffness, unthriftiness
Lithium (Li)	Deficiency	NR
Lithium (Li)	Toxicity	NR
Lutetium (Lu)	Toxicity	NR
Magnesium (Mg)	Deficiency	Collapse, convulsions, death, hypersalivation, hypocalcemia, hypomagnesemia, limb movements when collapsed, tetany
Magnesium (Mg)	Toxicity	Death, diarrhea, reduced food intake, lethargy, locomotion disturbed, reduced nutrient utilization, poor performance
Manganese (Mn)	Deficiency	Appetite loss, balance difficulty, death, depressed growth joint pain, locomotion impaired, reproductive failure
Manganese (Mn)	Toxicity	Reduced appetite, reduced growth rate, interferes with Fe metabolism
Mercury (Hg)	Toxicity	Alopecia, anorexia, blindness, death, depression, diarrhea, convulsions, acute hemorrhagic gastroenteritis, incoordination, stiff legged gait, paresis. Polioencephalomalacia, pruritus, scabby skin lesions (anus and vulva), tooth loss, acute renal failure
Molybdenum (Mo)	Deficiency	Xanthine calculi
Molybdenum (Mo)	Toxicity	*Oral intake:* Alkaline phosphatase increased, connective tissue changes (subperiosteal hemorrhages on the humerus, spontaneous fractures), central nervous system degeneration, motor neurone disease, diarrhea, reduced weight gain, depigmentation of skin and hair, delayed puberty, loss of crimp in wool
Neodymium (Nd)	Toxicity	NR
Nickel (Ni)	Deficiency	NR
Nickel (Ni)	Toxicity	NR
Niobium (Nb)	Toxicity	NR
Palladium (Pd)	Toxicity	NR
Phosphorus (P)	Deficiency	Bone undermineralized, conception rate low, reduced food intake, reduced milk production, estrus suppression, pica, skeletal deformities, reduced weight gain
Phosphorus (P)	Toxicity	Bones undermineralized *Lambs:* urolithiasis
Potassium (K)	Deficiency	poor growth rate, reduced feed intake, weight loss, emaciation, listlessness, loss of wool, rough hair, hypoglycemia, increased liver glycogen, urinary calculi, impaired responses, stiffness, death. Histological changes in many tissues
Potassium (K)	Toxicity	Hypocalcemia (milk fever), hypomagnesemia (grass tetany)

Praseodymium (Pr)	Toxicity	NR
Promethium (Pr)	Toxicity	NR
Rhodium (Rh)	Toxicity	NR
Rubidium (Rb)	Toxicity	NR
Samarium (Sm)	Toxicity	NR
Scandium (Sc)	Toxicity	NR
Selenium (Se)	Deficiency	Arrhythmia—cardiac changes (frequency, rhythm), muscular dystrophy, early embryonic loss/failure to implant, reduced growth rate recumbency, stiffness, tachycardia, high P wave, shorter PR, ST, and QT interval, short T wave weakness, white muscle disease, muscular dystrophy (white muscle disease), reluctance to move
Selenium (Se)	Toxicity	Ataxia, labored breathing, congenitally deformed eyes, death, dyspnea growth impairment, kidney damage, high tissue Se content, tetanic spasms
Silicon (Si)	Deficiency	NR
Silicon (Si)	Toxicity	NR
Silver (Ag)	Toxicity	NR
Sodium (Na)	Deficiency	Reduced feed intake, impaired growth, pica
Sodium (Na)	Toxicity	NR
Strontium (Sr)	Toxicity	NR
Sulfur (S)	Deficiency	Death, emaciation, reduced feed intake, reduced dietary digestibility and utilization, hypersalivation, lachrimation, poor production, high blood urea, wool loss
Sulfur (S)	Toxicity	Death, diarrhea, dyspnea, halitosis (hydrogen sulfide smell on breath), muscle tremors
Tellurium (Te)	Toxicity	NR
Terbium (Tb)	Toxicity	NR
Thallium (Tl)	Toxicity	NR
Thorium (Th)	Toxicity	NR
Thulium (Tm)	Toxicity	NR
Tin (Sn)	Toxicity	NR
Titanium (Ti)	Toxicity	NR
Tungsten (W)	Toxicity	NR
Uranium (U)	Toxicity	NR
Vanadium (V)	Toxicity	Anorexia, ataxia, death, diarrhea, reduced feed intake, poor growth, small intestine mucosal hemorrhage, renal hemorrhage
Ytterbium (Yb)	Toxicity	NR
Yttrium (Y)	Toxicity	NR
Zinc (Zn)	Deficiency	Deaths (postnatal), growth retardation, Impaired development and function of male sex organs, aspermatogenesis. Impaired reproduction (estrus through pregnancy to parturition, Wool abnormalities: loose, brittle, loses crimp, alopecia, skin: thick inflamed and wrinkled, Lack of wool growth. Horns soft, deformed, lack striations. Low serum Zn
Zinc (Zn)	Toxicity	Ataxia, death (postnatal), diarrhea, pica, poor growth, poor food intake, poor food utilization
Zirconium (Zr)	Toxicity	NR
Not Applicable (Element is not essential in this species)	Deficiency	Al, Sb, Ba, Be, Bi, Br, Cd, Cs, Ce, Dy, Er, Eu, Gd, Ga, Ge, Au, Ho, In, La, Pb, Li, Lu, Hg, Nd, Nb, Pd, Pr, Pm, Rh, Rb, Sm, Sc, Ag, Sr, Te, Tb, Tl, Th, Tm, Sn, Ti, W, U, V, Yb, Y, Zr

References

Knowles and Grace (2014), NRC (2014).

Hindgut fermentors

This group of animals have specially adapted hindgut anatomy (cecum, colon) to accommodate microorganisms that can ferment food material in the gut lumen into a form that can be absorbed across the intestinal wall.

Elephants

Asian and African elephants (*Elephas maximus* and *Loxodonta Africana*) have gastrointestinal tracts similar to the horse and they are monogastric hindgut fermenters belonging to a group called megaherbivores. They eat grasses, foliage, twigs, branches, and bark.

The natural vegetation eaten by wild elephants is often low in mineral content but intake can also be obtained from water supplies—some waterholes are especially high in mineral content, and elephants, like many species, indulge in geophagy which also provides minerals.

Deficiency

Mineral deficiencies have been rarely reported in elephants, however, Ca, Fe, and Zn deficiency case reports exist and I deficiency is suspected. Tusk growth significantly increases Ca requirements as does lactation and subclinical hypocalcemia have been reported in lactating cows—Van der Kolk et al. (2008).

Element	Clinical signs	Comments
Calcium deficiency	Hypocalcemia, metabolic bone disease (nutritional secondary hyperparathyroidism, osteodystrophy)	
Iodine deficiency	Reduced reproductive performance	*Based on observation of higher reproductive performance in elephants fed high compared to low I content foods*
Iron deficiency	Anemia	
Zinc deficiency	Immune deficiency, skin lesions (hyperkeratosis)	Caused by excess Ca intake

Toxicity

Element	Clinical signs	Comments
Calcium toxicity	Immune deficiency, skin lesions (hyperkeratosis)	Due to secondary Zn deficiency

References

Sach et al. (2019): Dierenfeld (2008), Van Der Kolk et al. (2008), Hatt and Clauss (2006), Ensley et al. (1994), Schmidt (1989), Kuntze and Hunsdorff (1978).

Van Der Kolk JH, Van Leeuwen JPTM, Van Den Belt AJM, Van Schaik RHN, Schaftenaar W. Subclinical hypocalcaemia in captive elephants (Elephas maximus). Vet Rec 2008;162(15):475–9. doi:10.1136/vr.162.15.475.

Horses

Equids (horses, donkeys, Persian onagers) develop hepatic hemosiderosis and hemochromatosis due to iron overload, as do other hindgut fermenters Rhino and Tapirs.

Mineral	Deficiency/ toxicity	Clinical signs reported
Aluminium (Al)	Toxicity	equine granulomatous enteritis
Antimony (Sb)	Toxicity	NR
Arsenic (As)	Deficiency	NR
Arsenic (As)	Toxicity	NR
Barium (Ba)	Toxicity	NR
Beryllium (Be)	Toxicity	NR
Bismuth (Bi)	Toxicity	NR
Boron (B)	Deficiency	NR
Boron (B)	Toxicity	NR
Bromine (Br)	Toxicity	NR
Cadmium (Cd)	Toxicity	Chronic interstitial nephritis
Cesium (Cs)	Toxicity	NR
Calcium (Ca)	Deficiency	*Foals:* rickets [poor mineralization of osteoid, enlarged joints, crooked long bones (angular deformity)] *Adults:* osteomalacia, lameness
Calcium (Ca)	Toxicity	NR
Cerium (Ce)	Toxicity	NR
Chlorine (Cl)	Deficiency	NR
Chlorine (Cl)	Toxicity	NR
Chromium (Cr)	Deficiency	NR
Chromium (Cr)	Toxicity	NR
Cobalt (Co)	Deficiency	Not reported—even when on pastures so low in Co that ruminants die
Cobalt (Co)	Toxicity	NR
Copper (Cu)	Deficiency	Osteochondrosis (Cu deficient grazing or secondary to high Zn) osteodysgenesis, hemorrhage at parturition (due to ruptured uterine artery)
Copper (Cu)	Toxicity	NR
Dysprosium (Dy)	Toxicity	NR
Erbium (Er)	Toxicity	NR
Europium (Eu)	Toxicity	NR
Fluorine (F)	Deficiency	NR
Fluorine (F)	Toxicity	Bones thick, rough and irregular surfaced, dental abscesses, food intake reduced, hair coat dry and rough, hypersalivation, lameness, reluctant to move, shifting body weight, skin lost normal elasticity, swollen jaw (due dental abscesses), teeth discoloration, teeth wear excessively, teeth traumatize the cheek/gums, exposure of dental pulp cavity, unthriftiness
Gadolinium (Gd)	Toxicity	NR
Gallium (Ga)	Toxicity	NR
Germanium (Ge)	Toxicity	NR
Gold (Au)	Toxicity	NR

Holmium (Ho)	Toxicity	NR
Indium (In)	Toxicity	NR
Iodine (I)	Deficiency	Goiter, abnormal estrus cycles, male infertility (lack libido and poor semen quality), stillborn foals, weak neonates, neonatal death
Iodine (I)	Toxicity	Alopecia, goiter (foals)
Iron (Fe)	Deficiency	Anemia, labored breathing, reduced growth rate, diarrhea, pneumonia, pallor, weakness, lethargy, rough haircoat, tires easily
Iron (Fe)	Toxicity	*Foals:* coma, death, dehydration, diarrhea, jaundice, jejunal villi erosion, pulmonary hemorrhage, hepatic accumulation of Fe, liver degeneration, reduced serum and liver Zn *Acute:* anorexia, coagulopathy, depression, jaundice, neurological signs *Chronic:* hemosiderosis: asymptomatic, hemochromatosis: liver failure
Lanthanum (La)	Toxicity	NR
Lead (Pb)	Toxicity	Erythrocyte stippling, loud respiratory noises (roaring) due pharyngeal and laryngeal paralysis
Lithium (Li)	Deficiency	NR
Lithium (Li)	Toxicity	NR
Lutetium (Lu)	Toxicity	NR
Magnesium (Mg)	Deficiency	Ataxia, aortic mineralization, collapse, convulsions, death. Hyperpnoea, hypomagnesemia, muscle tremors, nervousness, sweating, with limb movements, soft tissue mineralization and tissue degeneration at postmortem examination, tetany
Magnesium (Mg)	Toxicity	NR
Manganese (Mn)	Deficiency	NR
Manganese (Mn)	Toxicity	NR
Mercury (Hg)	Toxicity	*Acute:* colic, death, dermatitis (ulcerative), edema (ventral), stomatitis, acute renal failure, tubular necrosis, shock *Chronic*: ataxia, depression, dermatitis, gingival swelling and necrosis, head bobbing, hypermetria, masseter muscle atrophy, trembling, neurologic signs. Interstitial renal fibrosis, chronic renal failure, weakness, weight loss
Molybdenum (Mo)	Deficiency	NR
Molybdenum (Mo)	Toxicity	Rickets in foals
Neodymium (Nd)	Toxicity	NR
Nickel (Ni)	Deficiency	NR
Nickel (Ni)	Toxicity	NR
Niobium (Nb)	Toxicity	NR
Palladium (Pd)	Toxicity	NR
Phosphorus (P)	Deficiency	*Foals:* Rickets *Adults:* lameness, osteomalacia, osteofibrosis, fibrous osteodystrophy "big head," jaw enlargement
Phosphorus (P)	Toxicity	Poor bone calcification, nutritional secondary hyperparathyroidism, enlarged facial bones (called "big head" or "bran disease")
Potassium (K)	Deficiency	Reduced growth rate, reduced appetite, hypokalemia, unthriftiness
Potassium (K)	Toxicity	NR
Praseodymium (Pr)	Toxicity	NR
Promethium (Pr)	Toxicity	NR
Rhodium (Rh)	Toxicity	NR
Rubidium (Rb)	Toxicity	NR
Samarium (Sm)	Toxicity	NR

Scandium (Sc)	Toxicity	NR
Selenium (Se)	Deficiency	Anorexia, aspartate aminotransferase activity increased, increased aspartic-pyruvic transaminase, blood urea nitrogen (BUN) increased, cardiac function impairment, creatine kinase activity increased, dysphagia, fever, gamma-glutamyl transferase increased, hematocrit low, hyperkalemia, lactate dehydrogenase high locomotor difficulties myopathy, neck pain, edema (ventral), recumbency, respiratory distress, steatitis (yellow fat disease), stiffness, suckling difficulty, weakness, weight loss, white muscle disease
Selenium (Se)	Toxicity	*Acute:* Abdominal pain, blindness, colic, diarrhea, head pressing, lameness, lethargy, sweating, tachycardia, tachypnea *Chronic:* alopecia (mane and tail especially), cracking of hooves around coronary band, emaciation *Deaths reported following injection of Se Vitamin E preparations*
Silicon (Si)	Deficiency	NR
Silicon (Si)	Toxicity	NR
Silver (Ag)	Toxicity	NR
Sodium (Na)	Deficiency	Anorexia, chewing uncoordinated, fatigue, reduced rate of feed intake, gait unsteady, decreased milk production, muscle spasms, pica (licking materials, e.g., mangers, rocks, skin turgor reduced, reduced thirst
Sodium (Na)	Toxicity	NR
Strontium (Sr)	Toxicity	NR
Sulfur (S)	Deficiency	NR
Sulfur (S)	Toxicity	Convulsions, cyanosis, death, dullness, jaundice, labored breathing, lethargy, nasal discharge (yellow), respiratory failure
Tellurium (Te)	Toxicity	NR
Terbium (Tb)	Toxicity	NR
Thallium (Tl)	Toxicity	NR
Thorium (Th)	Toxicity	NR
Thulium (Tm)	Toxicity	NR
Tin (Sn)	Toxicity	NR
Titanium (Ti)	Toxicity	NR
Tungsten (W)	Toxicity	NR
Uranium (U)	Toxicity	NR
Vanadium (V)	Toxicity	NR
Ytterbium (Yb)	Toxicity	NR
Yttrium (Y)	Toxicity	NR
Zinc (Zn)	Deficiency	*Foals:* alopecia, alkaline phosphatase reduced, enlarged epiphyses, reduced food intake, reduced growth rate, lameness, skin lesions (crusting, desquamation, flaking, parakeratosis, roughness, hoof, face, muzzle, nose, legs, stiffness *Adults*: effusions on legs
Zinc (Zn)	Toxicity	NR
Zirconium (Zr)	Toxicity	NR
Not Applicable (Element is not essential in this species)	Deficiency	Al, Sb, Ba, Be, Bi, Br, Cd, Cs, Ce, Dy, Er, Eu, Gd, Ga, Ge, Au, Ho, In, La, Pb, Li, Lu, Hg, Nd, Nb, Pd, Pr, Pm, Rh, Rb, Sm, Sc, Ag, Sr, Te, Tb, Tl, Th, Tm, Sn, Ti, W, U, V, Yb, Y, Zr

References

Pearce et al. (1998), Bridges and Ghagan Moffitt (1990), Knight et al. (1990), NRC (1989), Bridges and Harris (1988), Glade and Belling (1986), Bridges et al. (1984).

Rabbits

Rabbits are strict herbivores and hindgut fermenters relying on the microbiome of their enlarged cecum to digest fermentable plant material. Unfermentable fibers are excreted in the feces and not diverted into the cecum from the proximal colon. Soft cecotrophs are passed and reingested and these are an essential source of cobalt-containing cobalamin (vitamin B12) for the rabbit.
Unusually, in rabbits, serum total calcium concentrations vary widely and are directly related to dietary intake (usually they are maintained within a very narrow reference range irrespective of dietary intake). Ca is passively absorbed and vitamin D may not be necessary in rabbits.

Mineral	Deficiency/ toxicity	Clinical signs reported
Aluminium (Al)	Toxicity	Reduced testes weights, reduced sperm quality Neurotoxic: ataxia, seizures, motor inhibition, neurofibrils form
Antimony (Sb)	Toxicity	Interstitial pneumonia
Arsenic (As)	Deficiency	NR
Arsenic (As)	Toxicity	NR
Barium (Ba)	Toxicity	NR
Beryllium (Be)	Toxicity	*Intravenous Be metal in water:* Osteosarcomas *Intravenous: Be salts:* Bone tumors *Intra-medullary bone injection or subperiosteal implantation*: convulsions, death, hypoglycemia, liver necrosis
Bismuth (Bi)	Toxicity	NR
Boron (B)	Deficiency	NR
Boron (B)	Toxicity	*Fetuses (44 mg boron/kg/day on gestation days 6–19):* Decreases in the number of live fetuses and litters, increase in resorptions, decreases in body weight, and increases in the occurrence of external, visceral, and cardiovascular malformation *Adults:* conjunctivitis
Bromine (Br)	Toxicity	NR
Cadmium (Cd)	Toxicity	Glomerular sclerosis, tubular-cell degeneration, and interstitial fibrosis
Cesium (Cs)	Toxicity	NR
Calcium (Ca)	Deficiency	NR
Calcium (Ca)	Toxicity	NR *Rabbits appear to be very tolerant of high Ca intake*
Cerium (Ce)	Toxicity	NR
Chlorine (Cl)	Deficiency	NR
Chlorine (Cl)	Toxicity	NR
Chromium (Cr)	Deficiency	Aortic plaques
Chromium (Cr)	Toxicity	NR
Cobalt (Co)	Deficiency	NR
Cobalt (Co)	Toxicity	NR
Copper (Cu)	Deficiency	Achromotrichia, alopecia, Anemia, abnormal cartilage formation, bone abnormalities, dermatitis, hypopigmentation (graying) of dark hairs
Copper (Cu)	Toxicity	NR
Dysprosium (Dy)	Toxicity	*Application of $DyCl_3$:* conjunctivitis, dermatitis, Reduced ileal contractility, severe irritation to abraded skin
Erbium (Er)	Toxicity	Conjunctivitis, severe irritation on abraded skin
Europium (Eu)	Toxicity	Conjunctivitis, inflammatory, and fibrotic responses in the skin, abraded skin irritation
Fluorine (F)	Deficiency	NR

Fluorine (F)	Toxicity	NR
Gadolinium (Gd)	Toxicity	Severe irritation to abraded skin
Gallium (Ga)	Toxicity	NR
Germanium (Ge)	Toxicity	NR
Gold (Au)	Toxicity	NR
Holmium (Ho)	Toxicity	Conjunctivitis, severe irritation on abraded skin
Indium (In)	Toxicity	NR
Iodine (I)	Deficiency	NR
Iodine (I)	Toxicity	Death (neonates)
Iron (Fe)	Deficiency	*Rare:* Anemia (microcytic, hypochromic)
Iron (Fe)	Toxicity	NR
Lanthanum (La)	Toxicity	NR
Lead (Pb)	Toxicity	NR
Lithium (Li)	Deficiency	NR
Lithium (Li)	Toxicity	NR
Lutetium (Lu)	Toxicity	Conjunctivitis, severe irritation to abraded skin
Magnesium (Mg)	Deficiency	alopecia, blanching of ears, cardiac changes (endocardial, subendocardial, interstitial, and perivascular fibrosis), convulsions, fur chewing, poor growth, hyperexcitability
Magnesium (Mg)	Toxicity	NR
Manganese (Mn)	Deficiency	Angular deformity (bowing) forelegs, bone poorly mineralized, bone growth retarded, libido lack, otolith abnormalities, shortened forelimbs, skull deformity, sperm lack, sterility (male), tubular degeneration testes
Manganese (Mn)	Toxicity	NR
Mercury (Hg)	Toxicity	NR
Molybdenum (Mo)	Deficiency	NR
Molybdenum (Mo)	Toxicity	*Oral:* Alopecia, anemia, anorexia, death, dermatosis, death, epiphyseal cartilage widening, fractures, growth impairment, limb deformity (weakness) of front legs, metaphyseal cartilage irregularly calcified, muscle degeneration, thyroxine T4 decline, weight loss *Instilled into eye:* conjunctivitis (mild)
Neodymium (Nd)	Toxicity	Conjunctivitis, severe irritation to abraded skin
Nickel (Ni)	Deficiency	NR
Nickel (Ni)	Toxicity	Reduced feed intake *Inhalation:* pneumonia *Intramuscular implants*: rhabdomyosarcomas
Niobium (Nb)	Toxicity	NR
Palladium (Pd)	Toxicity	Anorexia, death, reduced defecation, dermatitis: eschar, erythema, and edema in intact and abraded skin. Eye irritation, bone damage, heart damage, kidney damage, liver damage, sluggishness, reduced urine output, water intake reduced
Phosphorus (P)	Deficiency	Hypophosphatemia, increased calciuria
Phosphorus (P)	Toxicity	Anorexia
Potassium (K)	Deficiency	Muscular dystrophy
Potassium (K)	Toxicity	Nephritis
Praseodymium (Pr)	Toxicity	Conjunctivitis, severe irritation on abraded skin
Promethium (Pm)	Toxicity	NR

Rhodium (Rh)	Toxicity	NR
Rubidium (Rb)	Toxicity	NR
Samarium (Sm)	Toxicity	Increased blinking but no conjunctivitis, severe irritation to abraded skin
Scandium (Sc)	Toxicity	NR
Selenium (Se)	Deficiency	NR *Se may not be essential in rabbits*
Selenium (Se)	Toxicity	Death
Silicon (Si)	Deficiency	NR
Silicon (Si)	Toxicity	NR
Silver (Ag)	Toxicity	NR
Sodium (Na)	Deficiency	Poor weight gain
Sodium (Na)	Toxicity	NR
Strontium (Sr)	Toxicity	NR
Sulfur (S)	Deficiency	NR
Sulfur (S)	Toxicity	*Carbon disulfide exposure in atmosphere*: reduced weight gain, loss of muscle control, neurological damage, increased Cu in tissues (thyroid, pancreas, spinal cord). Death (sodium tetrathionate)
Tellurium (Te)	Toxicity	NR
Terbium (Tb)	Toxicity	NR
Thallium (Tl)	Toxicity	NR
Thorium (Th)	Toxicity	NR
Thulium (Tm)	Toxicity	NR
Tin (Sn)	Toxicity	NR
Titanium (Ti)	Toxicity	NR
Tungsten (W)	Toxicity	NR
Uranium (U)	Toxicity	NR
Vanadium (V)	Toxicity	NR
Ytterbium (Yb)	Toxicity	NR
Yttrium (Y)	Toxicity	NR
Zinc (Zn)	Deficiency	Alopecia, dermatitis, reduced food intake, low hematocrit, hair pigment change (dark to gray), reproductive failure (nonreceptive to male, failure of ovulation), sores around mouth, wet matted hair on lower jaw and ruff, weight loss
Zinc (Zn)	Toxicity	Grayish areas with pulmonary congestion, peribronchial leukocytic infiltration, bronchial exudate composed of polymorphonuclear leukocytes
Zirconium (Zr)	Toxicity	Pneumonia, peribronchial granulomas
Not Applicable (Element is not essential in this species)	Deficiency	Al, Sb, Ba, Be, Bi, Br, Cd, Cs, Ce, Dy, Er, Eu, Gd, Ga, Ge, Au, Ho, In, La, Pb, Li, Lu, Hg, Nd, Nb, Pd, Pr, Pm, Rh, Rb, Sm, Sc, Ag, Sr, Te, Tb, Tl, Th, Tm, Sn, Ti, W, U, V, Yb, Y, Zr

References

Varga (2014), NRC (1977).

Rhinoceros

Rhinos belong to the Order Perissodactyla and they are hindgut fermenters. Nutritional data for rhinos is limited, but from what is known the horse (which also belongs to the Perissodactyla) appears to be a close nutritional model for Indian and White Rhinos (Clauss et al., 2007) but Black Rhinos have significantly higher absorption coefficients for Ca and Mg and lower ones for Na and K than horses. Several zoo diets are deficient in Cu, Mn, or Zn, and most contained excessive levels of Fe compared to horse requirements.

Hemosiderosis occurs in captive Rhinos (especially browsing species) and may be associated with mineral imbalance. Ulcerative dermatitis, hemolytic anemia, impaired immune function, and peripheral vasculitis may also be linked to mineral status.

Hemosiderosis results in cellular degeneration, necrosis and fibrosis, and in affected animals bone marrow is hypocellular and often fibrotic and neonatal leukoencephalomalacia has been suggested to be linked to iron overload in the mother.

Due to free radicle formation iron overload disease may predispose black rhinoceros to hemolytic anemia, mucocutaneous ulcerative diseases, stress intolerance, and increased susceptibility to infections (Kock et al., 1992; Paglia and Dennis, 1999; Smith et al., 1995).

Captive Black and Sumatran rhinoceros get Fe overload with high blood and tissue concentrations and organ damage, however wild Rhino do not get this disorder.

Element	Deficiency/ toxicity	Clinical signs
Iron	Toxicity	Anemia (hemolytic), hypocellular bone marrow, hemosiderosis, mucocutaneous ulcerative disease, stress intolerance

Treatments for Fe overload in Rhino include phlebotomy, Fe chelators, and selection of dietary components low in Fe content.

References

Radcliffe and Khairani (2019), Duncan (2018), Olias et al. (2012), Dennis et al. (2007), Clauss et al. (2007), Dierenfeld et al. (2005), Paglia and Dennis (1999), Smith et al. (1995); Kock et al. (1992).

Tapirs

Tapirs are members of the Order Perissodactyla like Horses and Rhinoceroses however their alimentary tract anatomy is different in that the cecum is the most voluminous section of the tract rather than the colon (Hagen et al., 2014). They are browsing herbivores consuming a wide range of different parts of numerous plants including fruits, grasses, leaves, shrubs, stems, and twigs.

Element	Clinical signs	Comments
Copper deficiency	Stillbirths, light haircoat, low serum Cu	Janssen et al. (1999), Janssen (2003)
Iron deficiency	Anemia	In Malayan tapir neonates (Helmick and Milne, 2012)
Iron toxicity	Death, hemochromatosis, hemosiderosis, iron overload	In Baird's, Brazilian and Malayan tapirs (Bonar et al., 2006)

Hagen et al. (2014), Helmick and Milne (2012), Bonar et al. (2006), Janssen (2003), Janssen et al. (1999).

Marine mammals

There are 33 species in the Pinnipeds which are fin-footed aquatic mammals including true seals (family Phocide), fur seals, sea lions (family Otariide), and walrus (family Odobenide). They have simple stomachs.
Cetaceans include Whales, Dolphins, and Porpoises and they have a complex chambered stomach with a forestomach but, like simple stomached animals they rely on acid and digestive enzymes to break down foods prior to absorption.

Deficiencies

Element	Clinical signs	Comments
Copper deficiency		Dugongs in Australia due to seagrass diet
Iron deficiency	Anemia	Fish-induced anemia (due to interreference with Fe absorption) suspected in cetaceans which develop anemia that is unresponsive to oral Fe
Sodium deficiency	Anorexia, collapse (Addisonian crisis), convulsions, death, hypochloridemia, hyponatremia, incoordination, lethargy, tremor, weakness	In captive Pinnipeds—usually kept in fresh water—also in phocid seals, otariids and others

Toxicities

Environmental pollution from Industries is a major issue for the health of all marine life, especially due to heavy metals. The Arctic is heavily contaminated with Cd and the Mediterranean with Hg and these differences are reflected in the body tissue content of dolphins from these different areas. Marine mammals are high up the food chain and so their food (e.g., bivalves) is often already heavily contaminated and further accumulation leads to clinical disease and death.
Routes of exposure include food, inhalation, seawater, via milk (suckling young), transplacental and transdermal.
Accumulation of toxic metals is associated with an increased risk of infection, but studies have shown there are usually several metals involved, not individual elements. In some cases, the relative amounts of different metals (such as Hg:Se:Br ratio) have been found to be a factor leading to ill health. Apparently in vitro dolphin lymphocytes are more resistant to the genotoxic and cytotoxic effects of methylmercury compared to human or rat cells.
Captive Bottlenose dolphins get iron overload and Dolphins in collections have higher iron, ferritin, and transferrin saturation values than free-ranging dolphins.

Element	Clinical signs	Comments
Cadmium toxicity	NR	Despite high accumulation in their bodies
Copper toxicity	Copper accumulation in liver, death	Florida manatees
Iron toxicity	Hemochromatosis, high postprandial insulin, hyperglobulinemia, hyperglycemia, leucocytosis, fatty liver disease	Dolphins
Lead toxicity	Death, liver damage	Bottle-nose dolphin
Mercury toxicity	Anorexia, death, hepatitis, lethargy, renal failure, uremia, weight loss	Seals
	lipofuscin accumulation in liver	Bottlenose dolphins
	Damage to organ of Corti sensory cells	Harp Seals
Nickel toxicity	Stillbirths	Airborne Ni. In ringed seals (Finland)

References

Mazzaro et al. (2012), Venn-Watson et al. (2012), Kakuschke and Prange (2007), Das et al. (2003), Hÿvarinen and Sipilä (1984).

Marsupials

Limited information is available.

This group of animals including kangaroos, opossums, sugar gliders and wallabies. They have special adaptations for minerals, Quokka need 50% less copper and cobalt than sheep while the hairy nose-wombats get copper toxicity indicating a minute requirement for copper in marsupials (Hume, 2005).

Kangaroo

Kangaroos belong to the family Macropodide which comprises 40 species including wallaroos, wallabies, hare-wallabies, nail-tail wallabies, rock-wallabies, pademelons, quokka, tree-kangaroos and swamp wallaby that are grazing and/or browsing herbivores with sacculated stomachs adapted to ferment plant material, so most digestion occurs in the forestomach similar to ruminants.

Kangaroos are adapted to survive in extreme arid conditions. They have physiological and micromorphological adaptations to conserve sodium and other minerals. They actively seek out natural rock licks which contain significantly higher concentrations of magnesium, sodium, and sulfur than surrounding soils.

Little information is available.

Mineral	Deficiency/ toxicity	Clinical signs reported
Selenium (Se)	Deficiency	Myopathy (White muscle disease)
Sodium (Na)	Deficiency	Salt hunger, adrenal hypertrophy
Not Applicable (Element is not essential in this species)	Deficiency	Al, Sb, Ba, Be, Bi, Br, Cd, Cs, Ce, Dy, Er, Eu, Gd, Ga, Ge, Au, Ho, In, La, Pb, Li, Lu, Hg, Nd, Nb, Pd, Pr, Pm, Rh, Rb, Sm, Sc, Ag, Sr, Te, Tb, Tl, Th, Tm, Sn, Ti, W, U, V, Yb, Y, Zr

Koala

Koala (Phascolarctos cinereusare) are not bears they are arboreal herbivorous marsupials. They mainly live in eucalyptus (gum tree) forests in Australia and so, not surprisingly, they mainly eat eucalyptus leaves which are relatively high in fiber and of low nutritional value.

Koalas have a large 2-m long cecum and the microflora there digest the cellulose in the leaves. Whilst some basic work has been done on minerals in Koalas (Schmid et al., 2013) we still know very little about their requirements and mineral-related disease have yet to be reported. Koalas living in regions where soil and eucalyptus are low in sodium content have been observed eating bark which is higher in sodium presumably to meet daily requirements (Au et al., 2017).

In a retrospective review of 519 Koala postmortems conducted between 1997 and 2016 (Gonzalez-Astudillo et al., 2019) there were no nutrition-related diseases identified, although poor body condition was a common finding.

Element	Deficiency/Toxicity	Comments
Aluminium	Toxicity	Accumulation associated with renal failure but cause and effect relationship to be confirmed
Calcium—phosphorus imbalance	Deficiency/toxicity Ca: P ratio	Metabolic bone disease—cause not identified, may be mineral-related or vitamin D deficiency

References

Gonzalez-Astudillo et al. (2019), Au et al. (2017), Pye et al. (2013), Schmid et al. (2013), Haynes et al. (2004)

Opossums

Opossums are omnivores with a very wide spectrum of foods including amphibians, carrion, crayfish, eggs, fish, insects (earwigs, flies), fruit, grass, mice, slugs, snails, vegetables, worms.
The most common nutritional disease in Virginia Opossums is due to Ca deficiency.

Element	Deficiency/toxicity	Clinical signs
Calcium,	Deficiency, inappropriate Ca:P ratio	Nutritional secondary hyperparathyroidism
Phosphorus	Toxicity inappropriate Ca:P ratio	Nutritional secondary hyperparathyroidism

References

Johnson-Delaney (2014b), Hume (2005).

Sugar gliders

Sugar gliders diet is quite specific and includes: acacia gum, arthropods, sap from eucalyptus, nectar, manna, honeydew, and lerp.

Element	Deficiency/toxicity	Clinical signs
Calcium	Deficiency	Osteodystrophy, periodontal disease, tetany
Phosphorus	Deficiency/toxicity	Osteodystrophy

References

Johnson-Delaney (2014a, b), Dierenfeld (2009).

Wombats

Wombats are large herbivorous burrowing marsupials native to Australia.

Element	Deficiency/ toxicity	Clinical signs
Calcium	Toxicity	*Associated with Hypervitaminosis D:* osteodystrophy, soft tissue calcification
Copper	Toxicity	Copper accumulation in liver, death, hemoglobinuria, intravascular hemolysis, hepatic congestion, jaundice
Sodium	Toxicity	*Sodium fluoroacetate:* Heart failure

References

Hume (2005), Barboza and Vanselow (1990), Obendorf (1989), Vanselow and Barboza (1988).

Avian

Birds

With the exception of poultry there is a lack of detailed information about the nutritional requirements of many species of bird.

Nutritional diseases are common in captive birds kept as pets or in zoological collections. The essential nutrient requirements have not been established for all species and there are large differences between species. Captive birds should be fed a ration as close to their natural ration as possible.

Birds can be one of many types of feeder including omnivores, insectivores, carnivores, and some have highly specialized diets, e.g., nectar, pollen.

Flamingos for example are omnivores and eat blue-green and red algae, crustaceans, small fish, insects, larvae, and mollusks.

There are many possible sources of toxic materials in bird environments including batteries, ceramics, coins, glass (stained), jewelry, lead shot, metal toys (e.g., Pb), paints, galvanized, food bowels (Zn), galvanized wire cages (Zn), fishing weights (Pb).

Mineral	Deficiency/ toxicity	Clinical signs reported
Aluminium (Al)	Toxicity	NR
Antimony (Sb)	Toxicity	NR
Arsenic (As)	Deficiency	NR
Arsenic (As)	Toxicity (Acute)	NR
Arsenic (As)	Toxicity (Chronic)	NR
Barium (Ba)	Toxicity	NR
Beryllium (Be)	Toxicity	NR
Bismuth (Bi)	Toxicity	NR
Boron (B)	Toxicity	NR
Bromine (Br)	Toxicity	NR
Cadmium (Cd)	Toxicity	NR
Cesium (Cs)	Toxicity	NR
Calcium (Ca)	Deficiency	Angular deformity of limbs, anorexia, beak malformations, bone demineralization, carpal rotation (*waterfowl*) collapse, egg abnormalities (shape, soft, thin), egg binding, falling from perches, dermatitis, feather abnormalities (brittle, frayed), fractures, poor growth, hypocalcemia, incoordination, leukocytosis, metabolic bone disease (hyperparathyroidism), opisthotonos, osteomalacia, osteoporosis, paralysis, polydipsia, polyuria, regurgitation, reproductive problems, rickets, seizures, skeletal deformities, syncope, tetany, weakness, (especially African Gray Parrots; Timneh parrots)
Calcium (Ca)	Toxicity	Visceral gout (uric acid crystal deposition in organs), nephrocalcinosis, nephropathy, renal fibrosis
Cerium (Ce)	Toxicity	NR
Chlorine (Cl)	Deficiency	NR
Chlorine (Cl)	Toxicity	NR
Chromium (Cr)	Deficiency	NR
Chromium (Cr)	Toxicity	NR
Cobalt (Co)	Deficiency	NR
Cobalt (Co)	Toxicity	NR
Copper (Cu)	Deficiency	Aortic rupture (*?ratites*), convulsions, pale colored feathers, muscle weakness, polydipsia, weakness

Copper (Cu)	Toxicity	NR
Dysprosium (Dy)	Toxicity	NR
Erbium (Er)	Toxicity	NR
Europium (Eu)	Toxicity	NR
Fluorine (F)	Toxicity	NR
Gallium (Ga)	Deficiency	NR
Germanium (Ge)	Toxicity	NR
Gold (Au)	Toxicity	NR
Holmium (Ho)	Toxicity	NR
Indium (In)	Toxicity	NR
Iodine (I)	Deficiency	Circulatory collapse (due compression), crop distension, death, dyspnea, hypothyroidism, goiter, regurgitation, increased expiratory respiratory noise (chirp, wheeze, thyroid follicular hyperplasia, thyroid microfollicular adenoma, thyroiditis (white granulomatous), weight loss
Iodine (I)	Toxicity	Central nervous system signs, depressed growth
Iron (Fe)	Deficiency	?Anemia
Iron (Fe)	Toxicity	Ascites
Lanthanum (La)	Toxicity	NR
Lead (Pb)	Toxicity	Decreased ALAD activity (blood, brain, liver), anemia, anorexia, balance loss, death, dehydration, depression, loss of depth perception, diarrhea (green), dropped wings, emaciation, bright green feces, gastrointestinal signs, hemoglobinuria (Amazon parrots), lethargy, locomotion impaired, neurological signs (seizures), edema (cephalic), esophageal impaction, renal failure, elevated protoporphyrin IX levels (blood, brain, liver), skeletal abnormalities, urinary signs (polydipsia, polyuria), tremors, weakness (neck and limb), weight loss (gradual)
Lithium (Li)	Toxicity	NR
Lutetium (Lu)	Toxicity	NR
Magnesium (Mg)	Deficiency	Angular deformity long bones, convulsions, death (chicks), poor growth, lethargy
Magnesium (Mg)	Toxicity	Reduced egg production, thin egg shells, diarrhea, dermatitis, feather abnormalities (brittle, frayed), irritability
Manganese (Mn)	Deficiency	Ataxia, bone deformities (chondrodystrophy), death (embryos), dermatitis, reduced egg production, feather abnormalities (brittle, frayed), growth impairment, reduced hatchability, perosis, slipped hock tendon, tetanic spasms (chicks), tibiometatarsal joint enlargement, tibial deformity (twisting), slipping of gastrocnemius muscle from condyle (*especially young cranes, gallinaceous and ratites, also cockatiels, raptors, rosellas*).
Manganese (Mn)	Toxicity	NR
Mercury (Hg)	Toxicity	NR
Molybdenum (Mo)	Deficiency	NR
Molybdenum (Mo)	Toxicity	NR
Neodymium (Nd)	Toxicity	NR
Nickel (Ni)	Deficiency	NR
Nickel (Ni)	Toxicity	NR
Niobium (Nb)	Toxicity	NR
Palladium (Pd)	Toxicity	NR
Phosphorus (P)	Deficiency	Angular deformity of limbs, beak malformations, bone demineralization, egg abnormalities (shape), egg binding, fractures, poor growth, metabolic bone disease (hyperparathyroidism), osteomalacia, polydipsia, polyuria, reproductive problems, rickets, skeletal deformities, (especially African Gray Parrots; Timneh parrots)
Phosphorus (P)	Toxicity	Bone fragility, poor eggshell
Potassium (K)	Deficiency	NR

Potassium (K)	Toxicity	NR
Praseodymium (Pr)	Toxicity	NR
Promethium (Pr)	Toxicity	NR
Rhodium (Rh)	Toxicity	NR
Rubidium (Rb)	Toxicity	NR
Samarium (Sm)	Toxicity	NR
Scandium (Sc)	Toxicity	NR
Selenium (Se)	Deficiency	Dermatitis, encephalomalacia, exudative diathesis, feather abnormalities (brittle, frayed), malabsorption, maldigestion, muscular dystrophy, subcutaneous edema, oral paralysis (*cockatiels*), ventricular muscle degeneration
Selenium (Se)	Toxicity	Abdominal distension, ascites, coughing, death, depression, dyspnea, hemochromatosis (iron storage disease) Hemosiderosis (liver) hemochromatosis (iron storage disease—*most commonly passarines,* e.g., *toucans* (*often asymptomatic*)*mynah birds, birds of paradise*) weakness.
Silicon (Si)	Toxicity	NR
Silver (Ag)	Toxicity	NR
Sodium (Na)	Deficiency	Cannibalism, convulsions, death, reduced egg production, feather plucking due dry pruritic skin, gout (visceral), kidney damage, paralysis, pecking behavior, seizures
Sodium (Na)	Toxicity	Diarrhea, muscle weakness, polydipsia, polyuria
Strontium (Sr)	Toxicity	NR
Sulfur (S)	Deficiency	NR
Sulfur (S)	Toxicity	NR
Tellurium (Te)	Toxicity	NR
Terbium (Tb)	Toxicity	NR
Thallium (Tl)	Toxicity	NR
Thorium (Th)	Toxicity	NR
Thulium (Tm)	Toxicity	NR
Tin (Sn)	Toxicity	NR
Titanium (Ti)	Toxicity	NR
Tungsten (W)	Toxicity	NR
Uranium (U)	Toxicity	NR
Vanadium (V)	Toxicity	NR
Ytterbium (Yb)	Toxicity	NR
Yttrium (Y)	Toxicity	NR
Zinc (Zn)	Deficiency	Chondrodystrophy, deformities (fetal), dermatitis (exfoliative legs and face), reduced egg production, feather abnormalities (brittle, frayed), reduced hatchability, skin scaly, swollen hocks
Zinc (Zn)	Toxicity	Anemia, anorexia, cyanosis, death, diarrhea, embryo mortality, feather plucking, gastrointestinal perforation, gastrointestinal ulceration, hematuria, hemoglobinuria, lethargy, peritonitis, polydipsia, polyuria, regurgitation, sepsis, weakness, weight loss
Zirconium (Zr)	Toxicity	NR
Not Applicable (Element is not essential in these species)	Deficiency	Al, Sb, Ba, Be, Bi, Br, Cd, Cs, Ce, Dy, Er, Eu, Gd, Ga, Ge, Au, Ho, In, La, Pb, Li, Lu, Hg, Nd, Nb, Pd, Pr, Pm, Rh, Rb, Sm, Sc, Ag, Sr, Te, Tb, Tl, Th, Tm, Sn, Ti, W, U, V, Yb, Y, Zr

References

Hernández (2014), Markowski et al. (2013), Sanchez-Migallon (2012), Chow and Pollock (2018), Puschner and Poppenga (2009), Lightfoot and Yeager (2008), Richardson (2006), Schoemaker et al. (1999), Bauck (1995), Macwhirter (1984).

Poultry

The domestic chicken has been selectively bred to maximize the production of eggs or meat and the latter breeds have extreme growth rates with appropriate nutritional requirements. Minerals are critically important for the production of healthy eggs and shells, and also to avoid skeletal deformities during rapid growth.

Mineral	Deficiency/ toxicity	Clinical signs reported
Aluminium (Al)	Toxicity	Chicks: Cardiac teratogenesis in embryonic chick heart: defects in ventricular septation and ventricular myocardium Signs of P deficiency as Al binds to P. See P
Antimony (Sb)	Toxicity	NR
Arsenic (As)	Deficiency	Increased arginase activity in kidney, depressed hematocrit (PCV), depressed hemoglobin, reduced plasma uric acid
Arsenic (As)	Toxicity	Increased erythrocyte formation, reduced leukocyte production (Turkeys): reduced weight gain, death
Barium (Ba)	Toxicity	NR
Beryllium (Be)	Toxicity	NR
Bismuth (Bi)	Toxicity	NR
Boron (B)	Deficiency	Impaired growth chicks: (*low Br diet—not confirmed as an essential nutrient yet*)
Boron (B)	Toxicity	Chicks: curled-toe paralysis, skeletal abnormalities. *Due riboflavin deficiency*
Bromine (Br)	Toxicity	NR
Cadmium (Cd)	Toxicity	NR
Cesium (Cs)	Toxicity	NR
Calcium (Ca)	Deficiency	NR
Calcium (Ca)	Toxicity	*Growing:* Food intake reduced, death, gout (visceral), hypercalcemia, hypophosphatemia, nephrosis, sexual maturity delayed, urate urolithiasis (calcium urate), weight gain poor *Laying:* Conflicting studies about effects on egg laying
Cerium (Ce)	Toxicity	NR
Chlorine (Cl)	Deficiency	Dehydration, poor growth rate, high mortality rate, hypochloridemia, tetany
Chlorine (Cl)	Toxicity	NR
Chromium (Cr)	Deficiency	*Turkeys:* Growth disorders
Chromium (Cr)	Toxicity	*Egg embryos:* short and twisted limbs, microphthalmia, exencephaly, short and twisted neck, everted viscera, edema and reduced body size
Cobalt (Co)	Deficiency	reduced hatchability, embryonic death
Cobalt (Co)	Toxicity	NR
Copper (Cu)	Deficiency	Anemia, poor growth rate, death (chicks), internal hemorrhage due ruptured vessels, thickened aorta walls, aneurisms (chicks), cardiomegaly, lameness (chicks), fragile bones, angular deformity metatarsals, thickened epiphyseal cartilage, failure of cartilage cross-linking, enlarged hocks, poor bone mineralization, reduced egg production, reduced hatchability, retarded growth, reduced monoamine oxidase activity, low plasma Cu, shell-less eggs, abnormal egg shape, abnormal egg texture, hypopigmentation
Copper (Cu)	Toxicity	NR
Dysprosium (Dy)	Toxicity	NR
Erbium (Er)	Toxicity	NR
Europium (Eu)	Toxicity	NR

Fluorine (F)	Deficiency	NR
Fluorine (F)	Toxicity	Reduced growth rate and hatchability
Gadolinium (Gd)	Toxicity	NR
Gallium (Ga)	Toxicity	NR
Germanium (Ge)	Toxicity	NR
Gold (Au)	Toxicity	NR
Holmium (Ho)	Toxicity	NR
Indium (In)	Toxicity	NR
Iodine (I)	Deficiency	Rare. *Experimental* poor growth, poor feather condition poor egg production, poor egg size, thyroid enlargement in embryos, weight gain, small testes, lack of sperm, decreased size of comb in males
Iodine (I)	Toxicity	reduced egg production, reduced egg size, reduced hatchability
Iron (Fe)	Deficiency	*Rare*: anemia, abnormal erythrocytes, reduced growth rate, low hemoglobin, low hematocrit, lack of pigment in feathers (achromatrichia—seen in Chickens and Turkeys), poor feathers, embryonic mortality, Chicks: high serum triglycerides, low muscle myoglobin
Iron (Fe)	Toxicity	NR
Lanthanum (La)	Toxicity	*Newborn Chicks:* reduced activity of Ca2+-ATPase, Mg2+-ATPase, and cholinesterase, inhibition of calcium binding to brain synaptosomal membrane
Lead (Pb)	Toxicity	Death, erythrocyte changes, liver damage, renal intranuclear inclusions
Lithium (Li)	Deficiency	NR
Lithium (Li)	Toxicity	NR
Lutetium (Lu)	Toxicity	NR
Magnesium (Mg)	Deficiency	Rare. Poor growth, lethargy, ataxia, convulsions, coma, death, panting, gasping, loss of muscle tone, poor feathering, poor survivability, reduced egg production, poor egg quality, hypomagnesemia, hypocalcemia.
Magnesium (Mg)	Toxicity	Decrease egg production, thin egg shells, depressed growth, diarrhea, increased mortality
Manganese (Mn)	Deficiency	*Chickens:* Embryo: chondrodystrophy Chicks and poults: ataxia (neonatal), perosis (slipped tendon—enlarged tibiometatarsal joint; distortion of tibia, thickening and shortening long bones, gastrocnemious tendon slips off condyles, reluctant to walk, walk on hocks, die Chondrodystrophy: abnormal skull development (globular) short thick limbs and wings, short lower beak "parrot beak," death, adults: reduced egg production, poor eggshells, poor hatchability *Ducks:* perosis (slipped tendon)—enlarged tibiometatarsal joint; distortion of tibia, thickening and shortening long bones, gastrocnemious tendon slips off condyles, reluctant to walk, walk on hocks, die. Chondrodystophy: abnormal skull development (globular) short thick limbs and wings, short lower beak "parrot beak," death
Manganese (Mn)	Toxicity	NR
Mercury (Hg)	Toxicity	NR
Molybdenum (Mo)	Deficiency	Abnormal growth and development (impaired ossification of long bones; changes in articular cartilage with inability to move, feather defects and loss, high embryonic mortality, reduced viability of chicks
Molybdenum (Mo)	Toxicity	*Chicks:* Anemia, growth retardation
Neodymium (Nd)	Toxicity	NR
Nickel (Ni)	Deficiency	Low hematocrit, liver lesions
Nickel (Ni)	Toxicity	Reduced feed intake
Niobium (Nb)	Toxicity	NR

Palladium (Pd)	Toxicity	NR
Phosphorus (P)	Deficiency	NR
Phosphorus (P)	Toxicity	NR
Potassium (K)	Deficiency	*Experimental:* muscle weakness, ileus, cardiac failure, respiratory failure. Chicks: reduced feed intake, impaired growth, high mortality, reduced bone mineralization, reduced protein metabolism, weakness, tetanic seizures, paralysis, death, polyuria, renal and ureteral lesions *Laying hens*: reduced egg production, egg weight, shell thickness, albumin content, weakness, collapse, death
Potassium (K)	Toxicity	NR
Praseodymium (Pr)	Toxicity	NR
Promethium (Pm)	Toxicity	NR
Rhodium (Rh)	Toxicity	NR
Rubidium (Rb)	Toxicity	NR
Samarium (Sm)	Toxicity	NR
Scandium (Sc)	Toxicity	NR
Selenium (Se)	Deficiency	NR
Selenium (Se)	Toxicity	Failure of eggs to hatch due embryo deformities, increased mortality rate, reduced food intake, weight loss
Silicon (Si)	Deficiency	*Chicks:* retarded growth, reduced glycosaminoglycans and collagen concentrations in tibia, impaired bone and cartilage formation, pathological changes in epiphyseal cartilage
Silicon (Si)	Toxicity	NR
Silver (Ag)	Toxicity	*Chicks*: Reduced aortic elastin, cardiomegaly, growth depression, low hemoglobin, death *Ducks, Turkeys:* Signs of Vit E/Se deficiency—see Se
Sodium (Na)	Deficiency	poor growth, undermineralized (soft) bones, corneal keratinization, gonadal inactivity, adrenal hypertrophy, impaired feed utilization, decreased plasma volume, inappetence increased water consumption, impaired protein, and energy metabolism. Laying hens: weight loss, cannibalism, reduced egg production and egg size, reduced hatchability
Sodium (Na)	Toxicity	NR
Strontium (Sr)	Toxicity	NR
Sulfur (S)	Deficiency	NR
Sulfur (S)	Toxicity	NR
Tellurium (Te)	Toxicity	NR
Terbium (Tb)	Toxicity	NR
Thallium (Tl)	Toxicity	NR
Thorium (Th)	Toxicity	NR
Thulium (Tm)	Toxicity	NR
Tin (Sn)	Toxicity	NR
Titanium (Ti)	Toxicity	NR
Tungsten (W)	Toxicity	NR
Uranium (U)	Toxicity	NR
Vanadium (V)	Toxicity	Depressed growth rate, death
Ytterbium (Yb)	Toxicity	NR
Yttrium (Y)	Toxicity	NR

Zinc (Zn)	Deficiency	Anorexia, breathing labored, comb color, dermatitis, embryonic deformities, the curvature of the spine, death, growth impairment, lower feed efficacy, hyperkeratosis, low egg production, delayed egg production, posture abnormal, short thick leg bones, hock enlargement, micromelia, scaly skin, poor feathering, broken feathers, stiff gait, reproductive failure, tachypnea, vertebral fusion, toes missing, agenesis of skeleton or limbs, weakness
Zinc (Zn)	Toxicity	*Chicks:* depressed growth rate, depressed growth, death
Zirconium (Zr)	Toxicity	NR
Not Applicable (Element is not essential in this species)	Deficiency	Al, Sb, Ba, Be, Bi, Br, Cd, Cs, Ce, Dy, Er, Eu, Gd, Ga, Ge, Au, Ho, In, La, Pb, Li, Lu, Hg, Nd, Nb, Pd, Pr, Pm, Rh, Rb, Sm, Sc, Ag, Sr, Te, Tb, Tl, Th, Tm, Sn, Ti, W, U, V, Yb, Y, Zr

References

Scott et al. (1982).

Reptiles

There are over 8000 species of reptile and the most common mineral-related disease is nutritional secondary hyperparathyroidism due to Ca deficiency. One reason is that commonly used foods notably meat, pinkies, day old chicks, crickets and mealworms have low Ca content and an inverse Ca:P ratio. A ratio of 2:1 is recommended for growing reptiles, 1.5:1 for adults and during egg laying periods 10:1 may be needed. Feeding prey with a Ca supplement for a few days to gut load them, or dusting them with Ca powder (less reliable) can improve Ca status.
For vegetarian species calcium containing greens like dandelions, kale, or lettuce should be fed, and high oxalate containing plants avoided: spinach, rhubarb, cabbage, peas, potatoes and beet greens as this binds Ca. Celery, cucumber and lettuce are naturally low in Ca content.
Lack of dietary vitamin D, or lack of exposure to sunlight (or UV light sources) can cause low calcium absorption from the gut.
Toxicities have been reported from ingestion of common household items:
Pb—paint, fishing line sinkers, ammunition.
Zn—ointments, pennies, thermometer weights, zinc supplements.
Limited information is available.

Mineral	Deficiency/ toxicity	Clinical signs reported
Calcium (Ca)	Deficiency *Common*	Anorexia, poor bone mineralization, prolapsed cloaca, death (due asphyxiation respiratory failure), dystocia, eggs poorly mineralized, hypocalcemia, limb fractures (spontaneous), muscle twitches (toes and legs). neuromuscular signs, osteopenia, hind limb paresis, limb deformity (bowing due weakness, healing fractures), paralysis respiratory muscles, poor shell development (Chelonia), shell deformity, swollen limbs, mandibular swelling, muscles flabby, seizures, spinal deformity, tetany *Can be secondary to vitamin D deficiency (insufficient in the diet or in renal or liver disease), excess P, hypoparathyroidism, due parathyroid glands injury or their removal after surgery.*
Calcium (Ca)	Toxicity	abdominal pain, cardiac disorders, fatigue, muscle weakness, nausea, acute pancreatitis, renal hypertension, renal urolithiasis *In addition to dietary excess causes include: hyperthyroidism, hyperparathyroidism activation of vitamin D3 excess absorption of calcium* via *the GI tract and excessive bone resorption.* *Excessive calcium also reduces the use of proteins, fats, phosphorus, magnesium, zinc, iron, iodine and can lead to soft tissue mineralization if in conjunction with inadequate or excessive Vitamin D3 levels.*
Dysprosium (Dy)	Deficiency	NR
Dysprosium (Dy)	Toxicity	NR
Erbium (Er)	Deficiency	NR
Erbium (Er)	Toxicity	NR
Europium (Eu)	Deficiency	NR
Iodine (I)	Deficiency	Goiter, myxedema *Common in tortoises (especially Giant Land Tortoises), reptiles fed brassica plants (broccoli, brussels sprouts, cabbage, cauliflower, kale. Rape, swede, turnip) high in goitrogens.*
Iodine (I)	Toxicity	Goiter
Iron (Fe)	Deficiency	Anemia (hypochromic), anorexia, depression, lethargy, pallor, prolonged hibernation or brumation
Iron (Fe)	Toxicity	NR
Lead (Pb)	Toxicity	Anemia (hemolytic), anorexia, depression, gastrointestinal signs, lethargy, neurological signs, pallor, prolonged hibernation or brumation

Phosphorus (P)	Deficiency	Metabolic bone disease, poor bone calcification, renal damage, urolithiasis
Phosphorus (P)	Toxicity	nutritional secondary hyperparathyroidism. Bone demineralization, renal damage, urolithiasis
Potassium (K)	Deficiency	NR
Potassium (K)	Toxicity	NR
Praseodymium (Pr)	Deficiency	NR
Praseodymium (Pr)	Toxicity	NR
Promethium (Pr)	Deficiency	NR
Promethium (Pr)	Toxicity	NR
Rhodium (Rh)	Deficiency	NR
Rhodium (Rh)	Toxicity	NR
Rubidium (Rb)	Deficiency	NR
Rubidium (Rb)	Toxicity	NR
Samarium (Sm)	Deficiency	NR
Samarium (Sm)	Toxicity	NR
Scandium (Sc)	Deficiency	NR
Scandium (Sc)	Toxicity	NR
Selenium (Se)	Deficiency	Myopathy, impaired immune function, neurological signs, impaired vision,
Selenium (Se)	Toxicity	abnormalities to the nails and skin and softening of the bones, selenosis
Silicon (Si)	Deficiency	NR
Silicon (Si)	Toxicity	NR
Silver (Ag)	Deficiency	NR
Silver (Ag)	Toxicity	NR
Sodium (Na)	Deficiency	NR
Sodium (Na)	Toxicity	NR
Strontium (Sr)	Deficiency	NR
Strontium (Sr)	Toxicity	NR
Sulfur (S)	Deficiency	NR
Sulfur (S)	Toxicity	NR
Tellurium (Te)	Deficiency	NR
Tellurium (Te)	Toxicity	NR
Terbium (Tb)	Deficiency	NR
Terbium (Tb)	Toxicity	NR
Thallium (Tl)	Deficiency	NR
Thallium (Tl)	Toxicity	NR
Thorium (Th)	Deficiency	NR
Thorium (Th)	Toxicity	NR
Thulium (Tm)	Deficiency	NR
Thulium (Tm)	Toxicity	NR
Tin (Sn)	Deficiency	NR

Tin (Sn)	Toxicity	NR
Titanium (Ti)	Deficiency	NR
Titanium (Ti)	Toxicity	NR
Tungsten (W)	Deficiency	NR
Tungsten (W)	Toxicity	NR
Uranium (U)	Deficiency	NR
Uranium (U)	Toxicity	NR
Vanadium (V)	Deficiency	NR
Vanadium (V)	Toxicity	NR
Ytterbium (Yb)	Deficiency	NR
Ytterbium (Yb)	Toxicity	NR
Yttrium (Y)	Deficiency	NR
Yttrium (Y)	Toxicity	NR
Zinc (Zn)	Deficiency	NR
Zinc (Zn)	Toxicity	Anemia (hemolytic), anorexia, depression, gastrointestinal signs, lethargy, neurological signs, pallor, prolonged hibernation, or brumation
Zirconium (Zr)	Deficiency	NR
Zirconium (Zr)	Toxicity	NR

Amphibians

Good scientific data are lacking on the nutritional requirements for amphibian species. In their natural habitat, many are insectivores but the transdermal supply of nutrients from water is also important. In general, most commercially available insect sources fed to amphibians are nutritionally adequate for minerals except they are low in calcium content. As a result gut feeding Ca to the insects or dusting them with Ca is commonly practized in an attempt to increase intake. Insects like crickets can remove dust from their surfaces within minutes of application so this is less reliable than feeding the insects for a short period before they are given.
Amphibians have highly adapted skin with a thin stratum corneum layer making it permeable to fat-soluble substances, water, and water-soluble substances. Amphibians require a consistent electrolyte balance in their environment especially Na concentration with some adapted to saline water others to fresh water habitats. Like Na, Ca is transported by active transport systems transdermally and stored in paravertebral lime sacs for use when need, such as during metamorphosis. High F or P content in water supplies may impair Ca availability. Potassium is mainly obtained from the diet.
The causes of common diseases in amphibians, for example, skin disease and kidney disease have not been determined and nutritional factors including mineral imbalances may be involved in some of them.
Heavy metal contamination is recognized as a serious health risk to amphibians and there is now growing concerned about the health risk posed by rare earth metal accumulation in the environment, as reported in marine life in two systematic reviews (Blinova et al., 2020; Malhotra et al., 2020).

Mineral	Deficiency/ toxicity	Clinical signs reported
Calcium (Ca)	Deficiency	Poor bone mineralization, limb fractures, metabolic bone disease, spinal deformities (e.g., scoliosis) tetany, weakness
Chromium (Cr)	Toxicity	Ovarian dysfunction, oestradiol and progesterone levels depressed; follicle stimulating hormone (FSH) and luteinizing hormone levels were found significantly elevated.
Fluorine (F)	Toxicity	Metabolic bone disease (Frogs)
Iodine (I)	Toxicity	Excess iodine added to water may cause species such as the axolotl to shed its external gills
Lead (Pb)	Toxicity	*Frogs:* death, erythrocytopenia, excitation, hypersalivation, leukocytopenia, metamorphosis delayed (tadpoles), monocytopenia, loss of muscle tone, muscle twitches, neutrocytopenia, skin sloughing, sluggishness
Nickel (Ni)	Toxicity	*Salamander:* melanoma (after intraocular injection)

References

Blinova et al. (2020), Malhotra et al. (2020), AZA Amphibian Husbandry Guide (2020), Ferrie et al. (2014), Shaw et al. (2012).

Fish

Fish have simple gastrointestinal tracts but some lack acid production in the stomach reducing their ability to digest bone and so reduced Ca and P bioavailability. In plants P is often bound to phytic acid rendering it nonavailable.
Many essential minerals are absorbed from environmental water so, for example, Mg deficiency has not been reported in seawater fish, whereas it has in freshwater fish.
Fishbone homeostasis is not regulated by parathyroid hormone as it is in terrestrial animals, instead, hypocalcium and hypercalcium hormones are involved and calcium is regulated by the gills, fins, and oral mucosa. Fish obtain their Ca requirement from the water, but P has to be supplied in food material.
Fishbone is either cellular similar to other vertebrates (e.g., Clupeideor, Cyprinade, Salmonide) or acellular (e.g., Perciformes) which lacks osteocytes, osteoblasts move away from the bone once formed so the bone cannot remodel.

Deficiency

Mineral or trace element deficiency can occur due to:

a. Inadequate amounts of an essential nutrient in the food supply
b. Reduced bioavailability of the essential nutrient due to:
  a. Dietary imbalances.
  b. The source and form of the element
  c. Stored reserves in the body
  d. Antagonism from other elements in the gastrointestinal tract or in body tissues. High Ca reduces Fe, Mn, P, and Zn availability.

Toxicity

Maximum tolerable levels (for example, Mg, Se) can vary between fish species and probably reflects different gastrointestinal tract anatomy and physiology, such as the presence or absence of gastric acid.
The higher up a food chain the fish is the more likely it is to ingest high levels of minerals and other toxic elements that have accumulated in prey lower down the food chain. Bottom feeders in rivers and estuaries are most likely to be exposed to high levels of pollutants especially heavy metals from industry and farming.
For captive fish hard water (which is common in domestic water supplies in the UK and many other parts of the World) contains high levels of calcium and magnesium which contributes to total intake and may have a role in disease.
There is little evidence about the role of trace elements and rare earth elements in fish diseases. Accumulation of elements seen in fish include: As, Cu, Cd, F, Pb, Hg. Mo, Se, and there has been a recent systematic review of minerals in fish (Prabhu et al., 2016).

Mineral	Deficiency/ toxicity	Clinical signs reported
Aluminium (Al)	Toxicity	NR
Antimony (Sb)	Toxicity	NR
Arsenic (As)	Deficiency	Death (increased mortality rate), reduced food utilization impaired growth, increased liver enzymes activity—SGOT and SGPT
Arsenic (As)	Toxicity	Impaired growth, erythrocytopenia, low hematocrit, low hemoglobin
Barium (Ba)	Toxicity	NR
Beryllium (Be)	Toxicity	NR
Bismuth (Bi)	Toxicity	NR
Boron (B)	Deficiency	NR
Boron (B)	Toxicity	NR
Bromine (Br)	Toxicity	NR

Cadmium (Cd)	Toxicity	Death (increased mortality rate), reduced food utilization impaired growth, increased liver enzymes activity—SGOT and SGPT
Cesium (Cs)	Toxicity	NR
Calcium (Ca)	Deficiency	Food intake reduced; growth depressed
Calcium (Ca)	Toxicity	Cataracts, dwarfism, reduced growth
Cerium (Ce)	Toxicity	NR
Chlorine (Cl)	Deficiency	NR
Chlorine (Cl)	Toxicity	NR
Chromium (Cr)	Deficiency	NR
Chromium (Cr)	Toxicity	Death (increased mortality rate), reduced food utilization impaired growth, increased liver enzymes activity—SGOT and SGPT
Cobalt (Co)	Deficiency	NR
Cobalt (Co)	Toxicity	NR
Copper (Cu)	Deficiency	Depressed growth
Copper (Cu)	Toxicity	Death, reduced feed efficiency, reduced growth, high liver copper levels
Dysprosium (Dy)	Toxicity	NR
Erbium (Er)	Toxicity	NR
Europium (Eu)	Toxicity	NR
Fluorine (F)	Deficiency	NR
Fluorine (F)	Toxicity	Reduced hatching time
Gadolinium (Gd)	Toxicity	NR
Gallium (Ga)	Toxicity	NR
Germanium (Ge)	Toxicity	NR
Gold (Au)	Toxicity	NR
Holmium (Ho)	Toxicity	NR
Indium (In)	Toxicity	NR
Iodine (I)	Deficiency	Goiter (due to thyroid hyperplasia), mortality increased
Iodine (I)	Toxicity	NR
Iron (Fe)	Deficiency	Microcytic anemia, low hematocrit, low hemoglobin, low plasma iron, low transferrin
Iron (Fe)	Toxicity	Diarrhea, reduced growth, hepatocellular damage increased mortality
Lanthanum (La)	Toxicity	Embryotoxic and developmental effects reduced survival and hatching rates, tail malformations.
Lead (Pb)	Toxicity	Death (increased mortality rate), reduced food utilization impaired growth, increased liver enzymes activity—SGOT and SGPT
Lithium (Li)	Toxicity	NR
Lutetium (Lu)	Toxicity	NR
Magnesium (Mg)	Deficiency	Anorexia, cataracts, convulsions, death, gill filament degeneration, poor growth, lethargy, reduced tissue Mg content, muscle fiber degeneration, nephrocalcinosis, pyloric ceca epithelium degeneration, spinal curvature
Magnesium (Mg)	Toxicity	NR

Manganese (Mn)	Deficiency	Reduced growth and skeletal abnormalities, short body dwarfism (rainbow trout) *Rainbow trout*: reduced copper-zinc superoxide dismutase activities, reduced manganese-superoxide dismutase in cardiac muscle and liver and suppressed manganese and calcium concentrations of the vertebrae poor hatchability and low manganese concentration in the eggs
Manganese (Mn)	Toxicity	NR
Mercury (Hg)	Toxicity	NR
Molybdenum (Mo)	Deficiency	NR
Molybdenum (Mo)	Toxicity	Death of fertilized eggs (Rainbow trout),
Neodymium (Nd)	Toxicity	NR
Nickel (Ni)	Deficiency	NR
Nickel (Ni)	Toxicity	Death (increased mortality rate), reduced food utilization impaired growth, increased liver enzymes activity—SGOT and SGPT
Niobium (Nb)	Toxicity	NR
Palladium (Pd)	Toxicity	NR
Phosphorus (P)	Deficiency	Alkaline phosphatase increases, reduced feed utilization, poor growth, poor bone mineralization, low hematocrit, skeletal abnormalities (Atlantic salmon: bone weakness, spinal curvature; Carp: cephalic deformities frontal bones; Haddock, Halibut: scoliosis Atlantic halibut and haddock juveniles: twisted spines)
Phosphorus (P)	Toxicity	Poor growth, skeletal deformities
Potassium (K)	Deficiency	*Chinook Salmon:* anorexia, convulsions, death, tetany
Potassium (K)	Toxicity	NR
Praseodymium (Pr)	Toxicity	NR
Promethium (Pm)	Toxicity	NR
Rhodium (Rh)	Toxicity	NR
Rubidium (Rb)	Toxicity	NR
Samarium (Sm)	Toxicity	NR
Scandium (Sc)	Toxicity	NR
Selenium (Se)	Deficiency	Growth depression (rainbow trout) (Hilton et al., 1980) and catfish—with vitamin E deficiency) muscular dystrophy in Atlantic salmon, exudative diathesis in rainbow trout Glutathione peroxidase activity in plasma and liver
Selenium (Se)	Toxicity	Reduced activity, reduced growth, poor feed efficiency, high mortality, renal calcinosis, teratogenic malformations
Silicon (Si)	Deficiency	NR
Silicon (Si)	Toxicity	NR
Silver (Ag)	Toxicity	NR
Sodium (Na)	Deficiency	NR
Sodium (Na)	Toxicity	*Coho Salmon and Rainbow Trout*: Reduced growth and feed efficiency
Strontium (Sr)	Toxicity	NR
Sulfur (S)	Deficiency	NR
Sulfur (S)	Toxicity	NR
Tellurium (Te)	Toxicity	NR

Terbium (Tb)	Toxicity	NR
Thallium (Tl)	Toxicity	NR
Thorium (Th)	Toxicity	NR
Thulium (Tm)	Toxicity	NR
Tin (Sn)	Toxicity	Death (increased mortality rate), reduced food utilization impaired growth, increased liver enzymes activity—SGOT and SGPT
Titanium (Ti)	Toxicity	Death (increased mortality rate), reduced food utilization impaired growth, increased liver enzymes activity—SGOT and SGPT
Tungsten (W)	Toxicity	NR
Uranium (U)	Toxicity	NR
Vanadium (V)	Toxicity	NR
Ytterbium (Yb)	Toxicity	*Zebrafish:* embryotoxic and developmental effects reduced survival and hatching rates, and induced tail malformations *Goldfish:* Liver catalase activity was significantly decreased
Yttrium (Y)	Toxicity	NR
Zinc (Zn)	Deficiency	*Rainbow trout:* Cataracts, death, short body dwarfism, reduced growth, fin and skin erosions *Carp:* cataracts, poor growth, loss appetite, high mortality, skin and fin erosions *Catfish:* reduced appetite, reduced growth rate, low bone calcium, low bone zinc, low serum zinc, reduced egg production and hatchability
Zinc (Zn)	Toxicity	Death (increased mortality rate), reduced food utilization impaired growth, reduced hemoglobin, hematocrit, and hepatic copper concentrations, increased liver enzymes activity—SGOT and SGPT
Zirconium (Zr)	Toxicity	
Not Applicable (Element is not essential)	Deficiency	Al, Sb, Ba, Be, Bi, Br, Cd, Cs, Ce, Dy, Er, Eu, Gd, Ga, Ge, Au, Ho, In, La, Pb, Li, Lu, Hg, Nd, Nb, Pd, Pr, Pm, Rh, Rb, Sm, Sc, Ag, Sr, Te, Tb, Tl, Th, Tm, Sn, Ti, W, U, V, Yb, Y, Zr

References

Baeverfjord et al. (2019), Han et al. (2019), Gheorghe et al. (2017), Prabhu et al. (2016), NRC (2011, 1993), Lall and Lewis-McCrea (2007).

Hilton JW, Hodson PV, Slinger SJ. The requirement and toxicity of selenium in rainbow trout (Salmo gairdneri). J Nutr 1980;110(12):2527–35. https://doi.org/10.1093/jn/110.12.2527

Appendix 1

Treatments suggested for managing mineral/trace element/rare earth element deficiencies or toxicities

Please note: For human/veterinary use drugs/doses please consult your local formulary and/or licensing authority.

This list only includes treatments to help in cases of deficiency or toxicity and does not include medications that may be required to manage specific clinical signs.

Product	Action	Use(s)	Dose suggestion for small animals (Gfeller and Messonnier, 1998)
Aluminum hydroxide	Phosphate (P) binder	Reduces P absorption—used with a low P ration	30–90 mg/kg body weight every 8–24 h orally with meals
Apomorphine	Emetic	Remove ingested toxins	Dogs: 0.04 mg/kg body weight IV or 0.08 mg/kg SC
Ascorbic acid		Used for copper hepatotoxicity	Dogs: 25 mg/kg body weight orally
Calcitonin	Hormone stimulates osteoblast activity	Hypercalcaemia	Dogs: 4–6 IU/kg body weight SC or IM every 2–12 h
Calcium carbonate or calcium lactate	Calcium supplement	Hypocalcaemia	Carbonate: 100–150 mg/kg body weight orally in 2–3 doses per day Lactate: 130–200 mg/kg body weight orally every 8 h
Calcium chloride (10%) or calcium gluconate	Calcium supplement	Hypocalcaemia, hyperkalaemia	**Must monitor for bradycardia** Chloride: 0.1–0.3 mL/kg body weight IV over 10–30 min Gluconate: 0.25–1.5 mL/kg body weight IV slowly 5–30 min
Calcium EDTA		Lead poisoning	25 mg/kg body weight SC every 6 h for 20 doses
Castor oil	Cathartic	Remove toxins in gastrointestinal tract by defaecation	Cats: 4–10 mL orally Dogs: 8–30 mL orally
Cholestyramine	Iron-binding resin		Dog: 200–300 mg/kg body weight orally every 12 h
Deferoxamine (desferoxamine)	FDA licensed Iron chelator See note below	Iron toxicosis Aluminum toxicity (Smith, 2013)	15 mg/kg body weight/hour IV Adverse effects; hypersensitivity reactions, hypotension, acute respiratory distress syndrome (ARDS), renal failure

Continued

Product	Action	Use(s)	Dose suggestion for small animals (Gfeller and Messonnier, 1998)
Diethylenetriaminepenta acetate (DTPA),	Iron-chelating agent See below	Smith (2013): cadmium toxicity ? cobalt toxicity	1 g Ca-DTPA IV
Dimercaprol (2,3-dimercaptopropanol) also British anti-Lewisite (BAL)	It is a chelator and in the USA is licensed for arsenic, gold, and mercury poisoning and also licensed for use in acute lead poisoning when given concomitantly with edetate calcium disodium and in Wilson's disease in people in which Cu accumulates in the liver See note below	In the UK Dimercaprol Injection indicated in adults and children for the treatment of acute poisoning by antimony, arsenic, bismuth, gold, mercury and possibly thallium. There is evidence that when used in conjunction with sodium calcium edetate, but not by itself, it can be successful in the treatment of lead poisoning, particularly in children. ***Contraindications:*** cadmium, iron, or selenium toxicity	**a. Arsenic toxicity**: 3–4 mg/kg body weight IM every 8 h, up to 6–7 mg/kg if severe acute toxicosis **b. Lead toxicity:** 2.5 mg/kg body weight IM every 4 h days 1–2, every 8 h day 3, then every 12 h. Up to 5 mg/kg on day 1 only if very severe acute toxicosis *Common side-effects:* fever, hypertension, pain at injection site, vomiting. It can also be nephrotoxic Refer to the product's local SPC in your country for further details https://www.medicines.org.uk/emc/product/6702
DMPS 2,3-Dimercapto-1-propanesulfonic acid (Unithiol)	A chelating agent See note below	Not licensed for use in UK	**Not recommended**
DMSA meso-dimercaptosuccinic acid (chemet, succimer)	A chelating agent See note below	FDA licensed for Pb poisoning in children Used to treat As, Pb, and Hg toxicities. Not licensed for use in UK Smith (2013) says cadmium toxicity FDA has not found this to be safe and effective, and labeling claims have not been approved. (www.drugs.com 2nd July 2020)	**Not recommended** 10 mg/kg body weight every 8 h orally or parenterally *Common side effects*: neutropenia, skin rash, vomiting, diarrhea, rash, and low blood neutrophil levels. Also allergic reactions and liver problems.
Diphenylthiocarbazone (dithizone)	FDA approved chelator	Thallium toxicosis	50–70 mg/kg body weight orally every 8 h
EDTA (calcium disodium salt) Ethylenediaminetetraacetic acid	Approved chelator by FDA for treatment of lead poisoning by parenteral injection	Smith (2013): Cadmium toxicity ? cobalt toxicity	Adverse effects: acute renal failure, increase in hepatic transaminases, hypotension, cardiac arrhythmias, and allergic reactions
EDTA (Edetate disodium	Was FDA approved chelator for hypercalcaemia but not now	Withdrawn due to deaths caused by hypocalcaemia	No longer available in US

Continued

Product	Action	Use(s)	Dose suggestion for small animals (Gfeller and Messonnier, 1998)
Ferrous sulfate	Iron supplement	Iron deficiency	Cat: 50–100 mg/day orally Dog: 100–300 mg/day orally
Glyceryl monoacetate		Sodium fluoroacetate toxicity	0.55 mg/kg body weight IM every hour to maximum dose of 2–4 mg/kg body weight
Insulin (Regular, crystalline)	Hormone	Hyperkalaemia	0.5–1 Unit/kg body weight IV **with** concurrent 2 g dextrose per Unit of Insulin over 30 min
Magnesium chloride	Magnesium supplement	Hypomagnesaemia	**a.** 1 mEq/kg body weight/day in IV fluids **b.** If life-threatening: 0.15–0.3 mEq/kg IV slowly over 5–15 min
Magnesium sulfate	**a.** Cathartic **b.** Magnesium supplement	**a.** Toxicosis—encourage removal of toxin by defaecation **b.** Hypomagnesaemia—magnesium supplement	**a.** Cat: 200 mg/kg body weight orally Dog: 250–500 mg/kg body weight orally **b.** Life-threatening: 0.15–0.3 mEq/kg body weight IV over 5–15 min (sterile solution - NOT Epsom salts)
N-acetyl cysteine	Licensed (FDA) as an antidote for acetaminophen (paracetamol) overdose	Smith (2013) ? cobalt toxicity	
Oral colonic lavage solutions, e.g., polyethylene glycol	Lavage solutions (various)	Remove toxins from gastrointestinal tract	22–33 mL/kg body weight orally, or 25–40 mL/kg body weight orally, followed by nasogastric tube administration of 0.5 mL/kg/h
D-Penicillamine	Licensed by FDA as a chelator	Licensed for Wilson's disease copper (Cu) storage disease Used off label for lead (Pb) toxicosis *Adverse effects include*: aplastic anemia, haematuria, leukopenia, nephrotic syndrome, proteinuria, thrombocytopenia	Dog: Cu: 10–15 mg/kg body weight orally every 12 h Pb: 8 mg/kg body weight orally every 6 h or 10–55 mg/kg body weight orally every 12 h
Potassium chloride	Potassium supplement	Hypokalaemia	0.1–0.2 mL/kg body weight orally every 8 h, or 0.5 mEq/kg body weight/hour IV (maximum dose)
Potassium gluconate	Potassium supplement	Hypokalaemia	5–8 mEq/kg orally every 12–24 h
Prussian blue	Chelator	FDA licensed for treatment of thallium and cesium poisoning	

Continued

Product	Action	Use(s)	Dose suggestion for small animals (Gfeller and Messonnier, 1998)
Sodium sulfate	Cathartic	Removal of toxin by defaecation	Cat: 200 mg/kg body weight orally Dog: 250–500 mg/kg body weight orally
Sorbitol	Cathartic	Removal of toxin by defaecation	4 g/kg body weight orally
Tetramine		Copper hepatotoxicity	Dog: 10–15 mg/kg body weight orally every 12 h
Thiamine B_1	Vitamin	Lead poisoning	1–2 mg/kg body weight IM, or 2 mg/kg body weight orally every 24 h
Zinc acetate	Competitor for Copper	Copper hepatotoxicosis	Dog: 5–10 mg/kg body weight SC every 12 h

Chelators

There are 11 licensed chelators by the FDA in the US: dimercaprol (BAL); edetate calcium disodium (calcium EDTA); succimer (DMSA); penicillamine; trientine hydrochloride; deferoxamine mesylate; deferiprone; deferasirox; pentetate calcium trisodium (Ca-DTPA); pentetate zinc trisodium (Ca-DTPA); prussian blue (Radiogardase).

According to Smith (2013): while a metal might "technically" be capable of chelation (and readily demonstrable in urine or feces), this is an insufficient endpoint. Clinical relevance must be affirmed.

Aluminum toxicity: Deferoxamine is an accepted chelator for appropriately documented aluminum toxicity.

Cadmium toxicity: small treatment window in order to address chelation in cadmium toxicity. In acute toxicity, while no definitive chelation benefit is described, succimer (DMSA), diethylenetriaminepentaacetate (DTPA), and potentially ethylenediaminetetraacetic acid (EDTA) have been considered. In chronic toxicity, chelation is unsupported.

Chromium toxicity: There is little evidence to suggest that currently available chromium chelators are efficacious.

Cobalt toxicity scant human evidence exists with which to provide recommendation for cobalt chelation. DTPA has been recommended for cobalt radionuclide chelation, although DMSA, EDTA, and *N*-acetylcysteine have also been suggested.

Uranium toxicity: DTPA is unsupported for uranium chelation. Sodium bicarbonate is currently recommended, although animal evidence is conflicting.

Appendix 2

Managing acute exposures

General first aid recommendations to manage patients following exposure by various routes to toxic levels of minerals, trace elements, rare earth elements, or other metals.

When entering contaminated premises or handling contaminated patients wear appropriate personal protective equipment (PPE)—masks, gowns, chemical-resistant gloves, and shoes, avoid contact with a toxic substance, remove patient to a safe uncontaminated area.

Route of exposure	Action to take	Comments
Eye contact	**Immediately** flush eyes with copious amounts of water for at least 15 min, include under eyelids Seek immediate medical or veterinary attention if not present Ascertain exactly what substance has come into contact with the eyes	Prevent reexposure Wear splash resistant safety goggles with a face shield In a workplace an eyewash station should be available
Ingestion (In food, water or direct)	If possible, ascertain exactly what has been ingested, and how much Contact a poisons center immediately for instructions Do not induce vomiting, perform gastric lavage, or administer oral medications (e.g., activated charcoal) until confirmed that they are not contraindicated Give small amount of water to rinse out mouth Never give liquids to a patient that is unconscious or unaware If vomiting occurs, keep head lower than hips to prevent aspiration If not breathing, give artificial respiration by qualified personnel Seek immediate medical/veterinary attention if not already present	Caustic substances can cause very severe, life-threatening damage to the gastrointestinal tract if not treated appropriately and promptly
Inhalation	If adverse effects occur, move patient to an uncontaminated area If not breathing, give artificial respiration, or oxygen Seek immediate medical/veterinary attention if not present	Prevent reexposure If conditions warrant a respirator, a respiratory protection program that meets OSHA 29CFR 1910.134 should be followed. Refer to NIOSH 42 CFR 84 for applicable certified respirators or contact your local country equivalent
Skin contact	Wash skin with soap and water for at least 15 min while removing contaminated clothing and shoes or hair/wool Get immediate medical/veterinary attention if not present Ideally, destroy contaminated clothing or thoroughly clean and dry before reuse. Destroy contaminated shoes	Prevent reexposure Wear appropriate PPE

Appendix 3

Health risks from exposure to ionizing radiation

Most reports of clinical signs associated with exposure to Naturally Occurring Radioactive Material (NORM) is from human exposure as a result of occupational exposure involving burning coal, making and using fertilizers, and oil and gas production. Animals may also be exposed by ingestion of contaminated feeds or water, or direct exposure to the skin, or by inhalation. EU Council Directive 96/29/Euratom of 13 May 1996 specifies provisions relating to NORM.

Radioisotopes are the unstable form(s) of an element that emits radiation. The basic unit of radiation dose absorbed in tissue is the gray (Gy), where 1 Gy represents 1 J of energy deposited per kilogram of tissue and more than 1000 mGy is considered high exposure. A becquerel (Bq) is a measure of actual radioactivity in the material.

Naturally occurring isotopes are:

Barium-130; beryllium-1, bismuth-209; cadmium-113 cadmium-116, calcium-48, carbon-14, chlorine-36, europium-151, gadolinium-152, germanium-76, indium-115, iodine-129, lanthanum-138, lutetium-176, molybdenum-100, neodymium-144, neodymium-150, potassium-40, rubidium-87, samarium-147, samarium-148, selenium-82, silicon-32, sodium-22, tellurium-128, tellurium-130, tungsten-180, thorium-232, uranium-235 uranium-238, vanadium-50, zirconium-96

Other radioisotopes are artificially created in nuclear accelerators and reactors.

Of the naturally occurring radioisotopes U-238 and Th-232 series are of the most concern to human and animal health.

Radioactivity exposure is greater from their use in medical activities compared to natural background exposure:

1 kg Uranium ore (Canadian, 15%)	25 million Bq
Radioisotope for medical diagnosis	70 million Bq
A luminescent exit sign (1970s)	1,000,000 million Bq
Radioisotope source for medical therapy	100,000,000 million Bq (100 TBq)
1 kg of superphosphate fertilizer	5000 Bq

Radioactivity damages cells alters the genetic material in the body, and can cause the development of cancer. The specific health risks depend upon the type of isotope, the dose of exposed, the duration of exposure, and other factors including species-specific susceptibility. Exposure to alpha particles can do much more damage than the same dosage amount from beta particles or gamma rays.

In addition to naturally occurring radioactivity major nuclear incidents following nuclear fission such as Chernobyl, Fukushima, Hiroshima, Nagasaki, and Three Mile Island caused direct harm to exposed people, animals, and plants and also contributed to long-term environmental contamination resulting in soil and crop uptake.

Humans may be exposed due to occupation—working with mineral materials. Often mining exposes workers to multiple radioisotopes at the same time. Animals may be exposed through ingesting contaminated pastures, crops, or soil.

Radiation may be acquired by predators highest up the food chain consuming contaminated lower animals or plant.

Rapidly dividing cells in the body are most susceptible to acute radiation exposure, especially those in bone marrow, enterocytes, lymphatic tissue, and skin. Ironically, whilst potentially harmful these properties have also been used to advantage for example the use of low dose XRays in diagnostic imaging, but also the use of targeted radiotherapy to kill cancerous tissue.

Clinical signs of damage due to radiation toxicity may include:
Alopecia, brain hemorrhage, cancer (bladder, breast, leukemia, lung, multiple myeloma, esophagus, ovarian, stomach), cataracts, death, diarrhea, disorientation, dizziness, electrolyte imbalances, fatigue, fever, hematemesis, hematochezia, headache, hypotension, immunosuppression, increased risk of infections, nausea, neurological damage (confusion, coma, high intracranial pressure, incoordination), organ failure, skin lesions, skin burns, shock, tumors, vomiting, weakness

Following radiotherapy pets may develop local skin lesions including localized alopecia, erythema, moist eczema (dogs) or dry eczema (cats), pruritus.

Genetic changes caused by radiation damage are rarely transferred to offspring although exposure in utero does damage the fetus.

Beta rays are widely used in diagnostic medicine in positron emission tomography (PET) scanners, therapeutically in brachytherapy to prevent tissue growth, and in cancer therapy. In industry beta rays are used to measure the thickness of materials (paint coatings, paper, plastics, textiles). Radioisotopes such as carbon-14 can be used to trace activity in biological systems, iron-59 iron is used in the study of metabolism in the spleen and potassium-42 for assessing potassium in the blood.

Gamma rays induce free radical production which can cause cancer, but they are also used to kill bacteria in the sterilization of foods and equipment and cancer therapy. In industry, gamma cameras are used to detect rays passed through objects to inspect solid structures and identify defects or the character of materials in baggage or cargo for example. Cobalt 60 gamma rays are used to preserve foods.

Radioiodine isotope 123 emits gamma rays but it has such a short half-life that it does not harm tissues and can be used safely as a tracer to monitor thyroid iodine uptake. Flourine-18 is used as a tracer in oncology as the tracer. PET also uses carbon-11 in cardiac studies and nitrogen-13 in brain imaging.

Naturally occurring radioactive isotopes and their health risk to man and animals.

Element	Radioactive isotopes	Clinical signs	Treatment
Cesium	Chernobyl and Fukushima nuclear accidents		Prussian blue is the approved antidote for internal detoxification
Iodine	Chernobyl and Fukushima nuclear accidents		Prophylactic or immediate treatment with potassium iodide tablets is recommended
Strontium			Calcium salts chelators for gastrointestinal trapping and enhanced mobilization after strontium exposure

See also Appendix 2

References

Aaseth J, Nurchi VM, Andersen O. Medical therapy of patients contaminated with radioactive cesium or iodine. Biomolecules 2019;9 (12):856. Published 2019 Dec 11 https://doi.org/10.3390/biom9120856.

Abboud IA. Concentration effect of trace metals in Jordanian patients of urinary calculi. Environ Geochem Health 2008;30:11–20.

Abdulwahhab Y. Camels: Diseases & treatment. 1st ed. UAE: Amrit Advertising; 2003. ISBN–9948–03–059–1.

Abhilash KP, Arul JJ, Bala D. Fatal overdose of iron tablets in adults. Indian J Crit Care Med 2013;17(5):311–3. https://doi.org/10.4103/0972-5229.120326.

Acland HM, et al. Toxic hepatopathy in foals. Vet Pathol 1984;21(1):3–9.

Adam FH, Noble PJ, Swift ST, Higgins BM, Sieniawska CE (2010). Barium toxicosis in a dog. J Am Vet Med Assoc;237(5):547–50. https://doi.org/10.2460/javma.237.5.547. PMID: 20807132.

Adams GP, et al. A single treatment of yttrium-90-labeled CHX-A–C6.5 diabody inhibits the growth of established human tumor xenografts in immunodeficient mice. Cancer Res 2004;64(17):6200–6.

Agnew UM, Slesinger TL. Zinc toxicity. In: StatPearls. Treasure Island (FL): StatPearls Publishing; 2020. https://www.ncbi.nlm.nih.gov/books/NBK554548/. 2020 Jan, [Internet]. [Updated 2020 Feb 7]. Available from.

Ahmadi R, Ziaei S, Parsay S. Association between nutritional status with spontaneous abortion. Int J Fertil Steril 2017;10(4):337–42.

Ahmed WM, El-Khadrawy HH, Hanafi EM, El-Hameed ARA, Sabra HA. Effect of copper deficiency on ovarian activity in Egyptian buffalo-cows. World J Zool 2009;4(1):01–8. ref.64.

Alborough R, Grau-Roma L, de Brot S, et al. Renal accumulation of prooxidant mineral elements and CKD in domestic cats. Sci Rep 2020;10:3160. https://doi.org/10.1038/s41598-020-59876-6.

Albretsen JC. The toxicity of iron, an essential element. Vet Med 2006;101:82–90.

Al-Dissi AN, Blakley BR, Woodbury MR. Selenium toxicosis in a white-tailed deer herd. Can Vet J 2011;52:70–3.

Alfrey AC. Aluminium Chapter 9. In: Mertz W, editor. Trace elements in animal and human nutrition, vol. 2. Published by Academic Press; 1986. p. 399–409.

Al-Juboori A, Korian J. Technical report. Symptoms of nutritional deficiency in Arabian camels. Abu Dhabi Food Control Authority Development Sector Research & Development Division; 2011. تقرير ف.

Allen MJ, Sharma S. Magnesium. (2020) In: StatPearls [Internet]. Treasure Island (FL): StatPearls Publishing; https://www.ncbi.nlm.nih.gov/books/NBK519036/ [Updated 2020 Apr 20]. Available from.

Alsdorf R, Wyszynski DF. Teratogenicity of sodium valproate. Expert Opin Drug Saf 2005;4(2):345–53. https://doi.org/10.1517/14740338.4.2.345.

Altschul E, Grossman C, Dougherty R, Gaikwad R, Nguyen V, Schwimmer J, Merker E, Mandel S. Lithium toxicity: a review of pathophysiology, treatment, and prognosis. In: Practical neurology—online; 2016. Accessed 19th June 2020.

Ambrose, AM, Larson, PS, Borzelleca, JF, and Hennigar, GR Jr. Long-term toxicologic assessment of nickel in rats and dogs. J Food Sci Technol 1976;13:181–187.

Ames BN, Gold LS. Chemical carcinogenesis: too many rodent carcinogens. Proc Natl Acad Sci U S A 1990;87(19):7772–6. https://doi.org/10.1073/pnas.87.19.7772.

Anderson RA. Chromium Chapter 7. In: Merz W, editor. Trace elements in human and animal nutrition. 5th ed, vol. 1. London: Academic Press Inc; 1987a. p. 225–40.

Anderson LC. Guinea pig husbandry and medicine. Vet Clin North Am Small Anim Pract 1987b;17(5):1045–60.

Anderson RA. Chromium as an essential nutrient for humans. Regul Toxicol Pharmacol 1997;26:S35–41.

Andleeb S, Mahmood T, Khalid A. Environmental chromium from the tannery industry induces altered reproductive endpoints in the wild female small Indian mongoose (Urva auropunctatus). Toxicol Ind Health 2019;35(2):145–58. https://doi.org/10.1177/0748233718814975 [Epub 2019 Jan 17].

Anger F, Anger JP, Guillou L, Papillon A, Janson C, Sublet Y. Subchronic oral toxicity (six months) of carboxyethylgermanium sesquioxide [(HOOCCH2CH2Ge)2O3]n in rats. Appl Organomet Chem 1992;6 (3):267–72.

Anke M. Arsenic Chapter 6. In: Mertz W, editor. Trace elements in animal and human nutrition, vol. 2. Published by Academic Press; 1986. p. 347–63.

Anke M, et al. In 4 Spurenelement-Symposium p97 Abteilung Wissenschaftliche Publikationen. Jena, Germany: Freidrich Schiller Universitat; 1983.

Anke M, Muller M, Hoppe L. Recent progress in exploring the essentiality of the ultratrace element aluminium in the nutrition of animals and man. Biomed Res Trace Elem 2005a;16(3):185–7.

Anke M, Groppel B, Masaoka T. Recent progress in exploring the essentiality of the non-metallic ultratrace elements fluorine and bromine to the nutrition of animals and man. Biomed Res Trace Elem 2005b;16 (3):177–82.

Anke M, Dorn W, Muller M, Seifert M. Recent progress in exploring the essentiality of the ultratrace element cadmium to the nutrition of animals and man. Biomed Res Trace Elem 2005c;16(3):198–202.

Appleton DJ, Rand JSR, Sunvold GD, Priest J. Dietary chromium tripicolinate supplementation reduces glucose concentrations and improves glucose tolerance in normal-weight cats. J Feline Med Surg 2002;4 (1):13–25.

Ashby J. The genotoxicity of sodium saccharin and sodium chloride in relation to their cancer-promoting properties. Food Chem Toxicol 1985;23(4–5):507–19.

ATSDR. Agency for toxic substances and disease registry. Toxicological profile information sheet. Viewed online, June 29th 2020 at http://www.atsdr.cdc.gov/toxprofiles/index.asp; 2009.

ATSDR, 2020a, Available from: https://www.atsdr.cdc.gov/toxprofiles/tp212.pdf.

ATSDR, 2020b. Available from: https://www.atsdr.cdc.gov/toxprofiles/tp54.pdf.

ATSDR Barium. Agency for Toxic Substances and Disease Registry (2007) Toxicological profile for barium and barium compounds. U.S. Department of Health and Human Services Public Health Service; 2007. https://www.atsdr.cdc.gov/ToxProfiles/tp24.pdf. Accessed online 9th June 2020.

ATSDR Boron. Agency for toxic substances and disease registry toxicological profile for boron. Available online at https://www.atsdr.cdc.gov/toxprofiles/tp26-c2.pdf; 2020. Last Accessed 1st July 2020.

ATSDR Calcium. Calcium hypochlorite/sodium hypochlorite CAS ID #: 7778-54-3, 007681–52-9. Available online at https://www.atsdr.cdc.gov/substances/toxsubstance.asp?toxid=192; 2011. Accessed 15th September 2020.

ATSDR Cesium. Available online at https://www.atsdr.cdc.gov/toxprofiles/tp157.pdf; 2004. Last accessed 15th September 2020.

ATSDR Chlorine. ATSDR public health statement for chlorine CAS#: 7782-50-5 toxic substances. Agency for Toxic Substances and Disease Registry. Portal—Chlorine. Available online at https://www.atsdr.cdc.gov/phs/phs.asp?id=683&tid=36; 2010. Accessed 15th September 2020.

ATSDR Fluorides. Agency for toxic substances and disease registry. Fluorides, hydrogen fluoride, and fluorine CAS ID #: 7782-41-4. Accessed online on 15th September 2020 at https://www.atsdr.cdc.gov/substances/toxsubstance.asp?toxid=38; 2011.

ATSDR Lead. Agency for toxic substances and disease registry. Lead CAS ID #: 7439-92-1. Accessed online on 15th September 2020 at https://www.atsdr.cdc.gov/substances/toxsubstance.asp?toxid=22; 2011.

ATSDR Selenium. Selenium toxicity. Agency for Toxic Substances and Disease Registry (ATSDR) public health statement selenium CAS ID #: 7782-49-2; 2011.

ATSDR Silica. ToxGuideTM for silica SiO2 (numerous CAS#s). January 2020 available online at https://www.atsdr.cdc.gov/toxguides/toxguide-211.pdf; 2020. Accessed 15th September 2020.

ATSDR Silver. Agency for toxic substances and disease registry toxicological profile for silver. Available online at https://www.atsdr.cdc.gov/ToxProfiles/tp146.pdf; 1990. CAS#: 7440-22-4 Last reviewed 2020. Last Accessed 14th September 2020.

ATSDR Strontium. Agency for Toxic Agents & Disease Registry: Toxic substances portal—strontium, https://www.atsdr.cdc.gov/toxprofiles/tp.asp?id=656&tid=120; 2014. Accessed 08/06/2020.

ATSDR Strontium. Strontium toxicity. Agency for Toxic Substances and Disease Registry (ATSDR) public health statement strontium CAS#: 7440-24-6; 2020.

ATSDR Sulphur. Agency for Toxic Substances and Disease Registry Sulphur dioxide CAS#: 7446-09-5, https://www.atsdr.cdc.gov/MMG/MMG.asp?id=249&tid=46; 2014.

ATSDR Thorium. Agency for Toxic Agents & Disease Registry: Toxic substances portal—thorium, https://www.atsdr.cdc.gov/phs/phs.asp?id=658&tid=121; 2019.

ATSDR Tungsten. Available online at https://www.atsdr.cdc.gov/toxprofiles/tp186-c2.pdf; 2020. Accessed 13th September 2020.

ATSDR Uranium. Agency for Toxic Substances And Disease Registry: uranium toxicity how should patients exposed to uranium be treated and managed? Available online at https://www.atsdr.cdc.gov/csem/csem.asp?csem=16&po=13; 2020. Accessed 13th September 2020.

ATSDR Vanadium Toxicity. Agency for toxic agents & disease. available online at https://www.atsdr.cdc.gov/ToxProfiles/tp58-c2.pdf; 2020. Accessed 13th September 2020.

ATSDR Zinc Toxicity. Agency for toxic agents & disease. available online at https://www.atsdr.cdc.gov/toxprofiles/tp60-c3.pdf; 2020. Accessed online 22nd June 2020.

Au J, Youngentob KN, Clark RG, Phillips R, Foley WJ. Bark chewing reveals a nutrient limitation of leaves for a specialist folivore. J Mammal 2017;98(4):1185–92.

Aulerich RJ, Ringer RK, Iwamoto S. Effects of dietary mercury on mink. Arch Environ Contam Toxicol 1974;2:43–51. https://doi.org/10.1007/BF01985799.

Ávalos A, Haza AI, Mateo D, Morales P. In vitro and in vivo genotoxicity assessment of gold nanoparticles of different sizes by comet and SMART assays. Food Chem Toxicol 2018;120:81–8. https://doi.org/10.1016/j.fct.2018.06.061.

AZA Amphibian Husbandry Guide. Accessed on 30th September 2020 online at https://assets.speakcdn.com/assets/2332/amphibianhusbandryresourceguide.pdf; 2020.

Baeverfjord G, Prabhu PAJ, Fjelidal PG, et al. Mineral nutrition and bone health in salmonids. Rev Aquac 2019;11:740–65.

Bai Y, Hunt CD. Dietary boron increases serum antibody concentrations in rats immunized with heat-killed *Mycobacterium tuberculosis* (MT). FASEB J 1996;10:A819.

Bailly R, Lauwerys R, Buchet JP, Mahieu P, Konings J. Experimental and human studies on antimony metabolism: their relevance for the biological monitoring of workers exposed to inorganic antimony. Br J Ind Med 1991;48(2):93–7. https://doi.org/10.1136/oem.48.2.93.

Bakirdere S, Örenay S, Korkmaz M. Effect of boron on human health. Open Miner Process J 2010;3(1):54–9. https://doi.org/10.2174/1874841401003010054.

Barai M, Ahsan N, Paul N, et al. Amelioration of arsenic-induced toxic effects in mice by dietary supplementation of *Syzygium cumini* leaf extract. Nagoya J Med Sci 2017;79(2):167–77. https://doi.org/10.18999/nagjms.79.2.167.

Baranwal AK, Singhi SC. Acute iron poisoning: management guidelines. Indian Pediatr 2003;40(6):534–40.

Barboza PS, Vanselow BA. Copper toxicity in captive wombats (Marsupialia: vombatidae). In: Proceedings of the American Association of Zoo Veterinarians. South Padre Island, TX; 1990. p. 204–6.

Barlow RM, Butler EJ. An ataxic condition in red deer (*Cervus elaphus*). J Comp Pathol 1964;74:519–29.

Bauck L. Nutritional problems in pet birds. Semin Avian Exot Pet Med 1995;4(1):3–8.

Beaser SB. The anticoagulant effects in rabbits and man of intravenous injection of salts of the rare earths. J Clin Invest 1942;21(4):447–54.

Beckers-Trapp ME, Lanoue L, Keen CL, Rucker RB, Uriu-Adams JY. Abnormal development and increased 3-nitrotyrosine in copper-deficient mouse embryos. Free Radic Biol Med 2006;40:35–44.

Bermudez E, Mangum JB, Wong BA, Asgharian B, Hext PM, Warheit DB, Everitt JI. Pulmonary responses of mice, rats, and hamsters to subchronic inhalation of ultrafine titanium dioxide particles. Toxicol Sci 2004;77(2):347–57. https://doi.org/10.1093/toxsci/kfh019.

Bernhoft RA. Mercury toxicity and treatment: a review of the literature. J Environ Public Health 2012;2012:460508.

Bernhoft A, Waaler T, Mathiesen SD, Flåøyen A. Trace elements in reindeer from Rybatsjij Ostrov, north western Russia. Rangifer 2002; 22:67–73.

Bichu S, Tilve P, Kakde P, et al. Relationship between the use of aluminium utensils for cooking meals and chronic aluminum toxicity in patients on maintenance hemodialysis: a case control study. J Assoc Physicians India 2019;67(4):52–6.

Bilčíková J, Fialková V, Kováčiková E, Miškeje M, Tombarkiewicz B, Kňažická Z. Influence of transition metals on animal and human health: a review. Contemp Agric 2016;67(3–4):187–95. 2018.

Blinova I, Muna M, Heinlaan M, Lukjanova A, Andleeb AK. Potential hazard of lanthanides and lanthanide-based nanoparticles to aquatic ecosystems: data gaps, challenges and future research needs derived from bibliometric analysis. Nanomaterials 2020;10:328. https://doi.org/10.3390/nano10020328.

Blunden S, Wallace T. Tin in canned food: a review and understanding of occurrence and effect. Food Chem Toxicol 2003;41(12):1651–62.

Boffetta P, Fordyce TA, Mandel JS. A mortality study of beryllium workers. Cancer Med 2016;5(12):3596–605. https://doi.org/10.1002/cam4.918.

Bomhard EM. The toxicology of indium oxide. Environ Toxicol Pharmacol 2018;58:250–8.

Bomhard EM, Gelbke HP, Schenk H, Williams GM, Cohen SM. Evaluation of the carcinogenicity of gallium arsenide. Crit Rev Toxicol 2013;43(5):436–66. https://doi.org/10.3109/10408444.2013.792329.

Bonar CJ, Trupkiewicz JG, Toddes B, Lewandowski AH. Iron storage disease in tapirs. J Zoo Wildl Med 2006;37(1):49–52. https://doi.org/10.1638/03-032.1. [1 March 2006].

Borbinha C, Serrazina F, Salavisa M, et al. Bismuth encephalopathy- a rare complication of long-standing use of bismuth subsalicylate. BMC Neurol 2019;19:212. https://doi.org/10.1186/s12883-019-1437-9.

Bourke CA. Motor neuron disease in molybdenum-deficient sheep fed the endogenous purine xanthosine: a possible mechanism for Tribulus staggers. Aust Vet J 2012;90:272–4.

Bourke CA. Molybdenum deprivation, purine ingestion and an astrocyte associated motor neurone syndrome in sheep: the assumed clinical effects of inosine. Aust Vet J 2015;93:79–83.

Bourke CA. Molybdenum deficiency produces motor nervous effects that are consistent with amyotrophic lateral sclerosis. Front Neurol 2016;7:28.

Bourke CA. Astrocyte dysfunction following molybdenum-associated purine loading could initiate Parkinson's disease with dementia. NPJ Parkinsons Dis 2018;4:7. 2018 Mar 20.

Bradberry SM, Beer ST, Vale JA (1996) UK PID monograph—bismuth. National Poisons Information Service, West Midlands Poisons Unit, City Hospital NHS Trust Accessed online 1st July 2020.

Bradberry SM, Beer ST, Vale JA. UKPID Monograph Antimony National Poisons Information Service (Birmingham Centre),West Midlands Poisons. Accessed 19th June 2020 http://www.inchem.org/documents/ukpids/ukpids/ukpid40.htm; 2020.

Breazile JE, Brown EM. Anatomy. In: Wagner JE, Manning PJ, editors. Biology of the Guinea pig. New York: Academic Press; 1976. p. 53–62.

Bridges CH, Ghagan Moffitt P. Influence of variable content of dietary zinc on copper metabolism of weanling foals. Am J Vet Res 1990;51:275–80.

Bridges CH, Harris ED. Experimentally induced cartilaginous fractures (osteochondritis dissecans) in foals fed low-copper diets. J Am Vet Med Assoc 1988;193:215–21.

Bridges CH, Womack JE, Harris ED, et al. Considerations of copper metabolism in osteochondrosis of suckling foals. J Am Vet Med Assoc 1984;185:173–8.

Brieger H, Semisch 3rd CW, Stasney J, Piatnek DA. Industrial antimony poisoning. Ind Med Surg 1954;23:521–3.

Bromism. In: Parfitt K, editor. Martindale. 32nd ed. Pharmaceutical Press; 1999. p. 1620–3.

Burbano X, Miguez-Burbano MJ, McCollister K, Zhang G, Rodriguez A, Ruiz P, et al. Impact of a selenium chemoprevention clinical trial on hospital admissions of HIV-infected participants. HIV Clin Trials 2002;3:483–91.

Calellor D. Selenium. In: Goldfrank's toxicologic emergencies. 9th ed. New York: McGraw-Hill; 2011. p. 1316–20.

Campbell KS, Foster AJ, Dillon CT, Harding MM. Genotoxicity and transmission electron microscopy studies of molybdocene dichloride. J Inorg Biochem 2006;100:1194–8.

Canan F, Kaya A, Bulur S, Albayrak ES, Ordu S, Ataoglu A. Lithium intoxication related multiple temporary ecg changes: a case report. Cases J 2008; 1(1):156. 2008 Sep 17.

Cancelliere EC, DeAngelis N, Nkurunungi JB, Raubenheimer D, Rothman JM. Minerals in the foods eaten by mountain gorillas (Gorilla beringei). PLoS One 2014;9(11), e112117. Published 2014 Nov 5 https://doi.org/10.1371/journal.pone.0112117.

Carlisle EM. Silicon Chapter 7. In: Mertz W, editor. Trace elements in human and animal nutrition. 5th ed, Vol. 1. New York: Academic Press; 1986. p. 373–88.

Casteel SW. Metal toxicosis in horses. Vet Clin North Am Equine Pract 2001;17(3):520–2.

CDC. Centres for Disease Control and Prevention Mercury (elemental): lung damaging agent. The National Institute for Occupational Safety and Health (NIOSH); 2011. May 12, 2011.

CDC. Toxic substances portal: aluminum. Centers for Disease Control and Prevention; 2015. website Available at http://wwwatsdrcdcgov/toxfaqs/TFasp?id=190&tid=34. Updated March 12, 2015.

CDC. Thallium: systemic agent. Emergency response safety and health database. The National Institute for Occupational Safety and Health (NIOSH); 2020. https://www.cdc.gov/niosh/ershdb/emergencyresponsecard_29750026.html.

Chesney RW, Hedberg GE. Rickets in polar bear cubs: is there a lesson for human infants? Neonatology 2011;99:95–6. https://doi.org/10.1159/000315150.

Chitambar CR. Review medical applications and toxicities of gallium compounds. Int J Environ Res Public Health 2010;7:2337–61. https://doi.org/10.3390/ijerph7052337.

Choi R, Kim MJ, Sohn I, Kim S, et al. Serum trace elements and their associations with breast cancer subgroups in Korean breast cancer patients. Nutrients 2019;11(1):37. https://doi.org/10.3390/nu11010037.

Chow C, Pollock C. Heavy metal poisoning in birds. LafeberVet; 2018. Updated January 7, 2012 http://www.lafeber.com/vet/heavy-metal-poisoning-in-birds/. Accessed March 2018.

Chua AC, Klopcic B, Lawrance IC, Olynyk JK, Trinder D. Iron: an emerging factor in colorectal carcinogenesis. World J Gastroenterol 2010; 16(6):663–72. https://doi.org/10.3748/wjg.v16.i6.663.

Clark RG, Hepburn JD. Deer liver and serum copper levels. Dent Surv 1986;(13):11–4.

Clark P, Tugwell P, Bennett KJ, Bombardier C, Shea B, Wells GA, Suarez-Almazor ME. Injectable gold for rheumatoid arthritis. Cochrane Database Syst Rev 1997;(4). https://doi.org/10.1002/14651858.CD000520, CD000520.

Clarkson TW. Mercury Chapter 12. In: Mertz W, editor. Trace elements in human and animal nutrition. 5th ed, vol. 1. London: Academic Press Inc; 1987. p. 417–26.

Clauss M, Hänichen T, Hummel J, Ricker U, Block K, Grest P, Hatt J-M. Excessive iron storage in captive omnivores? The case of the coati (Nasuas pp). In: Zoo animal nutrition, vol. III. Fürth: Filander; 2006. p. 91–9.

Clauss M, Castell JC, Kienzle E, et al. Mineral absorption in the black rhinoceros (*Diceros bicornis*) as compared with the domestic horse. J Anim Physiol Anim Nutr (Berl) 2007;91(5–6):193–204. https://doi.org/10.1111/j.1439-0396.2007.00692.x.

Cohn LA, Dodam JR, McCaw DL, Tate DJ. Effects of chromium supplementation on glucose tolerance in obese and nonobese cats. Am J Vet Res 1999;60:1360–3.

Corbera JA, Morales M, Pulido, Montoya MJA, Gutierrez C. An outbreak of nutritional muscular dystrophy in dromedary camels. J Appl Anim Res 2003;23:117–22.

Cossellu G, Motta V, Dioni L, Angelici L, Vigna L, Farronato G, Pesatori ACA, Bollati V. Titanium and zirconium levels are associated with changes in MicroRNAs expression: results from a human cross-sectional study on obese population. PLoS One 2016;11(9), e0161916.

Cowan V, Blakley B. Characterizing 1341 cases of veterinary toxicoses confirmed in western Canada: a 16-year retrospective study. Can Vet J 2016; 57(1):53–8.

Cronin RE, Ferguson ER, Shannon WA, Knochel JP. Skeletal muscle injury after magnesium depletion in the dog. Am J Physiol 1982;243(2):F113–20. https://doi.org/10.1152/ajprenal.1982.243.2.F113.

Culver DA, Dweik RA. Chronic beryllium disease. Clin Pulm Med 2003;10:72–9.

Cummings KJ, Stefaniak AB, Virji MA, Kreiss K. A reconsideration of acute beryllium disease. Environ Health Perspect 2009;117 (8):1250–6. https://doi.org/10.1289/ehp.0800455.

Czarnek K, Terpilowska S, Siwicki AK. Genotoxicity and mutagenicity of nickel(II) and iron(III) and interactions between these microelements. Trace Elem Electrolytes 2018;36(01). https://doi.org/10.5414/TEX0154.

Damian P. Stand alone report 14 a literature review of the health and ecological effects of the rare earth elements. Prepared for 225 Union Blvd., Suite 250 Lakewood, Colorado 80228: Rare Element Resources, Inc; 2014. August 27, 2014. Accessed 10th June 2020 https://www.nrc.gov/docs/ML1513/ML15134A344.pdf.

Daniel MS, Chausow G, Forbes RM, Czarnecki GL, Corbin JE. Experimentally-induced magnesium deficiency in growing kittens. Nutr Res 1986; 6(4):459–68.

Das K, Debacker V, Pillet S, Bouquegneau J-M. Heavy metals in marine mammals. In: Toxicology of marine mammals. Taylor and Francis Publishers; 2003. p. 135–67. https://doi.org/10.1201/9780203165577.ch7.

Das BC, Thapa P, Karki R, et al. Boron chemicals in diagnosis and therapeutics. Future Med Chem 2013;5(6):653–76. https://doi.org/10.4155/fmc.13.38.

Davies NL. Lithium toxicity in two dogs. J S Afr Vet Assoc 1991;62:140–2.

Davies M, Alborough R, Jones L, et al. Mineral analysis of complete dog and cat foods in the UK and compliance with European guidelines. Sci Rep 2017;7:17107. https://doi.org/10.1038/s41598-017-17159-7.

Davis GK, Mertz W. Copper Chapter 10. In: Mertz W, editor. Trace elements in human and animal nutrition. 5th ed, vol. 1. London: Academic Press Inc; 1987. p. 301–50.

Davis J, Desmond M, Berk M. Lithium and nephrotoxicity: a literature review of approaches to clinical management and risk stratification. BMC Nephrol 2018;19(1):305. 2018 Nov 03.

De Boeck M, Kirsch-Volders M, Lison D. Cobalt and antimony: genotoxicity and carcinogenicity. Mutat Res 2003;533(1–2):135–52. https://doi.org/10.1016/j.mrfmmm.2003.07.012.

Dennert G, Zwahlen M, Brinkman M, Vinceti M, Zeegers MP, Horneber M. Selenium for preventing cancer. Cochrane Database Syst Rev 2011;2011, CD005195.

Dennis PM, Funk JA, RajalaSchultz PJ, Blumer ES, Miller RE, Thomas M, Wittum E, Saville WJA. A review of some of the health issues of captive black rhinoceroses (*Diceros bicornis*). J Zoo Wildl Med 2007;38(4):509–17.

Der Kolk JH, Van Leeuwen JPTM, Van Den Belt AJM, Van Schaik RHN, Schaftenaar W. Subclinical hypocalcaemia in captive elephants (*Elephas maximus*). Vet Rec 2008;162(15):475–9. https://doi.org/10.1136/vr.162.15.475.

Destefani CA, Motta LC, Costa RA, et al. Evaluation of acute toxicity of europium–organic complex applied as aluminescent marker for the visual identification of gunshot residue. Microchem J 2016;124:195–200.

Dierenfeld E. Biology, medicine, and surgery of elephants. In: Fowler M, Mikota SK, editors. Biology, medicine, and surgery of elephants. Iowa: Wiley-Blackwell Publishing; 2008. p. 229–307.

Dierenfeld S. Feeding behavior and nutrition of the sugar glider (*Petaurus breviceps*) animal practice. Vet Clin North Am Exot Anim Pract 2009; 12(2):209–15.

Dierenfeld ES, Atkinson S, Craig AM, Walker KC, Streich WJ, Clauss M. Mineral concentrations in blood and liver tissue of captive and free-ranging Rhinoceros species. Zoo Biol 2005;24:51–72.

Digitalfire Reference Library. Zirconium compounds toxicity. Accessed 8th June 2020 https://digitalfire.com/hazard/zirconium+compounds+toxicity; 2020.

DiPalma JR. Bismuth toxicity, often mild, can result in severe poisonings. Emerg Med News 2001;23(3):16. April 2001.

Doig AT. Baritosis: a benign pneumoconiosis. Thorax 1976;31(1): 30–9. https://doi.org/10.1136/thx.31.1.30. PMC 470358. PMID 1257935.

Domingo L, Paternain JL, Llobet JM, Corbella J. The developmental toxicity of uranium in mice. Toxicology 1989;55(1–2):143–52.

Dong J-Y, Xun P, He K, Qin L-Q. Magnesium intake and risk of type 2 diabetes: meta-analysis of prospective cohort studies. Diabetes Care 2011;34:2116–22.

Donnelly SM, Smith EK. The role of aluminum in the functional iron deficiency of patients treated with erythropoietin: case report of clinical characteristics and response to treatment. Am J Kidney Dis 1990;16(5):487–90.

Duncan M. Perissodactyls in pathology of wildlife and zoo animals; 2018. p. 2018.

Durnev AD, Solomina AS, Daugel-Dauge NO, et al. Evaluation of genotoxicity and reproductive toxicity of silicon nanocrystals. Bull Exp Biol Med 2010;149(4):445–9. https://doi.org/10.1007/s10517-010-0967-3.

Eapen JT, Kartha CC, Rathinam K, et al. Levels of cerium in the tissues of rats fed a magnesium-restricted and cerium-adulterated diet. Bull Environ Contam Toxicol 1996;56(2):178–82.

Eapen J. Elevated levels of cerium in tubers from regions endemic for endomyocardial fibrosis (EMF). Bull Environ Contam Toxicol 1998;60:168–70. https://doi.org/10.1007/s001289900606.

EC. European commission position paper on mercury. accessible online at https://ec.europa.eu/environment/archives/air/pdf/pp_mercury6.pdf; 2020. Last accessed 28th June 2020.

Edited by Karen A. Terio, Denise McAloose and Judy St. Leger Published by Academic Press, Cambridge Massachusetts USA p. 433–454.

Emsley J. Natures building bocks: an A-Z guide to the elements. Oxford University Press; 2011. p. 384–7.

Ensley PT, Anderson M, Osborn K, Bissonnette S, Deftos L. Osteodystrophy in an orphan Asian elephant. Pittsburgh: Proceedings of the American Association of Zoo Veterinarians; 1994. p. 12–143.

EPA. Gadolinium (CASRN 7440-54-2). United States: Environmental Protection Agency; 2007a. Accessed online on 15th September 2020 at https://cfpub.epa.gov/ncea/pprtv/documents/Gadolinium.pdf.

EPA. Promethium (CASRN 7440-12-2). Accessed 29th August 2020 at https://cfpub.epa.gov/ncea/pprtv/documents/Promethium.pdf; 2007b.

EPA. Toxicological review of cerium oxide and cerium compounds (CAS No. 1306-38-3) in support of summary information on the integrated risk information system (IRIS). September 2009, Washington, DC: U. S. Environmental Protection Agency; 2009a. https://cfpub.epa.gov/ncea/iris/iris_documents/documents/toxreviews/1018tr.pdf.

EPA. Provisional peer-reviewed toxicity values for stable (nonradioactive) samarium chloride (CASRN 10361-82-7) and stable (Nonradioactive) samarium nitrate (CASRN 10361-83-8). EPA/690/R-09/050F Final 9-17-2009. Accessed 110620 https://cfpub.epa.gov/ncea/pprtv/documents/SamariumChlorideStableNonradioactive.pdf; 2009b.

EPA. United States Environmental Protection Agency (EPA) provisional peer-reviewed toxicity values for soluble zirconium compounds (CASRN 7440-67-7); 2012. EPA/690/R-12/038F Final 12-27-2012.

EPA. Rubidium compounds (CASRN 7440-17-7, rubidium) (CASRN 7791-11-9, rubidium chloride) (CASRN 1310-82-3, rubidium hydroxide) (CASRN 7790-29-6, rubidium iodide). Accessed on line 15th September 2020 https://cfpub.epa.gov/si/si_public_record_report.cfm?Lab=NCEA&dirEntryId=339495; 2016.

EPA. United States environmental protection agency provisional peer-reviewed toxicity values for stable (nonradioactive) soluble lanthanum (CASRN 7439-91-0); 2018a. EPA/690/R-18/004 | September 27, 2018.

EPA. United States environmental protection agency. Provisional peer-reviewed toxicity values for stable (nonradioactive) soluble lutetium (CASRN 7439-94-3); 2018b. EPA/690/R-18/003 | August 16, 2018.

Erway LC, Fraser AS, Hurley LS. Prevention of congenital otolith defect in pallid mutant mice by manganese supplementation. Genetics 1971;67 (1):97–108.

ETN. Extension toxicology network sulfur (1995). Accessed online 14th September 2020 http://pmep.cce.cornell.edu/profiles/extoxnet/pyrethrins-ziram/sulfur-ext.html; 1995.

EU. European commission position paper on mercury. accessible online at https://ec.europa.eu/environment/archives/air/pdf/pp_mercury6.pdf; 2020. Last accessed 28th June 2020.

EURARE. Health and safety issues in REE mining and processing. An internal EURARE guidance report. Kemakta Konsult AB, Geological Survey of Finland, Institute of Geology & Mineral Exploration, Fen Minerals A/S; 2014. September 2014 http://www.eurare.eu/docs/internalGuidanceReport.pdf. Accessed 10th June 2020.

Exley C. Human exposure to aluminium. Environ Sci Processes Impacts 2013;2013(15):1807–16.

Falk SA, Klein R, Haseman JK, Sanders GM, Talley FA, Lim DJ. Acute methyl mercury intoxication and ototoxicity in Guinea pigs. Arch Pathol 1974; 97(5):297–305.

Farina LL, Heard DJ, LeBlanc DM, Hall JO, Stevens G, Wellehan JFX, Detrisac CJ. Iron storage disease in captive Egyptian fruit bats (*Rousettus aegyptiacus*): relationship of blood iron parameters to hepatic iro n concentrations and hepatic histopathology. J Zoo Wildl Med 2005;36(2):212–21. https://doi.org/10.1638/03-115.1.

Farrar WP, Edwards JF, Willard MD. Pathology in a dog associated with elevated tissue mercury concentrations. J Vet Diagn Invest 1994;6:511–4.

Faye B, Seboussi R. Selenium in camel—a review. Nutrients 2009;1:30–49. https://doi.org/10.3390/nu1010030.

FEDIAF Nutritional Guidelines, http://www.fediaf.org/self-regulation/nutrition.html; 2019. Accessed 26th September 2020.

Fei M, Li N, Ze Y, Liu J, Wang S, Gong X, Duan Y, Zhao X, Wang H, Hong F. The mechanism of liver injury in mice caused by lanthanoids. Biol Trace Elem Res 2011;140:317–29.

Felder CC, Robillard JE, Roy S, Jose PA. Severe chloride deficiency in the neonate: the canine puppy as an animal model. Pediatr Res 1987;21 (5):497–501. https://doi.org/10.1203/00006450-198705000-00015.

Feng L, Xiao H, He X, Li Z, Li F, Liu N, Zhao Y, Huang Y, Zhang Z, Chai Z. Neurotoxicological consequence of long-term exposure to lanthanum. Toxicol Lett 2006;165:112–20.

Ferrie GM, Alford VC, Atkinson J, et al. Nutrition and health in amphibian husbandry. Zoo Biol 2014;33(6):485–501. https://doi.org/10.1002/zoo.21180.

Fireman E, Shai AB, Lerman Y, et al. Chest wall shrapnel-induced beryllium-sensitization and associated pulmonary disease. Sarcoidosis Vasc Diffuse Lung Dis 2012;29(2):147–50.

Fischer M, Modder G. Radionuclide therapy of inflammatory joint diseases. Nucl Med Commun 2002;23(9):829–31.

Fontenot AP, Falta MT, Kappler JW, Dai S, McKee AS. Beryllium-induced hypersensitivity: genetic susceptibility and neoantigen generation. J Immunol 2016;196(1):22–7. https://doi.org/10.4049/jimmunol.1502011.

Fosmire GJ. Zinc toxicity. Am J Clin Nutr 1990;51(2):225–7. https://doi.org/10.1093/ajcn/51.2.225.

Franzolin R, Dehority BA. Comparison of protozoal populations and digestion rates between water buffalo and cattle fed an all forage diet. J Appl Anim Res 1999;16(1):33–46.

Friberg LP, Boston G, Piscator NM, Roberts KH. Molybdenum—a toxicological appraisal; 1975. U. S. Environ. Protection Agency Rep. 600/1-75-004. 142 pp.

Friedman BJ, Freeland-Graves JH, Bales CW, et al. Manganese balance and clinical observations in young men fed a manganese-deficient diet. J Nutr 1987;117(1):133–43.

Frost DV. Spurenelement-symposium. Jena, Germany: Abteilung Wissenschaftliche Publikationen, Friedrich Schiller Universitat; 1983. p. P89.

Fuller TJ, Nichols WW, Brenner BJ, Peterson JC. Reversible depression in myocardial performance in dogs with experimental phosphorus deficiency. J Clin Invest 1978;62(6):1194–200. https://doi.org/10.1172/JCI109239.

Fyffe JJ. Serum copper concentrations and clinical signs in red deer (*Cervus elaphus* Serum) during drought in Central Victoria. Aust Vet J 1996;73:188–91.

Gale TF. The embryotoxicity of ytterbium chloride in golden hamsters. Teratology 1975;11(3):289–95. https://doi.org/10.1002/tera.1420110308. PMID 807987.

Ganrot PO. Metabolism and possible health effects of aluminum. Environ Health Perspect 1986;65:363–441. https://doi.org/10.1289/ehp.8665363.

Gapat SM, Bhikane AU, Gangane GR. Aetiological and therapeutic investigations on leukoderma in Indian buffaloes (*Bubalus bubalis*). Buffalo Bull 2016;35(4):10.

Garriga RM, Sainsbury AW, Goodship AE. Bone assessment of free-living red squirrels (*Sciurus vulgaris*) from the United Kingdom. J Wildl Dis 2004;40(3):515–22. q Wildlife Disease Association 2004.

Garruto RM, Shankar SK, Yanagihara R, Salazar AM, Amyx HL, Gajdusek DC. Low-calcium, high-aluminum diet-induced motor neuron pathology in cynomolgus monkeys. Acta Neuropathol 1989; 78:210–9.

Gebhart E, Rossman TG. Mutagenicity, carcinogenicity, teratogenicity. In: Merian E, editor. Metals and their compounds in the environment. Weinheim, Germany: VCH; 1991. p. 617–40.

Gerber GB, Léonard A. Mutagenicity, carcinogenicity and teratogenicity of germanium compounds. Mutat Res 1997;387(3):141–6. https://doi.org/10.1016/s1383-5742(97)00034-3.

Gerber GB, Léonard A, Hantson PH. Carcinogenicity, mutagenicity and teratogenicity of manganese compounds. Crit Rev Oncol Hematol 2002; 42(1):25–34.

German A. Magnesium deficiency in cats. Published online by PetMd; 2010. https://www.petmd.com/cat/conditions/endocrine/c_ct_hypomagnesemia.

Gfeller RW, Messonnier SP. Handbook of small animal toxicology. Published by Mosby, Inc; 1998.

Gheorghe S, Stoica C, Vasile GG, Nita-Lazar M, Stanescu E, Lucaciu IE. Metals toxic effects in aquatic ecosystems: modulators of water quality, water quality. Hlanganani Tutu: IntechOpen; 2017. https://doi.org/10.5772/65744. Available from: https://www.intechopen.com/books/water-quality/metals-toxic-effects-in-aquatic-ecosystems-modulators-of-water-quality.

Ghosh A, Sarkar S, Pramanik AK, Chaudhary PS, Ghosh S. Selenium toxicosis in grazing buffaloes and its relationship with soil and plant of West Bengal. Indian J Anim Sci 1993;63(5):557–60.

Gianduzzo T, Colombo JR, Haber G-P, Hafron J, et al. Laser robotically assisted nerve-sparing radical prostatectomy: a pilot study of technical feasibility in the canine model. BJU Int 2008;102(5):598–602. 2008 Aug 5.

Glade MJ, Belling TH. A dietary etiology for osteochondrotic cartilage. J Equine Vet Sci 1986;6:151–5.

Golub MS, Germann SL. Long-term consequences of developmental exposure to aluminum in a suboptimal diet for growth and behavior of Swiss Webster mice. Neurotoxicol Teratol 2001;23(4):365–72. https://doi.org/10.1016/S0892-0362(01)00144-1.

Gómez-Aracena J, Riemersma RA, Guitiérrez-Bedmar M, et al. Toenail cerium levels and risk of a first acute myocardial infarction: the EURAMIC and heavy metals study. Chemosphere 2006;64:112–20.

Gonzalez-Astudillo V, Henning J, Valenza L, Knott L, McKinnon A, Larkin R, Allavena R. A necropsy study of disease and comorbidity trends in morbidity and mortality in the koala (*Phascolarctos cinereus*) in South-East Queensland, Australia. Sci Rep 2019;9, 17494.

Gordon MF, Abrams RI, Rubin DB, Barr WB, Correa DD. Bismuth subsalicylate toxicity as a cause of prolonged encephalopathy with myoclonus. Mov Disord 1995;10(2):220–2.

Gould L, Kendall NR. Role of the rumen in copper and thiomolybdate absorption. Nutr Res Rev 2011;24(2):176–82. https://doi.org/10.1017/S0954422411000059. PMID: 22296933; PMCID: PMC3269883.

Graca JG, Davison FC, Feavel JB. Comparative toxicity of stable rare earth compounds: III. Acute toxicity of intravenous injections of chlorides and chelates in dogs. Arch Environ Health 1964;8:555–64.

Grace ND, Wilson PR, Quinn AK. Impact of molybdenum on the copper status of red deer (*Cervus elaphus*). N Z Vet J 2005;53(2):137–41. https://doi.org/10.1080/00480169.2005.36491.

Greentree WF, Hall JO. Iron toxicosis. In: Bonagura JD, editor. Kirk's current veterinary therapy XII: small animal practice. Philadelphia, PA: WB Saunders; 1995. p. 240–2.

Guo F, Lu X-W, Xu Q-P. Diagnosis and treatment of organotin poisoned patients. World J Emerg Med 2010;1(2):122–5.

Gupta RC, Kwatra MS, Singh N. Chronic selenium toxicity as a cause of hoof and horn deformities in buffalo cattle and goat. Indian Vet J 1982;59:738–40.

Habib G, Jabbar G, Siddiqui MM, Shah Z. Paralytic disorders associated with phosphorus deficiency in buffaloes. Pak Vet J 2004;24(1):18.

Haenlein GFW. Mineral and vitamin requirements and deficiencies. In: Proc. IVth intern. conf. goats, Brasilia, Brazil, March 8-13; 1987. p. 1249.

Hagen K, Müller DWH, Wibbelt G. The macroscopic intestinal anatomy of a lowland tapir (*Tapirus terrestris*). Eur J Wildl Res 2014;61(1):171–6.

Hakki SS, Bozkurt BS, Hakki EE. Boron regulates mineralized tissue-associated proteins in osteoblasts (MC3T3-E1). J Trace Elem Med Biol 2010; 24(4):243–50.

Haley PJ. Pulmonary toxicity of stable and radioactive lanthanides. Health Phys 1991;61:809–20.

Haley TJ, Komesu N, Raymond K. Pharmacology and toxicology of niobium chloride. Toxicol Appl Pharmacol 1962;4(3):385–92.

Haley TJ, Komesu N, Flesher AM, Mavis L, Cawthorne J, Upham HC. Pharmacology and toxicology of terbium, thulium, and ytterbium chlorides. Toxicol Appl Pharmacol 1963;5(4):427–36. July 1963.

Haley TJ, Komesu N, Efros MG, Koste L, Upham HC. Pharmacology and toxicology of praseodymium and neodymium chlorides. Toxicol Appl Pharmacol 1964;6:614–20.

Haley PJ, Finch GL, Mewhinney JA, Harmsen AG, Hahn FF, Hoover MD, Muggenburg BA, Bice DE. A canine model of beryllium-induced granulomatous lung disease. Lab Invest 1989;61(2):219–27.

Hambidge KM, Casey CE, Krebs NF. Zinc Chapter 1. In: Mertz W, editor. Trace elements in animal and human nutrition, vol. 2. Published by Academic Press; 1986. p. 1–109.

Han J, Park H, Kim J, et al. Toxic effects of arsenic on growth, hematological parameters, and plasma components of starry flounder, Platichthys stellatus, at two water temperature conditions. Fish Aquat Sci 2019;22:3. https://doi.org/10.1186/s41240-019-0116-5.

Handeland K, Bernhoft A. Osteochondrosis associated with copper deficiency in a red deer herd in Norway. Vet Rec 2004;155:676–8.

Handeland K, Flåøyen A. Enzootic ataxia in a Norwegian red deer herd. Acta Vet Scand 2000;41:329–31.

Handeland K, Bernhoft A, Aartun MS. Copper deficiency and effects of copper supplementation in a herd of red deer (*Cervus elaphus*). Acta Vet Scand 2008;50(1):8.

Hanko E, Erne K, Wanntorp H, Borg K. Poisoning in ferrets by tissues of alkyl mercury-fed chickens. Acta Vet Scand 1970;11(2):268–82.

Hansard 2nd SL, Ammerman CB, Henry PR, Simpson CF. Vanadium metabolism in sheep. I. Comparative and acute toxicity of vanadium compounds in sheep. J Anim Sci 1982;55(2):344–9. https://doi.org/10.2527/jas1982.552344x.

Hansard II SL, Ammerman CB, Fick KR, Miller SM. Performance and vanadium content of tissues in sheep as influenced by dietary vanadium. J Anim Sci 1978;46(4):1091–5. https://doi.org/10.2527/jas1978.4641091x.

Harari F, Bottai M, Casimiro E, Palm B, Vahter M. Exposure to lithium and cesium through drinking water and thyroid function during pregnancy: a prospective cohort study. Thyroid 2015;25(11):1199–208. https://doi.org/10.1089/thy.2015.0280.

Harrison WN, Bradberry SM, Vale JA. UKPID monograph zirconium national poisons information service (Birmingham Centre). West Midlands Poisons Unit, City Hospital NHS Trust; 2020. http://www.inchem.org/documents/ukpids/ukpids/ukpid90.htm. Accessed online 20th June 2020.

Hatt JM, Clauss M. Feeding Asian and African elephants *Elephas maximus* and *Loxodonta africana* in captivity. Int Zoo Yearb 2006;40(1):88–95.

Hawk SN, Uriu-Hare JY, Daston GP, Jankowski MA, Kwik-Uribe C, Rucker RB, Keen CL. Rat embryos cultured under copper-deficient conditions develop abnormally and are characterized by an impaired oxidant defense system. Teratology 1998;57:310–20.

Hawk SN, Lanoue L, Keen CL, Kwik-Uribe CL, Rucker RB, Uriu-Adams JY. Copper-deficient rat embryos are characterized by low superoxide dismutase activity and elevated superoxide anions. Biol Reprod 2003;68:896–903.

Haynes JI, Askew MJ, Leigh C. Dietary aluminium and renal failure in the Koala (*Phascolarctos cinereus*). Histol Histopathol 2004; 19:777–84.

Healthline. online Accessed 19th June 2020 https://www.healthline.com/health/lithium-toxicity#symptoms; 2020.

Hedya SA, Avula A, Swoboda HD. Lithium toxicity. Updated 2019 Nov 13, In: StatPearls [Internet]. Treasure Island (FL): StatPearls Publishing; 2019. 2020 Jan. Available from: https://www.ncbi.nlm.nih.gov/books/NBK499992/.

Heinrich U, Fuhst R, Rittinghausen S, Creutzenberg O, Bellmann B, Koch W, Levsen K. Chronic inhalation exposure of Wistar rats and two different strains of mice to diesel engine exhaust, carbon black, and titanium dioxide. Inhal Toxicol 1995;7(4):533–56. https://doi.org/10.3109/08958379509015211.

Helmick KE, Milne VE. Iron deficiency anaemia in captive Malayan tapir calves (*Tapirus indicus*). J Zoo Wildl Med 2012;43(4):876–84.

Hernández AJC. Nutritional diseases, poultry and avian diseases. In: Van Alfen NK, editor. Encyclopedia of agriculture and food systems. Published by Academic Press; 2014.

Hetzel BS, Maberly GF. Iodine Chapter 2. In: Mertz W, editor. Trace elements in animal and human nutrition, vol. 2. Published by Academic Press; 1986. p. 139–97.

Hewlings S, Kalman D. Sulfur in human health. EC. Nutrition 2019;14 (9):785–91.

Higdon J, Drake VJ, Delage B, Mendel RR. Molybdenum. Linus Pauling Institute Oregon State University; 2013. Last updated 2013. Last accessed online on 24th June 2020 at http://lpi.oregonstate.edu/mic/minerals/molybdenum#inherited-Mo-cofactor-deficiency.

Hirano S, Suzuki KT. Exposure, metabolism, and toxicity of rare earths and related compounds. Environ Health Perspect 1996;104(Suppl 1):85–95. https://doi.org/10.1289/ehp.96104s185.

Hogan DB. Bismuth toxicity presenting as declining mobility and falls. Can Geriatr J 2018;21(4):307–9.

Hopkins L, Mohr HE. Vanadium as an essential nutrient. Fed Proc Fed Am Soc Exp Biol 1974;33:1773.

Horn S, Naidus E, Alper SL, Danziger J. Cesium-associated hypokalemia successfully treated with amiloride. Clin Kidney J 2015;8(3):335–8. https://doi.org/10.1093/ckj/sfv017.

Horovitz CT. Biochemistry of scandium and yttrium, part 2: biochemistry and applications. In: Biochemistry and physiology of scandium and yttrium. New York, NY, USA: Springer; 2000. p. 39–163.

Hosnedlova P, Kepinska M, Skalickova S, Fernandez C, Ruttkay-Nedecky B, Malevu TD, Sochor J, Baron M, Melcova M, Kizek R. Summary of new findings on the biological effects of selenium in selected animal species—a critical review. Int J Mol Sci Rev 2017;18:2209. https://doi.org/10.3390/ijms18102209.

Hosseini, et al. New mechanistic approach of inorganic palladium toxicity: impairment in mitochondrial Electron transfer. Metallomics 2016;8:252–9.

Hove EL, Herndon JF. Potassium deficiency in the rabbit as a cause of muscular dystrophy. J Nutr 1955;55:363–4.

HSE. Molybdenum deficiency in animals. Available online at https://www.imoa.info/HSE/environmental_data/experimental/deficiency_in_animals.php; 2020. Last accessed 29th June 2020.

Hui Y, Qing J, Xiran Z. Studies on effects of yttrium chloride and praseodymium chloride on frequency of micronucleus in human blood lymphocytes. Zhong Yuf Yix Zaz 1998;32(3):156–8 [Article in Chinese with an English abstract].

Hume ID. Nutrition of marsupials in captivity. Int Zoo Yearb 2005;39 (1):117–32.

Hurley LS, Keen CL. Manganese Chapter 5. In: Mertz W, editor. Trace elements in human and animal nutrition. 5th ed, vol. 1. London: Academic Press Inc; 1987. p. 185–215.

Hurwitz BE, Klaus JR, Llabre MM, Gonzalez A, Lawrence PJ, Maher KJ, et al. Suppression of human immunodeficiency virus type 1 viral load with selenium supplementation: a randomized controlled trial. Arch Intern Med 2007;167:148–54.

Hÿvarinen H, Sipilä T. Heavy metals and high pup mortality in the Saimaa ringed seal population in eastern Finland. Mar Pollut Bull 1984;15 (9):335–7.

IARC. IARC monographs on the evaluation of carcinogenic risks to humans. Beryllium, cadmium, mercury and exposures in the glass manufacturing industry. vol. 58. Published by the World Health Organisation, International Agency for Research on Cancer; 1993.

IARC. Welding, molybdenum trioxide, and indium tin oxide/IARC Working Group on the Evaluation of Carcinogenic Risks to Humans (2017) Lyon, France IARC monographs on the evaluation of carcinogenic risks to humans. vol. 118; 2017.

IARC. In: Kuempel ED, Ruder A, editors. Titanium dioxide; 2018. Available online https://monographs.iarc.fr/wp-content/uploads/2018/06/TR42-4.pdf Accessed 13th September 2020.

IARC. Gallium arsenide. Accessed on 15th September 2020 online at https://monographs.iarc.fr/wp-content/uploads/2018/06/mono86-8.pdf; 2020.

IARC Monograph. 100F Occupational exposures during iron and steel founding. Accessed on 15th September 2020 at https://monographs.iarc.fr/wp-content/uploads/2018/06/mono100F-34.pdf; 2006.

Igbokwe IO, Igwenagu E, Igbokwe NA. Aluminium toxicosis: a review of toxic actions and effects. Interdiscip Toxicol 2019;12(2):45–70. Published online 2020 Feb 20 https://doi.org/10.2478/intox-2019-0007.

Imai K, Nakamura M. In vitro embryotoxicity testing of metals for dental use by differentiation of embryonic stem cell test. Congenit Anom 2006; 46(1):34–8. 2003.

Imai K, et al. Effects on ES cell differentiation on the corroded surface of four silver alloys for dental use. J Oral Tissue Eng 2016;14(1):34–40.

IMFN. Institute of Medicine, Food and Nutrition Board. Dietary Reference Intakes: (2000) Vitamin C, vitamin E, selenium, and carotenoids. Washington, DC: National Academy Press; 2000.

IMOA. Molybdenum deficiency in animals. Accessed online 20th August 2020 https://www.imoa.info/HSE/environmental_data/experimental/deficiency_in_animals.php; 2020.

Inchem. Summaries & evaluations antimony trioxide and antimony trisulfide. Lyon, France: International Agency for Research on Cancer; 1989. Available online http://www.inchem.org/documents/iarc/vol47/47-11.html.

Inchem. Cadmium. Accessed on 30th June 2020 online at http://www.inchem.org/documents/jecfa/jecmono/v46je11.htm; 2020.

IRIS. Summary selenium and compounds; CASRN 7782-49-2. Integrated risk information system (IRIS). U.S. Environmental Protection Agency National Center for Environmental Assessment Chemical Assessment; 1991.

IRIS. Summary antimony trioxide; CASRN 1309-64-4 integrated risk information system (IRIS). U.S. Environmental Protection Agency, Chemical Assessment Summary National Center for Environmental Assessment; 1995. Last reviewed 2002- no changes made.

IRIS. Integrated risk information system IRIS chemical assessment summary cerium oxide and cerium compounds CASRN 1306-38-3U.S. Environmental Protection Agency; 2009. https://cfpub.epa.gov/ncea/iris/iris_documents/documents/subst/1018_summary.

pdf#nameddest=canceroral. Available online at Accessed 15th September 2020.

Irving F, Butler DG. Ammoniated mercury toxicity in cattle. Can Vet Jour 1975;16(9):280–4.

Ivanoff CS, Ivanoff AE, Hottela TL. Gallium poisoning: a rare case report. Food Chem Toxicol 2012;50(2):212–5.

Janssen DL. Tapiridae. In: Fowler ME, Miller RE, editors. Zoo & wild animal medicine. 5th ed. Philadelphia, PA: W.B. Saunders; 2003. p. 569–77.

Janssen DL, Rideout BR, Edwards MS. Tapir Medicine. In: Fowler ME, Miller RE, editors. Zoo and wildlife medicine. 4th ed. Philadelphia, PA: W.B. Saunders; 1999. p. 562–8.

Jha AM, Singh AC. Clastogenicity of lanthanides: induction of chromosomal aberration in bone marrow cells of mice in vivo. Mutat Res 1995;341(3):193–7. https://doi.org/10.1016/0165-1218(95)90009-8. 7529360.

Johnson LE. Molybdenum deficiency in MSD manual professional version molybdenum deficiency. Last full review/revision May 2020 | Content last modified May 2020. Available online at https://www.msdmanuals.com/professional/nutritional-disorders/mineral-deficiency-and-toxicity/molybdenum-deficiency; 2020. Last accessed 24th June 2020.

Johnson PE, Lykken GI. Manganese and calcium absorption and balance in young women fed diets with varying amounts of manganese and calcium. J Trace Elem Exp Med 1991;4:19–35.

Johnson CH, Van Tassell VJ. Acute barium poisoning with respiratory failure and rhabdomyolysis. Ann Emerg Med 1991;20:1138–42.

Johnson-Delaney CA. Ferret nutrition. Vet Clin Exot Anim 2014a;17:449–70. https://doi.org/10.1016/j.cvex.2014.05.008.

Johnson-Delaney CA. Captive marsupial nutrition. Vet Clin North Am Exot Anim Pract 2014b. https://doi.org/10.1016/j.cvex.2014.05.006. September 2014.

Jugdaohsingh R. Silicon and bone health. J Nutr Health Aging 2007;11 (2):99–110.

Kaiser C, Wernery U, Kinne J, Marker L, Liesegang A. The role of copper and vitamin a deficiencies leading to neurological signs in captive cheetahs (*Acinonyx jubatus*) and lions (*Panthera leo*) in the United Arab Emirates. Food Nutr Sci 2014;5:1978–90. Published Online October 2014 in SciRes http://www.scirp.org/journal/fns, https://doi.org/10.4236/fns.2014.520209.

Kakuschke A, Prange A. The influence of metal pollution on the immune system a potential stressor for marine mammals in the North Sea. Int J Comp Psychol 2007;20:179–93.

Kane E, Morris JG, Rogers QR, Ihrke PJ, Cupps PT. Zinc deficiency in the cat. J Nutr 1981;111(3):488–95. https://doi.org/10.1093/jn/111.3.488.

Kasprzak KS, Waalkes MP. The role of calcium, magnesium, and zinc in carcinogenesis. Adv Exp Med Biol 1986;206:497–515. https://doi.org/10.1007/978-1-4613-1835-4_35.

Keen CL, Hanna LA, Lanoue L, Uriu-Adams JY, Rucker RB, Clegg MS. Developmental consequences of trace mineral deficiencies in rodents: acute and long-term effects. J Nutr 2003;133:1477S–80S.

Kessler J. Mineral nutrition of goats. In: Morand-Fehr P, editor. Goat nutrition. Wageningen, Netherlands: Pudoc Publ; 1991. EAAP Publ. No. 46, 104.

Khanal DR, Knight AP. Selenium: its role in livestock health and productivity. J Agric Environ 2010;11:101–6.

Kibirige D, Luzinda K, Ssekitoleko R. Spectrum of lithium induced thyroid abnormalities: a current perspective. Thyroid Res 2013;6(1):3.

Kim KT, Eo MY, Nguyen TTH, Kim SM. General review of titanium toxicity. Int J Implant Dent 2019;5:10. Published online 2019 Mar 11 https://doi.org/10.1186/s40729-019-0162-x.

Klein GL. Aluminum toxicity to bone: a multisystem effect? The Journal title is Osteoporosis and Sarcopenia 2019;5(1):2–5. https://doi.org/10.1016/j.afos.2019.01.001.

Klevay LM. The ratio of Zinc to Copper of Diets in the United States. Nutr Rep Int 1975;11:237.

Klevay LM. Cardiovascular disease from copper deficiency–a history. J Nutr 2000;130:489S–92S.

Knight DA, Weisbrode SE, Schmall LM, et al. The effects of copper supplementation on the prevalence of cartilage lesions in foals. Equine Vet J 1990;22:426–32.

Knochel JP. Phosphorus. In: Shils ME, Olson JA, Shike M, Ross AC, editors. Modern nutrition in health and disease. 9th ed. Philadelphia: Lippincott Williams & Wilkins; 1999. p. 157–67.

Knochel JP, Barcenas C, Cotton JR, Fuller TJ, Haller R, Carter NW. Hypophosphatemia and rhabdomyolysis. J Clin Invest 1978;62 (6):1240–6. https://doi.org/10.1172/JCI109244.

Knowles SO, Grace ND. A recent assessment of the elemental composition of New Zealand pastures in relation to meeting the dietary requirements of grazing livestock. J Anim Sci 2014;92(1):303–10. https://doi.org/10.2527/jas.2013-6847 [Epub 2013 Nov 15].

Knox DP, Reid HW, Peters JG. An outbreak of selenium responsive unthriftiness in farmed red deer (Cervus elaphus). Vet Rec 1987;120:91–2.

Kock N, Foggin C, Kock MD, Kock R. Hemosiderosis in the black rhinoceros (*Diceros bicornis*): a comparison of free-ranging and recently captured with translocated and captive animals. J Zoo Wildl Med 1992;23(2):230–4.

Koltun P, Tharumarajah A. Life cycle impact of rare earth elements. Int Scholar Res Notice 2014;2014, 907536. 10 pages https://doi.org/10.1155/2014/907536.

Kostial K. Cadmium Chapter 5. In: Mertz W, editor. Trace elements in animal and human nutrition, vol. 2. Published by Academic Press; 1986. p. 319–37.

Krakoff IH, Newman RA. Clinical toxicologic and pharmacologic studies of gallium nitrate. Cancer 1979;44:1722–7.

Krewski D, Yokel RA, Nieboer E, Borchelt D, et al. Human health risk assessment for aluminium, aluminium oxide, and aluminium hydroxide. J Toxicol Environ Health B Crit Rev 2007;10 (Suppl 1):1–269. https://doi.org/10.1080/10937400701597766.

Krishnamachari KAVR. Fluorine Chapter 11. In: Mertz W, editor. Trace elements in human and animal nutrition. 5th ed, vol. 1. London: Academic Press Inc; 1987. p. 365–407.

Kuivenhoven M. Arsenic (Arsine) toxicity. Last updated 4/29/2019 available online at accessed online on 20th June 2020 https://www.statpearls.com/kb/viewarticle/17832; 2019.

Kumar A, Ali M, Pandey BM. Understanding the biological effects of thorium and developing efficient strategies for its decorporation and mitigation. Bhabha Atomic Research Centre Newsletter Issue 2013;335:p55–60. Nov-Dec 2013.

Kuntze A, Hunsdorff P. Haematological and biochemical findings (Ca, P, Mg, Fe, glucose, enzymes) in Asiatic female elephants. Erkran Zoo 1978;14:309–13.

Kuria SG, Tura IA, Mboga S, Walaga HK. Status of minerals in camels (*Camelus dromedarius*) in north eastern Kenya as evaluated from the blood plasma. Livestock Res Rural Develop 2013;25 (8):2013.

Kurokawa Y, Maekawa A, Takahashi M, Hayashi Y. Toxicity and carcinogenicity of potassium bromate—a new renal carcinogen. Environ Health Perspect 1990;87:309–35. https://doi.org/10.1289/ehp.9087309.

Kutty VR, Abraham S, Kartha CC. Geographical distribution of endomyocardial fibrosis in South Kerala. Int J Epidemiol 1996;25(6):1202–7.

Lall SP, Lewis-McCrea LM. Role of nutrients in skeletal metabolism and pathology in fish—an overview. Aquaculture 2007;267(2007):3–19.

Lamand M. Metabolism and requirements of microelements by goats. In: Proc. ITOVIC-INRA intern. symposium nutrition systems goat feeding, Tours, France, May 12-15, I; 1981. p. 210.

Lapresle J, Duckett S, Galle P, Cartier L. A case of aluminum encephalopathy in man. Seances Soc Biol Ses Fil 1975;169:282–5.

Larsson SC, Wolk A. Magnesium intake and risk of type 2 diabetes: a meta-analysis. J Intern Med 2007;262:208–14.

Laulicht F, Brocato J, Cartularo L, Vaughan J, Wu F, Kluz T, Sun H, Oksuz BA, Shen S, Paena M, Medici S, Zoroddu MA, Costa M. Tungsten-induced carcinogenesis in human bronchial epithelial cells. Toxicol Appl Pharmacol 2015;288(1):33–9. https://doi.org/10.1016/j.taap.2015.07.003. 2015 October 1.

Lavoie JP, Teuscher E. Massive iron overload and cirrhosis resembling haemochromatosis in a racing pony. Equine Vet J 1993;26(6):552–4.

Leggett R. Biokinetics of yttrium and comparison with its geochemical twin holmium. Office of Radiation and Indoor Air, U. S. Environmental Protection Agency (EPA), under Interagency Agreement DOE No. 1824-S581-A1, under contract No. DE-AC05-00OR22725 with UT-Battelle; 2017.

Lemus R, Venezia CF. An update to the toxicological profile for water-soluble and sparingly soluble tungsten substances. Crit Rev Toxicol 2015; 45(5):388–411. https://doi.org/10.3109/10408444.2014.1003422.

Lenntech. Health effects of neodymium. available online accessed 30th August 2020 at https://www.lenntech.com/periodic/elements/nd.htm#ixzz6a6LwzwGi; 2013.

Lenntech. Bismuth Bi—chemical properties of bismuth—health effects of bismuth—environmental effects of Bismuth. available online at https://www.lenntech.com/periodic/elements/bi.htm; 2020a. Last accessed 1st July 2020.

Lenntech. Promethium. Accessed 21st August 2020 online at https://www.lenntech.com/periodic/elements/pm.htm; 2020b.

Leon A, Bain S, Levick W. Hypokalaemic episodic polymyopathy in cats fed a vegetarian diet. Aus Vet J 1992;69:249–54.

Léonard A, Gerber GB. Mutagenicity, carcinogenicity and teratogenicity of aluminium. Mutat Res 1988;196(3):247–57. https://doi.org/10.1016/0165-1110(88)90009-7.

Léonard A, Gerber GB. Mutagenicity, carcinogenicity and teratogenicity of vanadium compounds. Mutat Res 1994;317(1):81–8.

Léonard A, Gerber GB. Mutagenicity, carcinogenicity and teratogenicity of antimony compounds. Mutat Res 1996;366(1):1–8. https://doi.org/10.1016/s0165-1110(96)90003-2.

Léonard A, Gerber GB. Mutagenicity, carcinogenicity and teratogenicity of thallium compounds. Mutat Res 1997;387(1):47–53.

Léonard A, Lauwery R. Mutagenicity, carcinogenicity and teratogenicity of beryllium. Mutat Res 1987;186(1):35–42. https://doi.org/10.1016/0165-1110(87)90013-3. PMID: 3600684.

Léonard A, Lauwerys RR. Carcinogenicity and mutagenicity of chromium. Mutat Res 1980;76(3):227–39.

Léonard A, Gerber GB, Jacquet P. Carcinogenicity, mutagenicity and teratogenicity of nickel. Mutat Res 1981;87(1):1–15. https://doi.org/10.1016/0165-1110(81)90002-6.

Léonard A, Gerber GB, Léonard F. Mutagenicity, carcinogenicity and teratogenicity of zinc. Mutat Res 1986;168(3):343–53. https://doi.org/10.1016/0165-1110(86)90026-6.

Levander O. Selenium. In: Mertz W, editor. Trace elements in human and animal nutrition. 5th ed. Orlando Fl: Academic Press; 1986. p. 209–79.

Lewis RC, Meeker JD. Biomarkers of exposure to molybdenum and other metals in relation to testosterone among men from the United States National Health and Nutrition Examination Survey 2011–2012. Fertil Steril 2015;103:172–8.

Lieberherr M, Grosse B, et al. Calcif Tissue Int 1982;34:280.

Lightfoot TL, Yeager JM. Bird toxicity and related environmental concerns. Vet Clin North Am Exot Anim Pract 2008;11(2):229–59.

Lindemann M, Lepine A, Hayes M. Evaluation of chromium supplementation on several insulin-controlled parameters in beagles. Recent advances in canine and feline nutrition. Iams Nutr Symp Proc 2000;3:243–54.

Linhart C, Talasz H, Morandi EM, Exley C, Lindner HH, Taucher S, Egle D, Hubalek M, Concin N, Ulmer H. Use of underarm cosmetic products in relation to risk of breast cancer: a case-control study. EBioMedicine 2017;21:78–85.

Lintzenich B, Ward A, Edwards M, Griffin M, Robbins C. Polar bear nutrition guidelines. Polar Bears International; 2006. www.polarbearsinternational.org. Accessed 19th August 2020.

Linus Pauling Institute. Magnesium. Linus Pauling Institute, Micronutrient Information Center. Oregon State University; 2020. Available online at https://lpi.oregonstate.edu/mic/minerals/magnesium#function. Accessed 2nd July 2020.

Long N. Arsenic toxicity. Last update August 25, 2019 available online at Life in the Fastlane https://litfl.com/arsenic-toxicity/; 2019.

López-Alfaro C, Coogan SCP, Robbins CT, Fortin JK, Nielsen SE. Assessing nutritional parameters of brown bear diets among ecosystems gives insight into differences among populations. PLoS One 2015;10(6). https://doi.org/10.1371/journal. pone.0128088, e0128088.

Lozoff B, et al. J Pediatr 1982;101:948.

Lyon AW, Mayhew WJ. Cesium toxicity: a case of self-treatment by alternate therapy gone awry. Ther Drug Monit 2003;25(1):114–6.

MacFarquhar JK, Broussard DL, Melstrom P, Hutchinson R, Wolkin A, Martin C, Burk RF, Dunn JR, Green AL, Hammond R, Schaffner W, Jones TF. Acute selenium toxicity associated with a dietary supplement. Arch Intern Med 2010;170(3):256–61. https://doi.org/10.1001/archinternmed.2009.495. 2010 February 8.

Mackintosh CG, Gill J, Turner K. Selenium supplementation of young red deer (*Cervus elaphus*). N Z Vet J 1989;37:143–5.

Macwhirter P. Malnutrition. In: Branson W Ritchie and Greg J Harrison. Avian medicine: principles and application Chapter 31; 1984. p. 842–861. Published by www.avianmedicine.net

Malhotra N, Hsu H-S, Liang S-T, Lee J-S, Ger T-R, Hsiao C-D. An updated review of toxicity effect of the rare earth elements (REEs) on aquatic organisms. Animals 2020;10(9):1663. https://doi.org/10.3390/ani10091663.

Malloy MH, Graubard B, Moss H, McCarthy M, Gwyn S, Vietze P, Willoughby A, Rhoads GG, Berendes H. Hypochloremic metabolic alkalosis from ingestion of a chloride-deficient infant formula: outcome 9 and 10 years later. Pediatrics 1991;87(6):811–22.

Marathe MR, Thomas GP. Embryotoxicity and teratogenicity of lithium carbonate in Wistar rat. Toxicol Lett 1986;34(1):115–20. https://doi.org/10.1016/0378-4274(86)90153-0.

Markowski M, Kaliński A, Skwarska J, et al. Avian feathers as bioindicators of the exposure to heavy metal contamination of food. Bull Environ Contam Toxicol 2013;91(3):302–5.

Masok FB, Masiteng PL, Mavunda RD, Maleka PP. Health effects due to radionuclides content of solid minerals within Port of Richards

Bay. South Africa Int J Environ Res Public Health 2016;2016 (13):1180. https://doi.org/10.3390/ijerph13121180.

Massari S, Ruberti M. Rare earth elements as critical raw materials: focus on international markets and future strategies. Resour Pol 2013;38 (1):36–43.

Mazzaro LM, Johnson SP, Fair PA, et al. Iron indices in bottlenose dolphins (*Tursiops truncatus*). Comp Med 2012;62(6):508–15.

McCall AS, Cummings CF, Bhave G, Vanacore R, Page-McCaw A, Hudson BG. Bromine is an essential trace element for assembly of collagen IV scaffolds in tissue development and architecture. Cell 2014;157(6):1380–92.

McDonagh M, et al. A systematic review of public water fluoridation. NHS Centre for Reviews and Dissemination, University of York; 2000. Available on line. Last accessed 12th June 2020 https://www.nhs.uk/conditions/fluoride/documents/crdreport18.pdf.

McDowell LR. Minerals in animal and human. San Diego California: Nutrition Academic Press Inc; 1992a. p. 92101.

McDowell LR (1992b) Calcium and phosphorus Chapter 2 p 26-73 in Minerals in animal and human nutrition Academic Press Inc San Diego California 92101.

McDowell LR (1992c) Sodium and chlorine Chapter 3 p 78-82 in Minerals in animal and human nutrition Academic Press Inc San Diego California 92101.

McDowell LR (1992d) Potassium Chapter 4 p 98-113 in Minerals in animal and human nutrition Academic Press Inc San Diego California 92101.

McDowell LR (1992e) Magnesium Chapter 5 p 115-135 in Minerals in animal and human nutrition Academic Press Inc San Diego California 92101.

McDowell LR (1992f) Sulfur Chapter 6 p 137-150 in Minerals in animal and human nutrition Academic Press Inc San Diego California 92101.

McDowell LR (1992g) Iron Chapter 7 p 152-174 in Minerals in animal and human nutrition Academic Press Inc San Diego California 92101.

McDowell LR (1992h) Copper and molybdenum Chapter 8 p 176-202 in Minerals in animal and human nutrition Academic Press Inc San Diego California 92101.

McDowell LR (1992i) Cobalt Chapter 9 p 205-222 in Minerals in animal and human nutrition Academic Press Inc San Diego California 92101.

McDowell LR (1992j) Iodine Chapter 10 p 224-244 in Minerals in animal and human nutrition Academic Press Inc San Diego California 92101.

McDowell LR (1992k) Manganese Chapter 11 p 246-263 in Minerals in animal and human nutrition Academic Press Inc San Diego California 92101.

McDowell LR (1992l) Zinc Chapter 12 p 265-292 in Minerals in animal and human nutrition Academic Press Inc San Diego California 92101.

McDowell LR (1992m) Selenium Chapter 13 p 294-330 in Minerals in animal and human nutrition Academic Press Inc San Diego California 92101.

McDowell LR (1992n) Fluorine Chapter 14 p 333-349 in Minerals in animal and human nutrition Academic Press Inc San Diego California 92101.

McDowell LR (1992o) Aluminium, arsenic, cadmium, lead and mercury Chapter 15 p 352-364 in Minerals in animal and human nutrition Academic Press Inc San Diego California 92101.

McDowell LR (1992p) Newly discovered and trace elements Chapter 16 p 366-379 in Minerals in animal and human nutrition Academic Press Inc San Diego California 92101.

McLachlan DRC, et al. Seventh international symposium on trace elements in man and animals (TEMA-7) Dubrovnick Yugoslavia (Abstr); 1990. p. 42.

McLaughlin AIG, Kazantzis G, King E, et al. Pulmonary Fibrosis and Encephalopathy associated with the inhalation of aluminium dust. Br J Ind Med 1962;19:253–63.

Melber C, Mangelsdorf I. Palladium toxicity in animals and in in vitro test systems—an overview. In: Zereini F, Alt F, editors. Palladium emissions in the environment. Berlin, Heidelberg: Springer; 2006.

Melber C, Keller D, Mangelsdor I. WHO environmental health criteria 226 palladium. Published under the joint sponsorship of the United Nations environment programme, the international labour and the World Health Organization, and produced within the framework of the inter-organization programme for the sound Management of Chemicals. Geneva: WHO; 2002.

Melnikov P, Zanoni LZ. Clinical effects of cesium intake. Biol Trace Elem Res 2009. https://doi.org/10.1007/s12011-009-8486-7.

Melnikov P, Zanoni LZ. Clinical effects of cesium intake. Biol Trace Elem Res 2010;135(1–3):1–9. https://doi.org/10.1007/s12011-009-8486-7. PMID 19655100.

Mendy A, Gasana J, Vieira ER. Urinary heavy metals and associated medical conditions in the US adult population. Int J Environ Health Res 2012;22:105–18.

Mertz M. Lithium Chapter 8. In: Mertz W, editor. Trace elements in animal and human nutrition, vol. 2. Published by Academic Press; 1986. p. 391–7.

Middleton D, Kowalski P. Advances in identifying beryllium sensitization and disease. Int J Environ Res Public Health 2010;7(1):115–24. https://doi.org/10.3390/ijerph7010115.

Mills CF, Davis GK. Molybdenum Chapter 13. In: Mertz W, editor. Trace elements in human and animal nutrition. 5th ed, vol. 1. London: Academic Press Inc; 1987. p. 429–57.

Mitsui S, Ogata A, Yanagie H, Kasano H, Hisa T, Yamase T, Eriguchi M. antitumor activity of polyoxomolybdate, [NH3Pri]6 [Mo7O24].3H2O, against, human gastric cancer model. Biomed Pharmacother 2006;60:353–8.

Moreira A, et al. Potential impacts of lanthanum and yttrium through embryotoxicity assays with *Crassostrea gigas*. Ecol Indic 2020;108:105687. https://doi.org/10.1016/j.ecolind.2019.105687.

Morris ER. Iron Chapter 4. In: Mertz W, editor. Trace elements in human and animal nutrition. 5th ed, vol. 1. London: Academic Press Inc; 1987. p. 79–126.

MSD Veterinary. available online https://www.msdvetmanual.com/toxicology/mercury-poisoning/overview-of-mercury-poisoning; 2020. Last accessed 26th June 2020.

Mude S, SyaamaSundar N. Vitiligo in buffaloes. Vet World 2009;2(7):282.

Mullaney TP, Brown CM. Iron toxicity in neonatal foals. Equine Vet J 1988;20(2):119–24.

Nagano K, Nishizawa T, Umeda Y, Kasai T, Noguchi T, Gotoh K, Ikawa N, Eitaki Y, Kawasumi Y, Yamauchi T, Heihachiro H, Fukushima S. Inhalation carcinogenicity and chronic toxicity of indium-tin oxide in rats and mice. J Occup Health 2011;53(3):175–87.

Naghii MR, Samman S. The role of boron in nutrition and metabolism. Prog Food Nutr Sci 1993;17(4):331–49.

Nalabotu SK, Kolli MB, Triest WE, Ma JY, Manne N, Katta A, Addagarla HS, Rice KM, Bough ER. Intratracheal instillation of cerium oxide nanoparticles induces hepatic toxicity in male Sprague-Dawley rats. Int J Nanomedicine 2011;6:2327.

Namagondlu G, Main N, Yates L, Mooney J, Sathyamurthy S, Daryanani I, Crowe A, Ledson T, Banerjee A. Lanthanum associated abnormal liver function tests in two patients on dialysis: a case report. J Med Case Reports 2009;3:9321.

NAS. Dietary reference intakes for vitamin A, vitamin K, arsenic, boron, chromium, copper, iodine, iron, manganese, molybdenum, nickel, silicon, vanadium, and zinc: a report of the Panel on Micronutrients. Standing Committee on the Scientific Evaluation of Dietary Reference Intakes. Washington DC: Food and Nutrition Board, Institute of Medicine. National Academy of Sciences; 2000.

Navia JM, Hunt CE. Nutrition, nutritional diseases and nutrition research application. In: Wagner JE, Manning PJ, editors. Biology of the Guinea pig. New York: Academic Press; 1976. p. 235–65.

NCBI. Toxicological profile for uranium, https://www.ncbi.nlm.nih.gov/books/NBK158798/; 2020. Accessed online July 2nd 2020.

NCI National Cancer Institute. Bioassay of titanium dioxide for possible carcinogenicity CAS No. 13463-67-7. Online https://ntp.niehs.nih.gov/ntp/htdocs/lt_rpts/tr097.pdf; 1979.

Newman LS, et al. The natural history of beryllium sensitization and chronic beryllium disease. Environ Health Perspect 1996a;5:937–43.

Newman LS, et al. Significance of the blood beryllium lymphocyte proliferation test. Environ Health Perspect 1996b;5:953–6.

Newnham RE. Seventh international symposium on trace elements in man and animals TEMA-7 Dubrovnik Yugoslavia (Abstr); 1990. p. 39.

Nichols CP, Gregory NG, Goode N, Gill RMA, Drewe JA. Regulation of bone mineral density in the grey squirrel, *Sciurus carolinensis*: bioavailability of calcium oxalate, and implications for bark stripping. J Anim Physiol Anim Nutr (Berl) 2018;102(1):330–6. https://doi.org/10.1111/jpn.12740.

Nielsen FH. Ultratrace elements in nutrition. Ann Rev Nutr 1984;4:21–41.

Nielsen FH. Other elements Chapter 10. In: Mertz W, editor. Trace elements in animal and human nutrition, vol. 2. Published by Academic Press; 1986. p. 415–54.

Nielsen FH. Nickel Chapter 4. In: Mertz W, editor. Trace elements in human and animal nutrition. 5th ed, vol. 1. London: Academic Press Inc; 1987a. p. 245–66.

Nielsen FH. Vanadium Chapter 9. In: Mertz W, editor. Trace elements in human and animal nutrition. 5th ed, vol. 1. London: Academic Press Inc; 1987b. p. 282–91.

Nielsen FH. Zirconium. In: Mertz W, editor. Trace elements in human and animal nutrition. 5th ed, vol. 2. London: Academic Press Inc; 1987c. p. 447–8.

Nielsen FH. In: Smith KT, editor. Trace minerals in foods. New York: Marcel Dekker; 1988. p. 357.

Nielsen FH. Ultratrace elements in nutrition : current knowledge and speculation. J Trace Elem Exp Med 1998;11:251–74.

Nigra AE, Ruiz-Hernandez A, Redon J, Navas-Acien A, Tellez-Plaza M. Environmental metals and cardiovascular disease in adults: a systematic review beyond lead and cadmium. Curr Environ Health Rep 2016;3(4):416–33. https://doi.org/10.1007/s40572-016-0117-9.

NIH. Selenium institute of medicine, food and nutrition. (updated 2020); Last accessed online on 25th July 2020 at https://ods.od.nih.gov/factsheets/Selenium-HealthProfessional/; 2020.

NIH Molybdenum. National institute of health office of dietary supplements molybdenum. Last Updated: June 3, 2020. Accessed on line on 24th June 2020 at https://ods.od.nih.gov/factsheets/Molybdenum-HealthProfessional/; 2020.

Nimni ME, Han B. Cordoba F (2007) Are we getting enough sulfur in our diet? Nutr Metab 2007;4:24. https://doi.org/10.1186/1743-7075-4-24.

NIOSH Aluminium. Aluminium in NIOSH pocket guide to chemical hazards. The National Institute for Occupational Safety and Health (NIOSH), Centres for Disease Control and Prevention (CDC); 2020. available online at https://www.cdc.gov/niosh/npg/npgd0022.html. Accessed 11th September 2020.

NIOSH Arsenic, 2020. Arsenic in NIOSH Pocket Guide to Chemical Hazards. The National Institute for Occupational Safety and Health (NIOSH), Centres for Disease Control and Prevention (CDC). Available from: https://www.cdc.gov/niosh/topics/arsenic/ (Accessed 11 September 2020).

NIOSH Antimony. Antimony in NIOSH pocket guide to chemical hazards. The National Institute for Occupational Safety and Health (NIOSH), Centres for Disease Control and Prevention (CDC); 2020. available online accessed 11th September 2020 at https://www.cdc.gov/niosh/npg/npgd0036.html.

NIOSH Boron. Boron in NIOSH pocket guide to chemical hazards. The National Institute for Occupational Safety and Health (NIOSH), Centres for Disease Control and Prevention (CDC); 2020. available online at https://www.cdc.gov/niosh/npg/npgd0060.html. Accessed 11th September 2020.

NIOSH Bromine. Bromine in NIOSH pocket guide to chemical hazards. The National Institute for Occupational Safety and Health (NIOSH), Centres for Disease Control and Prevention (CDC); 2020. available online at https://www.cdc.gov/niosh/npg/npgd0064.html. Accessed 11th September 2020.

NIOSH Cadmium. Cadmium in NIOSH Pocket Guide to Chemical Hazards. The National Institute for Occupational Safety and Health (NIOSH), Centres for Disease Control and Prevention (CDC); 2020. https://www.cdc.gov/niosh/npg/npgd0087.html. available online at Accessed 11th September 2020.

NIOSH Germanium. National institute for occupational safety and health NIOSH pocket guide to chemical hazards germanium. Accessed online on 15th September 2020 at https://www.cdc.gov/niosh/npg/npgd0300.html; 2019.

NIOSH Tellurium. National institute for occupational safety and health centers for disease control and prevention. NIOSH pocket guide to chemical hazards tellurium, https://www.cdc.gov/niosh/npg/npgd0587.html; 2020. Accesses 14th September 2020.

NIOSH Titanium. Occupational exposure to titanium dioxide. Titanium carcinogenicity. National Institute for Occupational Safety and Health (NIOSH); 2011. Available online https://www.cdc.gov/niosh/docs/2011–160/pdfs/2011–160.pdf Accessed 13th September 2020.

NIOSH Yttrium (2020) Yttrium in NIOSH pocket guide to chemical hazards. The National Institute for Occupational Safety and Health (NIOSH), Centres for Disease Control and Prevention (CDC) available online at https://www.cdc.gov/niosh/npg/npgd0673.html Accessed 12th September 2020.

NJDHH. New Jersey Department of health hazardous substance fact sheet tellurium, https://nj.gov/health/eoh/rtkweb/documents/fs/1777.pdf; 2009. accessed 14th September 2020.

NORD. Rare disease database Menkes disease. Accessed online at https://rarediseases.org/rare-diseases/menkes-disease/; 2020a. last accessed 20th August 2020.

NORD. Heavy metal poisoning. National Organization for Rare Disorders; 2020b. Available online at https://rarediseases.org/rare-diseases/heavy-metal-poisoning/. Accessed 13th September 2020.

NORD Beryllium. National organisation for rare diseases (NORD). Rare disease database—berylliosis. Available online at https://rarediseases.org/rare-diseases/berylliosis/; 2009. Last accessed 1st July 2020.

Norose N, Terai M, Norose K. Manganese deficiency in a child with very short bowel syndrome receiving long-term parenteral nutrition. J Trace Elem Exp Med 1992;5:100–1 [abstract].

Novikova B, Dobenecker E, Kienzle. Secondary alimentary hyperparathyroidism in meerkat (*Suricata suricatta*) Anna Lineva*. E. accessible online on 26th September 2020 at https://www.academia.edu/31514130/Secondary_alimentary_hyperparathyroidism_in_meerkat_suricata_suricatta_Clinical_case; 2020.

NRC. Nutrient requirements of rabbits. Washington, D.C.: National Research Council Published by National Academy of Sciences; 1977.

NRC. Mineral tolerance of domestic animals. Washington, DC: National Research Council. National Academy of Sciences; 1980.

NRC. Nutrient requirements of foxes and mink. Washington, D.C: National Research Council. National Academy Press; 1982.

NRC. Nutrient requirements of the horse. Washington: National Research Council. National Academy Press; 1989.

NRC. Nutrient requirements of fish. Washington, D.C: National Research Council. National Academies Press; 1993.

NRC. Nutrient requirements of laboratory animals: Fourth revised edition, National Research Council. Washington, DC: The National Academies Press; 1995.

NRC. Nutrient requirements of dairy cattle: Seventh revised edition, National Research Council 2001. Washington, DC: The National Academies Press; 2001.

NRC. Nutrient requirements of nonhuman primates: second revised edition national research council. Washington DC: The National Academies Press; 2003. p. 20001.

NRC. Mineral tolerance of animals second revised edition. Washington, DC: National Research Council. The National Academies Press; 2005.

NRC. Nutrient requirements of dogs and cats. Washington DC: National Research Council. The National Academies Press; 2006.

NRC. Nutrient requirements of fish and shrimp. Washington, D.C: National Research Council 2011 National Academies Press; 2011.

NRC. Nutrient requirements of small ruminants (sheep, goats, cervids and new world camelids). Washington DC: National Research Council 2014 National Academies Press; 2014.

NRC. Examining special nutritional requirements in disease states: proceedings of a workshop (2018). National Academies Press; 2018. http://nap.edu/25164.

Nuclear Radiation and Health Effects published by the World Nuclear Association (updated April 2020) n.d.Available online at: https://www.world-nuclear.org/information-library/safety-and-security/radiation-and-health/nuclear-radiation-and-health-effects.aspx Last accessed 5th September 2020.

Nuttall KL. Review: evaluating selenium poisoning. Ann Clin Lab Sci 2006;36(4):2006.

Obendorf DL. Wombats and vitamin D don't mix. Vet Pathol Rep 1989;26:31.

Ohnishi K, Usuda K, Nakayama S, Sugiura Y, Kitamura Y, Kurita A, Tsuda Y, Kimura M, Koichi K. Distribution, elimination, and renal effects of single oral doses of europium in rats. Biol Trace Elem Res 2011;143:1054–63.

OHS Database. Occupational Health Services, Inc. 1993 (August) MSDS for Sulfur. Secaucus, NJ: OHS Inc; 1993.

Olias P, Mundhenk I, Bothe M, Ochs A, Gruber AD, Klopfleisch R. Iron overload syndrome in the black rhinoceros (*Diceros bicornis*): microscopical lesions and comparison with other Rhinoceros species. J Comp Pathol 2012;147:542e549.

Olson OE. Selenium toxicity in animals with emphasis on Man. J Am Colleg Toxicol 1986;5(1):45–70.

Org M. Titanium allergy, https://melisa.org/titanium/?v=79cba1185463; 2020. accessed July 2nd 2020.

Oruç HH, Uzunoğlu. Suspected iron toxicity in dairy cattle. Uludag Univ J Fac Vet Med 2009;28(1):75–7.

OSHA Contributors. Occupational safety and health guideline for yttrium and compounds. United States Occupational Safety and Health Administration; 2007. Archived from the original on March 2.

Ott M, Stegmayr B, Salander Renberg E, Werneke U. Lithium intoxication: incidence, clinical course and renal function—a population-based retrospective cohort study. J Psychopharmacol (Oxford) 2016;30 (10):1008–19.

Packer M. Cobalt cardiomyopathy: a critical reappraisal in light of a recent resurgence. Circulation 2016;9:12. Available online at https://doi.org/10.1161/CIRCHEARTFAILURE.116.003604.

Paglia DE, Dennis P. Role of chronic iron overload in multiple disorders of captive black rhinoceroses (*Diceros bicornis*). Proc Am Assoc Zoo Vet 1999;163–71.

Patlolla AK, Hackett D, Tchounwou PB. Genotoxicity study of silver nanoparticles in bone marrow cells of Sprague-Dawley rats. Food Chem Toxicol 2015;85:52–60. https://doi.org/10.1016/j.fct.2015.05.005.

Patterson RE, Haut M, Montgomery CA, Lowensohn HS, McQuilken CT, Djuh YY, Huott A, Olsson RA. Natural history of potassium-deficiency myopathy in the dog: role of adrenocorticosteroid in rhabdomyolysis. J Lab Clin Med 1983;102(4):565–76.

Pavlakis N, Pollock CA, McLean G, Bartrop R. Deliberate overdose of uranium: toxicity and treatment. Nephron 1996;72:313–7. https://doi.org/10.1159/000188862.

Pearce SG, Firth EC, Grace ND, et al. Effect of copper supplementation on the evidence of developmental orthopaedic disease in pasture-fed New Zealand thoroughbreds. Equine Vet J 1998;30:211–8.

Pearson EG, Hedstrom OR, Poppenga RH. Hepatic cirrhosis and hemochromatosis in three horses. J Am Vet Med Assoc 1994;204(7):1053–6.

Pechova A, Pavlata L. Chromium as an essential nutrient: a review. Medicina 2007;52(1):1–18. review article.

Pederson H, Mow T. Hypomagnesemia and mitral valve prolapse in Cavalier King Charles Spaniels. J Vet Med A 1998;45:607–14.

Penland JG, Prohaska JR. Abnormal motor function persists following recovery from perinatal copper deficiency in rats. J Nutr 2004;134:1984–8.

Perez-D'Gregorio RE, Miller RK, Baggs RB. Maternal toxicity and teratogenicity of tellurium dioxide in the Wistar rat: relationship to pair-feeding. Reprod Toxicol 1988;2(1):55–61. https://doi.org/10.1016/s0890-6238(88)80009-1.

PigSite. The PigSite. online https://thepigsite.com/disease-guide/mercury; 2020. Last accessed 26th June 2020.

Pinto J, et al. Increased urinary riboflavin excretion resulting from boric acid ingestion. J Lab Clin Med 1978;92:126.

Pitt MA. Molybdenum toxicity: Interactions between copper, molybdenum and sulphate. Agents and Actions 1976;6:756–758.

Pizzorno L. Nothing boring about boron. Integr Med (Encinitas) 2015;14 (4):35–48.

Prá D, Franke SI, Giulian R, et al. Genotoxicity and mutagenicity of iron and copper in mice. Biometals 2008;21(3):289–97. https://doi.org/10.1007/s10534-007-9118-3.

Prabhu PAJ, Schrama JW, Kaushik SJ. Mineral requirements of fish: a systematic review. Rev Aquac 2016;8:172–219.

Puschner B, Poppenga RH. Lead and zinc intoxication in companion birds. Comp Cont Educ Vet 2009;31(1):E1–12.

Pye GW, et al. Metabolic bone disease in juvenile koalas (phascolartcos cinereus). J Zoo Wildl Med 2013;44(2):273–9.

Qu A, Wang CR, Bo J 2004, Research on the cytotoxic and genotoxic effects of rare-earth element holmium to Vicia fabia. Hereditas 26 (2) 195–201. PMID: 15639987.

Quansah R, Armah FA, Essumang DK, Luginaah I, Clarke E, Marfoh K, Cobbina SJ, Nketiah-Amponsah E, Namujju PB, Obiri S, Dzodzomenyo M. Association of arsenic with adverse pregnancy outcomes/

infant mortality: a systematic review and meta-analysis. Environ Health Perspect 2015;123:412–21. https://doi.org/10.1289/ehp.1307894.

Quarterman J. Lead Chapter 4. In: Mertz W, editor. Trace elements in animal and human nutrition, vol. 2. Published by Academic Press; 1986. p. 281–307.

Qureshi AS, et al. Calcium and phosphorus: backbone of camel health—a review. EC Vet Sci 2020;5(3):01–9.

Radcliffe RW, Khairani KO. Health of the forest rhinoceros of Southeast Asia in Fowler's zoo and wild animal medicine current therapy. vol. 9; 2019. p. 2019.

Radio-Analysis Quality Management System, n.d. Based on ISO/IEC Standard 17025 (SANAS)-Accreditation Schedule T0111. Available online: http://www.sanas.co.za/af-directory/calibration_list.php?s_lab_no=&s_lab_name=necsa&s_disciplines=&s_accreditation_status=Accredited&s_phys_city=&s_province (accessed on 16 July 2016).

Rahimzadeh MR, Rahimzadeh MR, Kazemi S, Moghadamnia A-A. Caspian J Intern Med 2017;8(3):135–45. https://doi.org/10.22088/cjim.8.3.135.

RAIS. Mercury risk assessment information system, https://rais.ornl.gov/tox/profiles/methyl_mercury_f_V1.html; 2020.

Rallis T, Spais AG, Papasteriadis A, Agiannidis A, Leondidis S. Iron toxicity in sheep. J Trace Elem Electrolyt Health Dis 1989;3(3):131. 01 Sep 1989.

Randhawa C, Randhawa S, Sood N. Effect of molybdenum induced copper deficiency on peripheral blood cells and bone marrow in Buffalo calves Asian-Australas. J Anim Sci 2002;15(4):509–15. https://doi.org/10.5713/ajas.2002.509.

Rashedy AH, Solimany AA, Ismail AK, Wahdan MH, Saban KA. Histopathological and functional effects of antimony on the renal cortex of growing albino rat. Int J Clin Exp Pathol 2013;6(8):1467–80. www.ijcep.com. ISSN:1936-2625/IJCEP1304038.

Ratnaike RN. Acute and chronic arsenic toxicity. Postgrad Med J 2003;79:391–6.

Redman HC, McClellan RO, Jones RK, Boecker BB, Chiffelle TL, Pickrell JA, Rypka EW. Toxicity of 137-CsCl in the beagle. Early biological effects. Radiat Res 1972;50(3):629–48.

Richardson JA. Implications of toxins in clinical disorders. In: Harrison GJ, Lightfoot TL, editors. Clinical avian medicine. Palm Beach, FL: Spix Publishing; 2006.

Rim KT, Koo KH, Park JS. Toxicological evaluations of rare earths and their health impacts to workers: a literature review. Saf Health Work 2013;4(1):12–26.

Roberts M, Kohn F. Habitat use, foraging behavior, and activity patterns in reproducing Western tarsiers, *Tarsius bancanus*, in captivity: a management synthesis. Zoo Biol 1993;12:217–32.

Rodriguez-Moran M, Simental Mendia LE, Zambrano Galvan G, Guerrero-Romero F. The role of magnesium in type 2 diabetes: a brief based-clinical review. Magnes Res 2011;24:156–62.

Rogers PAM, Arora SP, Fleming GA, Crinion RAP, McLaughlin JG. Selenium toxicity in farm animals: treatment and prevention. Irish Vet J 1990;43:151–3.

Roohani N, Hurrell R, Kelishadi R, Schulin R. Zinc and its importance for human health: an integrative review. J Res Med Sci 2013;18(2):144–57.

Rossman MD. Chronic beryllium disease: diagnosis and management. Environ Health Perspect 1996;5:945–7.

Roswag VA, Becker NI, Mohlbach E, Encarnaceo JA. Mangelerscheinungen bei Fledermaus-Pfleglingen und vorsorgliche Gegenma Onahmen Nyctalus (N.F.), Berlin. Heft 2011;3–4:239–45.

Roth. Europium ICP standard solution 10000 mg/l Eu safety data sheet version: 1.0 en, https://www.carlroth.com/; 2016.

Rothman JM, Dierenfeld ES, Hintz HF, Pell AN. Nutritional quality of gorilla diets: consequences of age, sex, and season. Oecologia 2008;155(1):111–22. https://doi.org/10.1007/s00442-007-0901-1 [Epub 2007 Nov 13].

Rucki M, Kejlova K, Vlkova A, Jirova D, Dvorakova MD, Svobodova L, Kandarova H, Letasiova S, Kolarova H, Mannerstrom M, Heinonene T. Evaluation of toxicity profiles of rare earth elements salts (lanthanides). J Rare Earths 2020. Available online 28th February 2020 https://doi.org/10.1016/j.jre.2020.02.011.

Sach F, Dierenfeld ES, Langley-Evans SC, Watts MJ, Yon L. African savanna elephants (*Loxodonta africana*) as an example of a herbivore making movement choices based on nutritional needs. Peer J 2019;7, e6260. Published 2019 Feb 1 https://doi.org/10.7717/peerj.6260.

Sanai T, Okuda S, Onoyama K, Oochi N, Takaichi S, Mizuhira V, Fujishima M. Chronic tubulointerstitial changes induced by germanium dioxide in comparison with carboxyethylgermanium sesquioxide. Kidney Int 1991;40(5):882–90.

Sanchez-Migallon D. Heavy metal toxicity. In: Tully TN, editor. Clinical veterinary advisor: birds and exotic pets. St. Louis, MO: Elsevier; 2012. p. 192–4.

Santa Cruz. Biotechnology, Inc. Terbium. Available online accessed 7th June 2020 http://datasheets.scbt.com/sc-224297.pdf; 2020.

Schaumberg DA, Mendes F, Balaram M, Dana MR, Sparrow D, Hu H. Accumulated lead exposure and risk of age-related cataract in men. JAMA 2004;292:2750–4.

Schauss AG. Nephrotoxicity and neurotoxicity in humans from organogermanium compounds and germanium dioxide. Biol Trace Elem Res 1991;29(3):267–80.

Schmid DA, Pye GW, Hamlin-Andrus CC, Ellis WA, Fercovitch FB, Ellersieck MR, Chen TC, Holick MF. Fat-soluble vitamin and mineral comparisons between zoo-based and free-ranging koalas (*Phascolarctos cinereus*). J Zoo Wildl Med 2013;44(4):1079–82. https://doi.org/10.1638/2012-0207R1.1.

Schmidt M. Zinc deficiency, presumptive secondary immune deficiency and hyperkeratosis in an Asian elephant: a case report. Greensboro, North Carolina: American Association of Zoo Veterinarians; 1989. p. 23–31.

Schoemaker NJ, Lumeij JT, Dorrestein GM, Beynen AC. Nutrition-related problems in pet birds. Tijdschr Diergeneeskd 1999;124(2):39–43. 1999 Jan 15.

Schroeder HA, Mitchener M. Scandium, chromium (VI), gallium, yttrium, rhodium, palladium, indium in mice: effects on growth and life span. J Nutr 1971;101(10):1431–7.

Schulze MB, Schulz M, Heidemann C, Schienkiewitz A, Hoffmann K, Boeing H. Fiber and magnesium intake and incidence of type 2 diabetes: a prospective study and meta-analysis. Arch Intern Med 2007;167:956–65.

Schwarz K. Silicon, fibre, and atherosclerosis. Lancet 1977;1(8009):454–7. https://doi.org/10.1016/s0140-6736(77)91945-6.

Schwarz K, Milne DB. Growth-promoting effects of silicon in rats. Nature 1972;239:333–4.

Scinicariello F, Buser MC, Feroe AG, Attanasio R. Antimony and sleep-related disorders: NHANES 2005–2008. Environ Res 2017;156:247–52. https://doi.org/10.1016/j.envres.2017.03.036.

Scott ML, Nesheim MC, Young RJ. Nutrition of the chicken. Itheca, New York: ML Scott and Associates; 1982.

Scott BL, McCleskey TM, Chaudhary A, Hong-Geller E, Gnanakaran S. The bioinorganic chemistry and associated immunology of chronic beryllium disease. Chem Commun (Camb) 2008;25:2837–47. https://doi.org/10.1039/b718746g.

Sedman AB, Alfrey AC, Miller NL, Goodman WG. Tissue and cellular basis for impaired bone formation in Aluminum-related Osteomalacia in the pig. J Clin Invest 1987;79:86–92. January 1987.

See KA, Lavercombe PS, Dillon J, Ginsberg R. Accidental death from acute selenium poisoning. Med J Aust 2006;185(7):388–9. https://doi.org/10.5694/j.1326-(2006) 5377.2006.tb0061.

Segev G, Bandt C, Francey T, Cowgill LD. Aluminum toxicity following administration of aluminum-based phosphate binders in 2 dogs with renal failure. J Vet Intern Med 2008;22:1432–5.

Segev GI, Naylor S, Cowgill LD. Hematological and neurological side effects associated with the use of aluminum based phosphate binders in dogs with chronic kidney disease. Israel J Vet Med 2016;71(1). March 2016.

Sessions D, Heard K, Kosnet M. Fatal Cesium chloride toxicity after alternative cancer treatment. J Altern Complement Med 2013;19 (12):973–5.

Sethy J, Chauhan NPS. Dietary preference of Malayan sun bear *Helarctos malayanus* in Namdapha Tiger Reserve, Arunachal Pradesh, India. Available online from https://biooneorg/journals/Wildlife-Biology; 2018. on 23 Jul 2020.

Sharma MC, Joshi C, Das G. Therapeutic management of copper deficiency in buffalo heifers: impact on immune function. Vet Res Commun 2008;32(1):49–63. https://doi.org/10.1007/s11259-007-9002-1.

Shaw SD, Bishop PJ, Harvey C, et al. Fluorosis as a probable factor in metabolic bone disease in captive New Zealand native frogs (*Leiopelma species*). J Zoo Wildlife Med 2012;43:549–65.

Shils ME. Magnesium. In: 9th ed., Shils ME, Olson JA, Shike M, Ross AC, editors. Modern nutrition in health and disease. Philadelphia: Lippincott Williams & Wilkins; 1999. p. 169–92.

Shimada H, Nagano M, Funakoshi T, Kojima S. Pulmonary toxicity of systemic terbium chloride in mice. J Toxicol Environ Health 1996;48(1):81–92. https://doi.org/10.1080/009841096161483. 8637060.

Shroeder HA, Nason AP. Interactions of trace metals in mouse and rat tissues; zinc, chromium, copper, and manganese with 13 other elements. JNutr 1976;106:198–203.

Shupe JL, Brunner RH, Seymour JL, Aldens CL. The pathology of chronic bovine fluorosis: a review. Toxicol Pathol 1992;20(2):274–88.

Simmons D, Joshi S, Shaw J. Hypomagnesaemia is associated with diabetes: not pre-diabetes, obesity or the metabolic syndrome. Diabetes Res Clin Pract 2010;87:261–6.

Singhi SC, Baranwal AK, Jayashree M. Acute iron poisoning: clinical picture, intensive care needs and outcome. Indian Pediatr 2003;40 (12):1177–82.

Smith RM. Cobalt Chapter 5. In: Mertz W, editor. Trace elements in human and animal nutrition. 5th ed, vol. 1. Academic Press Inc London; 1987. p. 143–76.

Smith SW. The role of chelation in the treatment of other metal poisonings. J Med Toxicol 2013;9:355–69. https://doi.org/10.1007/s13181-013-0343-6.

Smith JE, Chavey PS, Miller RE. Iron metabolism in captive black (*Diceros bicornis*) and white (*Ceratotherium simum*) rhinoceroses. J Zoo Wildl Med 1995;26(4):525–31.

Smith DJ, Anderson GJ, Bell SC, Reid DW. Elevated metal concentrations in the CF airway correlate with cellular injury and disease severity. J Cyst Fibros 2014;13:289–95.

Sources, Effects and Risks of Ionizing Radiation. United Nations Scientific Committee on the Effects of Atomic Radiation UNSCEAR. In: Report to the General Assembly, with Scientific Annexes UNITED NATIONS New York, 2018; 2017.

Spears J, Brown T, Sunvold G, Hayek M. Influence of chromium on glucose metabolism and insulin sensitivity. Recent advances in canine and feline nutrition. Iams Nutr Symp Proc 1998;2:103–12.

Spencer AJ, Wilson SA, Batchelor J, Reid A, Rees J, Harpur E. Gadolinium chloride toxicity in the rat. Toxicol Pathol 1997;25(3):245–55. https://doi.org/10.1177/019262339702500301.

Spiegl CJ, Calkins MC, DeVoldre JJ, Scott JK, Steadman LT, Stokinger HE. Inhalation toxicity of zirconium compounds. 1. Short-term studies. report, July 31, 1956, United States: University of North Texas Libraries, UNT Digital Library, crediting UNT Libraries Government Documents Department; 1956. https://digital.library.unt.edu/ark:/67531/metadc1024703/. accessed June 8, 2020 https://digital.library.unt.edu.

Srinivas A, Rao PJ, Selvam G, Murthy PB, Reddy PN. Acute inhalation toxicity of cerium oxide nanoparticles in rats. Toxicol Lett 2011;205:105–15.

Stalenberg E. Nutritional ecology of the Mumbulla koala. Spatial variation in habitat quality effects fine-scale resource use by a low-density koala population Honours thesis summary; 2010. June 2010.

Stoffaneller R, Morse NL. A review of dietary selenium intake and selenium status in Europe and the Middle East. Nutrients 2015;7:1494–537. https://doi.org/10.3390/nu7031494.

Strupp C. Beryllium metal II. A review of the available toxicity data. Ann Occup Hyg 2011a;55(1):43–56. https://doi.org/10.1093/annhyg/meq073.

Strupp C. Beryllium metal I. Experimental results on acute oral toxicity, local skin and eye effects, and genotoxicity. Ann Occup Hyg 2011b;55(1):30–42. https://doi.org/10.1093/annhyg/meq071.

Su Y-K, Mackey RV, Riaz A, Gates VL, Benson AB, Miller FH, Yaghmai V, Gabr A, Salem R, Lewandowski RJ. Long-term hepatotoxicity of Yttrium-90 Radioembolization as treatment of metastatic neuroendocrine tumor to the liver. J Vasc Interv Radiol 2017;28(11):1520–6.

Sundar S, Jaya Chakravarty J. Review antimony toxicity. Int J Environ Res Public Health 2010;7:4267–77. https://doi.org/10.3390/ijerph7124267.

Sunde RA. Selenium. In: Bowman B, Russell R, editors. Present knowledge in nutrition. 9th ed. Washington, DC: International Life Sciences Institute; 2006. p. 480–97.

Susanne R, Ann-Kathrin O. Husbandry and management of new world species: marmosets and tamarins. In: The Laboratory Primate; 2005. p. 145–62. https://doi.org/10.1016/B978-012080261-6/50010-6.

Tao S-H, Bolger PM. Hazard assessment of germanium supplements regulatory. Toxicol Pharmacol 1997;25(3):211–9.

Tao H, Man Y, Shi X, et al. Inconceivable hypokalemia: a case report of acute severe barium chloride poisoning. Case Rep Med 2016;2016:2743134. https://doi.org/10.1155/2016/2743134.

ThermoFisher Scientific. Safety datasheet for scandium. accessed online at https://www.fishersci.com/store/msds?partNumber=AA35769AD&productDescription=SCNDM+PLSMA+STDSOL+SC+1K+50ML&vendorId=VN00024248&countryCode=US&language=en; 2020. Accessed 15th September 2020.

Thompson KG, Audigè L, Arthur DG, Julian AF, Orr MB, McSporran KD, Wilson PR. Osteochondrosis associated with copper deficiency in young farmed red deer and wapiti × red deer hybrids. N Z Vet J 1994;42:137–43.

Tietz T, Lenzner A, Kolbaum AE, et al. Aggregated aluminium exposure: risk assessment for the general population. Arch Toxicol 2019;93:3503–21. https://doi.org/10.1007/s00204-019-02599-z.

Titenko-Holland N, Shao JS, Zhang LP, Xi LQ, Ngo HL, Shang N, Smith MT. Studies on the genotoxicity of molybdenum salts in human cells in vitro and in mice in vivo. Environ Mol Mutagen 1998;32:251–9.

Tomson FN, Lotshaw RR. Hyperphosphatemia and hypocalcemia in lemurs. J Am Vet Med Assoc 1978;173:1103–6.

Keith S, Faroon O, Roney N, et al. Toxicological Profile for Uranium. Atlanta (GA): Agency for Toxic Substances and Disease Registry (US); 2013 Feb. 3, HEALTH EFFECTS. Available from: https://www.ncbi.nlm.nih.gov/books/NBK158798/.

Toyokuni S. Role of iron in carcinogenesis: cancer as a ferrotoxic disease. Cancer Sci 2008;100(1):9–16. https://doi.org/10.1111/j.1349-7006.2008.01001.x. 2009.

Ulemale AH, Kulkarni MD, Yadav GB, Samant SR, Komatwar SJ, Khanvilkar AV. Fluorosis in cattle. Vet World 2010;3(11):526–7.

Underwood EJ, Suttle NF. The mineral nutrition of livestock. 3rd ed. Wallingford: CAB International; 1999.

Usami M, Ohno Y. Teratogenic effects of selenium compounds on cultured postimplantation rat embryos. Teratogen Carcinogen Mutagen 1996;16(1):27–36.

USEPA. US Environmental Protection Agency (USEPA) molybdenum—a toxicological appraisal, EPA-600/1–75-004. Research Park Triangle, NC: Health Effects Research Laboratory, Office of Research and Development; 1975.

USPA. United States Protection Agency Provisional Peer-Reviewed Toxicity Values for Stable (Nonradioactive) Praseodymium Chloride (CASRN 10361-79-2); 2009.

Valiathan MS, Kartha CC, Eapen JT, et al. A geochemical basis for endomyocardial fibrosis. ICMR Centre for research in cardiomyopathy, Trivandrum. India Cardiovasc Res 1989;23(7):647–8.

Van Den Brock A, Thoday K. Skin disease in dogs associated with zinc deficiency: a report of five cases. J Sm Anim Pract 1986;27:313–23.

Van der Voet GB, Todorov TI, Centeno JA, Jonas W, Ives J, Mullick FG. Metals and health: a clinical toxicological perspective on tungsten and review of the literature. Mil Med 2007;172(9):1002–5.

Van Dyke MV, Martyny JW, Mroz MM, et al. Exposure and genetics increase risk of beryllium sensitisation and chronic beryllium disease in the nuclear weapons industry. Occup Environ Med 2011;68(11):842–8. https://doi.org/10.1136/oem.2010.064220.

Van Vleet J. Experimentally induced vitamin E-selenium deficiency in the growing dog. J Am Vet Med Assoc 1975;166:769–74.

van Zelst M, Hesta M, Gray K, Staunton R, Du Laing G, Janssens GP. Biomarkers of selenium status in dogs. BMC Vet Res 2016;12:15. https://doi.org/10.1186/s12917-016-0639-2.

VanderSchee CR, Kuter D, Bolt AM, Lo F-C, Feng R, Thieme J, Chen-Wiegart YK, Williams G, Mann KK, Bohle DS. Accumulation of persistent tungsten in bone as in situ generated polytungstate. Commun Chem 2018;1(1). https://doi.org/10.1038/s42004-017-0007-6.

Vanselow BA, Barboza PS. Chronic copper poisoning in a captive (southern) hairy nosed wombat (*Lasiorhinus latifrons*). Vet Pathol Rep 1988;21:11.

Varga M. Rabbit basic science textbook of rabbit medicine. 2014;2014:3–108. Published online 2013 Oct 10 https://doi.org/10.1016/B978-0-7020-4979-8.00001-7. PMC7158370.

Velásquez JI, Aranzazu DA. An acute case of iron toxicity on newborn piglets from vitamin E/Se deficient sows. Rev Col Cienc Pec 2004;17:1.

Venn-Watson S, Benham C, Carlin K, DeRienzo D, St Leger J. Hemochromatosis and fatty liver disease: building evidence for insulin resistance in bottlenose dolphins (*Tursiops truncatus*). J Zoo Wildl Med 2012;43(3 Suppl):S35–47. https://doi.org/10.1638/2011-0146.1.

Vikøren T, Bernhoft A, Waaler T, Handeland K. Liver concentrations of copper, cobalt, and selenium in wild Norwegian red deer (*Cervus elaphus*). J Wildl Dis 2005;2005(41):569–79.

Waggoner DJ, Bartnikas TB, Gitlin JD. The role of copper in neurodegenerative disease. Neurobiol Dis 1999;6:221–30.

Wang Y, Zhang H, Shi L, Xu J, Duan G, Yang H. A focus on the genotoxicity of gold nanoparticles. Nanomedicine (Lond) 2020;15(4):319–23.

Ward JM, Ohshima M. The role of iodine in carcinogenesis. Adv Exp Med Biol 1986;206:529–42. https://doi.org/10.1007/978-1-4613-1835-4_37.

Wax PM. Current use of chelation in American health care. J Med Toxicol 2013;9:303–7.

Weaver CM, Heaney RP. Calcium. In: Shils ME, Olson JA, Shike M, Ross AC, editors. Modern nutrition in health and disease. 9th ed. Philadelphia: Lippincott Williams & Wilkins; 1999. p. 141–55.

Wells WH, Wells VL. The lanthanides, rare earth metals. In: Bingham E, Cohrssen B, Powell CH, editors. Pattys industrial hygiene and toxicology. 5th ed, vol. 3. New York, NY: John Wiley and Sons, Inc.; 2001. p. 423–58.

WHO. Cadmium (Environmental Health Criteria 134), Geneva, www.who.org; 1992. Accessed online June 10th 2020.

WHO. Worldwide prevalence of anaemia 1993–2005. In: de Benoist B, McLean E, Egli I, Cogswell M, editors. WHO global database on anaemia; 2008 [Accessed online] https://apps.who.int/iris/bitstream/handle/10665/43894/9789241596657_eng.pdf;jsessionid=178FA6D45E48695EEDFBE401D67C2EC9?sequence=1.

WikEM. Selenium, https://wikem.org/wiki/Selenium_toxicity; 2020. Accessed online 10th August 2020.

Williams JP, Brown SL, Georgesc GE, others. Animal Models for Medical Countermeasures to Radiation Exposure Radiat Res. 2010. April 2010;173(4):557–78. https://doi.org/10.1667/RR1880.1.

Wilson PR. Bodyweight and serum copper concentrations of farmed red deer stags following oral copper oxide wire administration. N Z Vet J 1989;37:94–7.

Wilson PR, Grace ND. A review of tissue reference values used to assess the trace element status of farmed red deer (*Cervus elaphus*). N Z Vet J 2001;49:126–32.

Wilson PR, Orr MB, Key EL. Enzootic ataxia in red deer. N Z Vet J 1979;27:252–4.

Winder C, Bonin T. The genotoxicity of lead. Mutat Res 1993;285(1):117–24. https://doi.org/10.1016/0027-5107(93)90059-o.

Winship KA. Toxicity of antimony and its compounds. Adverse Drug React Acute Poisoning Rev 1987;2:67–90.

Winship KA. Toxicity of tin and its compounds. Adverse Drug React Acute Poisoning Rev 1988;7(1):19–38. Spring 1988.

Wobeser G, Swift M. Mercury poisoning in a wild mink. J Wildl Dis 1976;12(3):335–40. https://doi.org/10.7589/0090-3558-12.3.335.

Wobeser G, Nielsen NO, Schiefer B. Mercury and mink. II. Experimental methyl mercury intoxication. Can J Comp Med 1976;40(1):34–45.

Wolfe BA, Sladky KK, Loomis MR. Obstructive urolithiasis in a reticulated giraffe (*Giraffa camelopardalis reticulata*). Vet Rec 2000;146(9):260–1.

Wong LCK, Downs WL. Renal effects of potassium niobite. Toxicol Appl Pharmacol 1966;9(3):561–70.

World Nuclear Association. (2020) Naturally-Occurring Radioactive Materials (NORM) available online at https://www.world-nuclear.org/information-library/safety-and-security/radiation-and-health/naturally-occurring-radioactive-materials-norm.aspx Last updated April 2020. Accessed 5th September 2020.

Wu C-C, Chen Y-C. Assessment of industrial antimony exposure and immunologic function for workers in Taiwan. Int J Environ Res Public Health 2017;14(7):689.

Yang XF, Xu J, Hou XH, et al. Developmental toxic effects of chronic exposure to high doses of iodine in the mouse. Reprod

Toxicol 2006; 22(4):725–30. https://doi.org/10.1016/j.reprotox.2006.05.010.

Yang CF, Duro D, Zurakowski LDM, Jaksic T, Duggan C. High prevalence of multiple micronutrient deficiencies in children with intestinal failure: a longitudinal study. J Pediatr 2011;159(1):39–44.

Yokoi K, Kawaai T, Konomi A, Uchida Y. Dermal absorption of inorganic germanium in rats. Regul Toxicol Pharmacol 2008;52(2):169–73.

Yongxing W, Xiaorong W, Zichung H. Genotoxicity of lanthanum (III) and gadolinium (III) in human peripheral blood lymphocytes. Bull Environ Contam Toxicol 2000;64:611–6.

Yuen HW, Becker W. Iron toxicity. In: StatPearls [Internet]. Treasure Island (FL): StatPearls Publishing; 2020. [Updated 2020 Jun 30]. Available from: https://www.ncbi.nlm.nih.gov/books/NBK459224/.

Zentec J, Meyer H. Normal handling of diets—are all dogs created equal? J Sm Anim Pract 1995;36:354–9.

Useful online information sources

ATSDR Agency for Toxic Substances and Disease Registry https://www.atsdr.cdc.gov/toxprofiledocs/index.html.

CDC Centers for Disease Control and Prevention https://www.cdc.gov/niosh/index.htm.

EPA United States Environmental Protection Agency https://iaspub.epa.gov/sor_internet/registry/substreg/LandingPage.do.

EPA United States Environmental Protection Agency Ecotox database https://cfpub.epa.gov/ecotox/.

EURARE Rare Earth Elements http://www.eurare.eu/.

IARC International Agency for Research on Cancer (WHO) https://www.iarc.fr/.

IRIS Integrated Risk Information System https://www.epa.gov/iris.

NCBI National Center for Biochtechnology Information https://www.ncbi.nlm.nih.gov/.

NIH National Institutes of Health https://search.nih.gov/.

NIOSH The National Institute for Occupational Safety and Health https://www.cdc.gov/niosh/index.htm.

NORD National Organisation for Rare Disorders https://rarediseases.org/about/.

NRC National Research Council.

PubMed, https://pubmed.ncbi.nlm.nih.gov/.

RAIS Risk Assessment Information System https://rais.ornl.gov/.

WHO World Health Organisation Vitamin and Mineral Nutrition Information System (VMNIS) https://www.who.int/vmnis/en/.

Index

Note: Page numbers followed by *t* indicate tables.

C

D

E

I

J

K

Q

R

S

U

V

W

X

Y

Z

Printed in the United States
by Baker & Taylor Publisher Services